# Bioinformatics—A Student's Companion

Kalibulla Syed Ibrahim
Guruswami Gurusubramanian
Zothansanga · Ravi Prakash Yadav
Nachimuthu Senthil Kumar
Shunmugiah Karutha Pandian
Probodh Borah · Surender Mohan

# Bioinformatics—A Student's Companion

Springer

Kalibulla Syed Ibrahim
Department of Biotechnology
Mizoram University
Aizawl, Mizoram
India

Nachimuthu Senthil Kumar
Department of Biotechnology
Mizoram University
Aizawl, Mizoram
India

Guruswami Gurusubramanian
Department of Zoology
Mizoram University
Aizawl, Mizoram
India

Shunmugiah Karutha Pandian
Department of Biotechnology
Alagappa University
Karaikudi, Tamil Nadu
India

Zothansanga
Department of Biotechnology
Mizoram University
Aizawl, Mizoram
India

Probodh Borah
College of Veterinary Science
Assam Agricultural University
Guwahati, Assam
India

Ravi Prakash Yadav
Department of Biotechnology
Mizoram University
Aizawl, Mizoram
India

Surender Mohan
Department of Biotechnology and Bioinformatics
SRM University
Sonepat, Haryana
India

ISBN 978-981-10-9463-7 ISBN 978-981-10-1857-2 (eBook)
DOI 10.1007/978-981-10-1857-2



Printed on acid-free paper

This Springer imprint is published by Springer Nature
The registered company is Springer Nature Singapore Pte Ltd.
The registered company address is: 152 Beach Road, #21-01/04 Gateway East, Singapore 189721, Singapore

# Foreword

Biological research has progressed tremendously in the recent years due in part to the increased automation in the process of generation of data. On the other hand, storing, managing and interpreting huge volumes of data have become one the most challenging tasks for the scientists. These have ultimately necessitated the application of computers paving the way for emergence of a highly interdisciplinary field of bioinformatics. Being interdisciplinary, it attracts people from mathematics, statistics, chemistry, physics, etc., but when it comes to learning bioinformatics, people often get confused as to where to start with.

Internet has revolutionized bioinformatics by offering various tools for free. Despite the richness of bioinformatics resources and methods, teaching bioinformatics to the beginners has still remained a daunting task for the biology teachers. It is therefore essential, in my opinion, that learners should be exposed to the practical lessons in bioinformatics using an easily explicable and 'how-to-do' approach. I wish to emphasize in particular the effort made by the authors for undertaking this challenging and exciting concept of developing a manual on bioinformatics, keeping in mind the readers' requirements from diverse backgrounds and also incorporating into it the syllabi of UG and PG courses of various Indian universities. This could serve not only as a model for teaching the practical component of the syllabi, but also for offering trainings to the beginners.

This manual is intended to introduce the basic algorithmic ideas and step-by-step procedures, with screenshots, in order to make better use of available tools to be able to develop novel computational ideas during their own careers later. Though the tools discussed over here are freely available over the internet along with the online manuals and help pages, originality of this work is on presenting the content on a self-study basis making it possible for the students to perform the basic analyses even in the absence of an instructor. This manual is carefully drafted with a blend of theory, practice and examples. I wish that the authors would come out with more publications of this kind in the future.

Dr. T. Madhan Mohan
Adviser—DBT, Government of India

# Preface

Bioinformatics is playing a significant role in the current revolution in modern biology. It attempts to unfurl the mystery of the biologically meaningful patterns of a sequence string of nucleotides. Teaching bioinformatics is now becoming an integral part of the core curriculum at most universities. We believe that it is only by understanding and wisely using these resources we will be able to push the frontier forward.

In this process we have come up with the following features, which should make this edition particularly helpful to you offering essential, practical answers to the day-to-day problems encountered in a successful career in modern biology. The target audience for this book is the biologists, especially the beginners, who want to learn how to analyze DNA or protein sequences in a simple but meaningful fashion. Readers are not expected to possess special background in statistics, mathematics, or computer science, rather just a basic knowledge of molecular biology and internet is all required.

This manual is expected to serve as a resource guide to some of the web-based analysis packages, including some standalone computing resources that are available free for academics to address some of the basic tasks of bioinformatics. In this manual, we focus on few theoretical aspects and emphasize more on applied concepts on PCR primer designing, sequence retrieval, sequence-similarity searching using BLAST, multiple-sequence alignment, phylogenetic analysis, RNA structure analysis, and protein sequence and structure analysis. Molecular docking is also included in this manual. This could be seen as an enriched task-oriented manual that will direct the reader through step-by-step instructions, completing specific bioinformatics analyses. Every chapter and exercise begins with an overview, followed by specific examples on how to perform and interpret the results, and concludes with exercises to test and explore some of the additional concepts. The manual finally concludes with the authors' recommendations on how to keep current with the ever-increasing information on bioinformatics through the appendix, providing links to the resources discussed and additional resources.

Bioinformatics—A Student's Companion presents a broad overview of the computational resources to help orient those just beginning to use computational tools to address biological problems. We strongly believe that even a novice can quickly tackle each computational problem and arrive at a satisfying result when guided by this unique step-by-step approach on the collection of software with illustrative examples. Most exercises could be used for practical courses in advanced undergraduate- or graduate-level courses in life sciences.

We would like to thank the funding agencies—Bioinformatics Infrastructural Facility under the Biotechnology Information System Network (BTISNeT) of Department of Biotechnology, New Delhi at Mizoram University, Alagappa University, College of Veterinary Science (CVSc), Assam Agricultural University and the Department of Science and Technology—Science and Engineering Research Board (SERB), New Delhi. The literature access facility provided by DBT—DeLCON and UGC—INFLIBNET have facilitated writing of the manual. The administrative support of Mizoram University, Alagappa University, CVSc—Assam Agricultural University and SRM University are acknowledged. Thanks are also due to the master's degree students who have been helping in testing the protocols and providing suggestions to improve them. Hopefully, the outcome has been a set of useful exercises that will be of help to students, researchers and the academic community.

Finally, we are grateful for the support from the publishers, Springer Nature, particularly Ms. Sivajothi and Ms. Madhurima Kahali, who have been incredibly patient with us.

Aizawl, India — Dr. Kalibulla Syed Ibrahim
Aizawl, India — Dr. Guruswami Gurusubramanian
Aizawl, India — Dr. Zothansanga
Aizawl, India — Mr. Ravi Prakash Yadav
Aizawl, India — Dr. Nachimuthu Senthil Kumar
Karaikudi, India — Dr. Shunmugiah Karutha Pandian
Guwahati, India — Dr. Probodh Borah
Sonepat, India — Mr. Surender Mohan

# Contents

# About the Authors

**Kalibulla Syed Ibrahim** is postdoctoral fellow at the Department of Biotechnology, Mizoram University. He received his Ph.D. in Biotechnology from Alagappa University, Karaikudi. His fields of interest include protein expression and purification, microbial ecology, high throughput sequencing and data analysis. Besides having publications in peer reviewed journals he is also a recipient of DST-Young Scientist Award (YSS/2014/000657) for working in Human Microbiome of gastric cancer patients of Mizoram, India.

**Guruswami Gurusubramanian** is Professor of Zoology at Mizoram University with more than 23 years of experience in teaching and research. He received his Ph.D. in Zoology from the University of Madras, Chennai and served as Post Doctoral Humboldt Fellow at the Institut für Biologischen Pflanzenschutz, Germany. His contributions have been recognized with awards like Fellow of International Society for Ecological Communication, Poland, the Business Excellence Award from the International Business Council, and the Life Time Achievement Award from the Indian Compendium Bureau, New Delhi. He has published over 109 articles in leading peer-reviewed International journals, has contributed to two books published by leading publishers and holds four patents. He has been successful in attracting funding from various bodies like the UGC, DBT, CSIR and Central Silk Board as a Principal Investigator.

**Zothansanga** is currently working as a postdoctoral researcher at the Department of Biotechnology, Mizoram University with 5 years of teaching experience. His fields of interest include microbial diversity and metagenomics, molecular phylogeny, bioinformatics and microbial control of pests. He is presently working on microbial community analysis using next-generation sequencing methods.

**Ravi Prakash Yadav** is Research Scholar at the Department of Biotechnology, Mizoram University. He received his M.Sc. Biotechnology from the Kalinga Institute of Industrial Technology, Bhubaneswar (2010–2012). His fields of interest include computational biology/bioinformatics and cancer genomics. Currently, he is working on the study of genomic alterations in cell cycle regulatory genes that result in tumorigenesis.

**Nachimuthu Senthil Kumar** is Professor of Biotechnology at Mizoram University with more than 12 years of experience in teaching and research. He completed his Ph.D. in Zoology at Bharathiar University, Coimbatore and worked as a postdoctoral fellow at Sun Yat-sen University, Guangzhou, China. His fields of interest include bioinformatics and molecular phylogeny, cancer genomics and human genetics. With over 50 publications in reputed international peer-reviewed journals and authoring two books in leading publications, his contributions have been recognized with various awards from the CSIR and DBT. As a Principal Investigator, he has successfully completed research projects funded by government funding bodies like the DBT, CSIR and DST.

**Shunmugiah Karutha Pandian** is Professor of Biotechnology at Alagappa University. He received his Ph.D. from the Forensic Sciences Department (Government of Tamil Nadu) affiliated with the University of Madras. After his Ph.D., he moved to the Centre for Cellular and Molecular Biology (CCMB), Hyderabad followed by the National Tsing Hua University, Taiwan for postdoctoral research. An established pioneer in drug discovery from natural products, he has published 127 articles in leading peer-reviewed international (SCI) journals and has contributed to five books and two laboratory manuals published by leading publishers like Springer. He has been successful in completing nine Major Research Projects (MRPs) funded by government bodies like the UGC, DBT, DCI, CSIR and ICMR as a Principal Investigator. In recognition to his contributions, he has been awarded with the 'Certificate of Achievement' under the 'Performance Linked Incentive Scheme' of the Alagappa University, the 'Shri. P.K. Das Memorial Best Faculty Award in Biotechnology' by the Nehru Group of Institutions, and the Tamil Nadu Scientist Award (TANSA) in Biological Sciences for the year 2012.

**Probodh Borah** is Professor of Microbiology/Animal Biotechnology at Assam Agricultural University with more than 29 years of experience in teaching and research. He completed his Ph.D. in Veterinary Science at Assam Agricultural University, Guwahati and served as postdoctoral fellow at the University of California, USA. He has over 63 publications in peer-reviewed journals and has authored six books, two laboratory manuals and contributed three book chapters. He is an active member of the DBT Committee for North Eastern States. As a Principal Investigator, he has received funding from government agencies including the UGC, DBT, CSIR, ICMR and DST. He is also the Coordinator, Bioinformatics Infrastructure Facility and State Biotech Hub (Assam).

**Surender Mohan** is Assistant Professor of Biotechnology and Bioinformatics at SRM University, Haryana, India with more than four years of experience in teaching and research activities. He is also pursuing his doctoral research at Jawaharlal Nehru University (JNU), New Delhi, India in specializing in Infectious diseases. He has been awarded International Students Fellowship by National University of Singapore (NUS). His contributions have been recognized with collaborates with premier institutes like International Centre for Genetic Engineering and Biotechnology (ICGEB), New Delhi, India in research activities and publications in peer-reviewed journals.

# Chapter 1
# Nucleotide Analysis

All living cells are made up of major biological molecules like carbohydrates, lipids, proteins and nucleic acids which need an understanding of their functional aspects to be studied at the molecular level. Though the genome encompasses all hereditary instructions, the nucleotide is the fundamental unit of genetic material containing four different bases: the purines, adenine (A) and guanine (G); the pyrimidines, cytosine (C) and thymine (T). Uracil (U) is a pyrimidine that replaces T in RNA. The phosphodiester bond links the adjacent nucleotides through DNA polymerase. The entire genome is made up of several nucleotides and DNA sequencing resolves the order of nucleotides and their primary structure. Databases store the huge amounts of data that is generated through sequencing projects. Finding homologues by searching the sequence databases is obviously the first stage in the analysis of any new sequence. Characterizing DNA starts with PCR. Objectives like amplifying a few or many targets, cloning the full-length coding region, quantifying the target, identifying unknown sequences or issues like analyzing SNPs, most likely effects of mutant sequences and characterizing pseudogenes can be met with PCR. A successful PCR depends on 'a good primer' and all one need to follow is just the rules for good primer design. Several software programs are available for designing primers. If the sequence of our gene of interest is unknown, it is desirable to design a degenerate primer by using DNA sequences of closely related species from the database which can have two or more bases at any given nucleotide position based on the conserved regions.

Restriction mapping besides useful in the characterization of a large DNA molecule, it has been used routinely in preparation for subcloning, analysis of restriction fragment length polymorphisms, construction of physical maps of chromosomes and checking sequence assembly. Further characterization like Gene prediction, defines the gene content is embedded within the primary structure. Evolutionary relationships assume that a certain percentage of the amino acid residues in a protein sequence are conserved and one easily assesses the relationships between two sequences by identifying the number of identical and similar amino acids. This is done by aligning the sequences. Sequence alignment is used



K. Syed Ibrahim et al., *Bioinformatics—A Student's Companion*,
DOI 10.1007/978-981-10-1857-2_1

for analyses of evolutionary relationships especially in predicting functional and tertiary structure information's from the amino acid sequence. The primary objective of Bioinformatics is to make sense of DNA/aminoacid sequence data through the development and implementation of sophisticated algorithms. Nevertheless, one should be always aware that any observation generated by a computer program represents only one of many possibilities.

## 1.1 Sequence Retrieval

**Introduction**

There are two sequence databases, i.e. primary (archival) and secondary (curated) databases. Every entry in these databases is generally stored as flat-files with a unique identifier (accession number) enabling the retrieval of information. Retrieval process is facilitated by interface for easy and user-friendly data access. The primary databases (NCBI, EBI) contain raw sequences of DNA, RNA or protein with minimum annotation. The secondary databases (Uniprot) consist of information about patterns, motifs and functional sites derived from the analyses of the sequences in the primary databases.

Global Cross-database NCBI search (GQuery, http://www.ncbi.nlm.nih.gov/sites/gquery) database and retrieval system was developed by National Centre for Biotechnology Information (NCBI) of National Institute of Health (NIH) to allow data retrieval and literature citations from its integrated databases. GQuery Search can be done by choosing databases like (i) Nucleotides for DNA sequences (GenBank, EMBL and DDBJ), (ii) Proteins for protein sequences (NCBI, PIR, SWISS-PROT) (iii) 3D structures for the 3D structure coordinates (PDB), (iv) Genome and chromosome mapping data and (v) the PubMed literature survey (MEDLINE). The search term may comprise either plain text or any of several more specialized searchable fields and also supports Boolean query. Search results are then viewed and saved in FASTA format (Fig. 1.1).

**Exploration Activity**

1. Go to NCBI homepage (http://www.ncbi.nlm.nih.gov/).
2. From the dropdown box, select **Nucleotide** option (Fig. 1.2).
3. In search text box, type the name of the organism, or gene, or its accession number, or gi number in the search field. For example, type "**bacillus**".
4. To execute search of your gene of interest, Click the "**Search**" button to check the results (Fig. 1.3).
5. In the right hand menu, the filter results show the number of nucleotide sequences entered in the database.

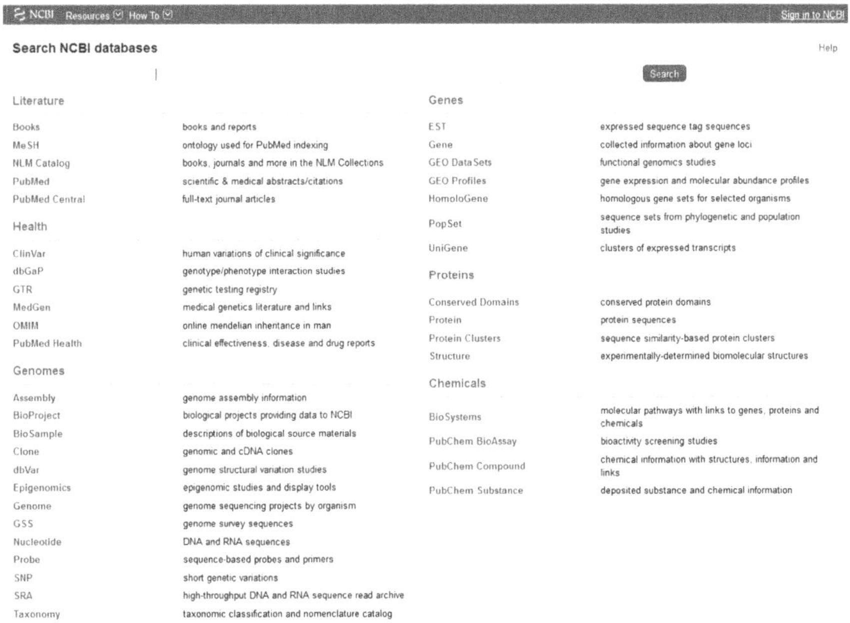


**Fig. 1.1** GQuery home page

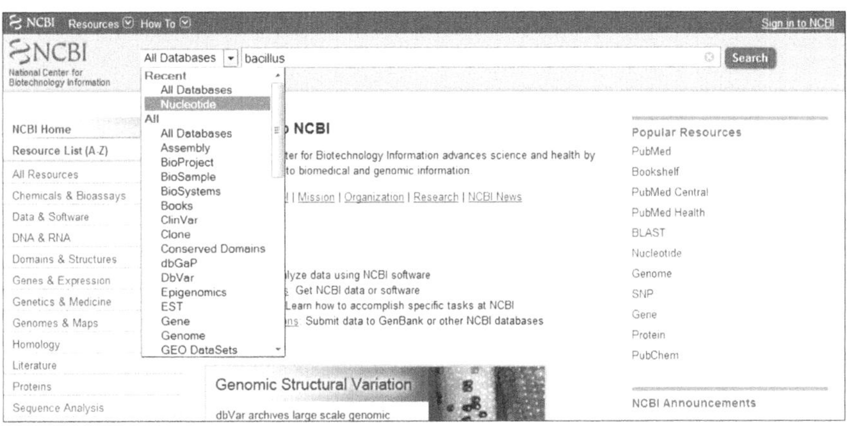


**Fig. 1.2** Selecting nucleotide database from NCBI

6. To narrow down the search hit, type the name of the species of your interest (***Bacillus pumilus***), and click "Search". The results will be specific to *Bacillus pumilus* only (Fig. 1.4).
7. To get a specific entry hit (gene/organism), use Boolean operators (*typed only in capital letters*):

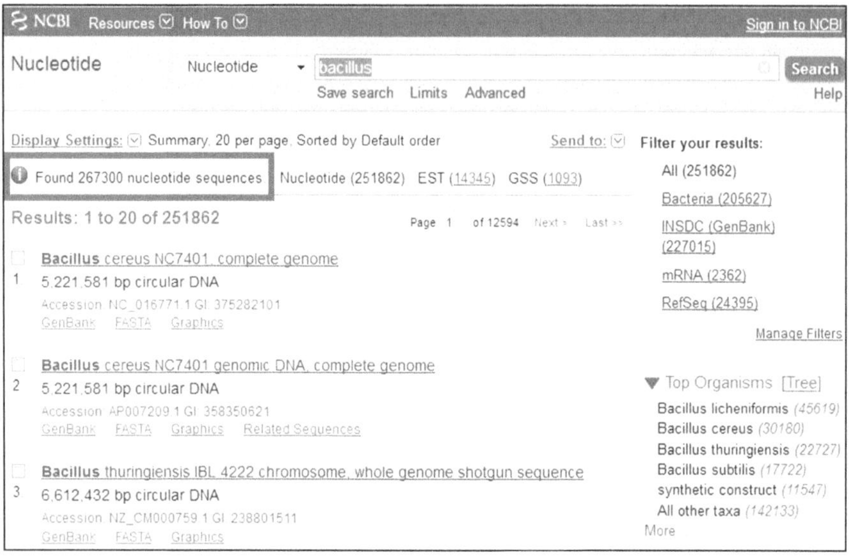

Fig. 1.3 Search results

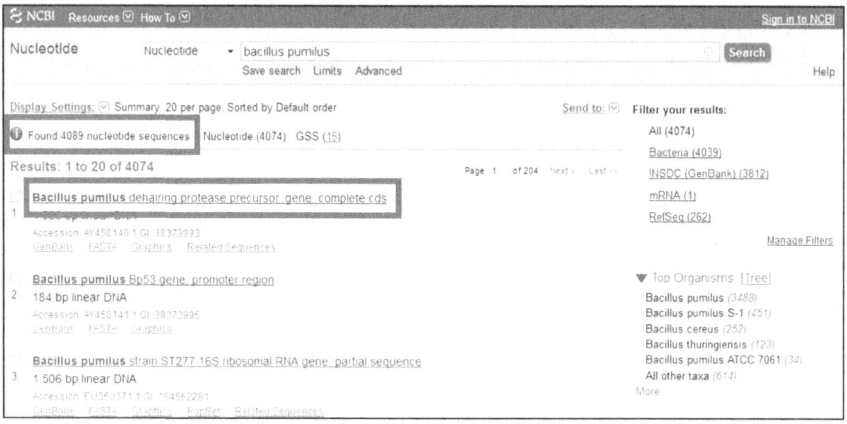

Fig. 1.4 Search results showing specific entry hits (*Bacillus pumilus*)

- Intersection (AND)—displays results that include selected entries,
- Union (OR)—displays results that include at least one of the selected entries.
- Difference (NOT)—excludes results with the selected entry.

8. To view the results, click on the link.
   By default the page displays the GenBank format of the sequence (Fig. 1.4).

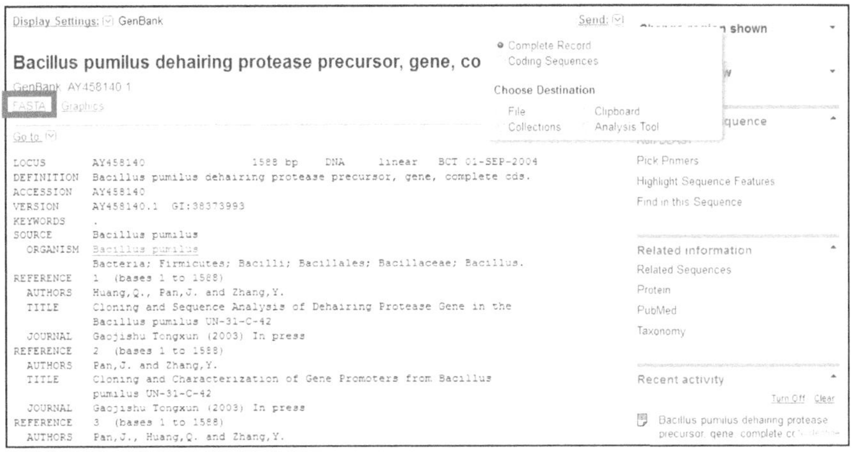


**Fig. 1.5** The GenBank record

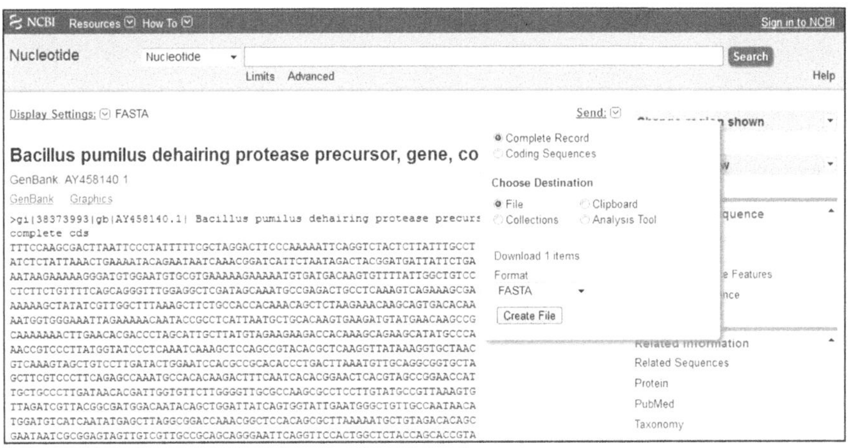


**Fig. 1.6** Downloading a sequence in FASTA format

9. To view the sequence in FASTA format click on the FASTA link (Fig. 1.5).
10. To download the sequence, choose Send > File and click 'Create File' button (Fig. 1.6).
11. Once you have found the sequence of interest, GenBank record helps to navigate to **perform analysis of the sequence** (Run BLAST, Pick Primers, Highlight Sequence Features, find in this sequence) and their **related information** (related sequences, bioproject, protein, taxonomy and trace archives) (Fig. 1.7).

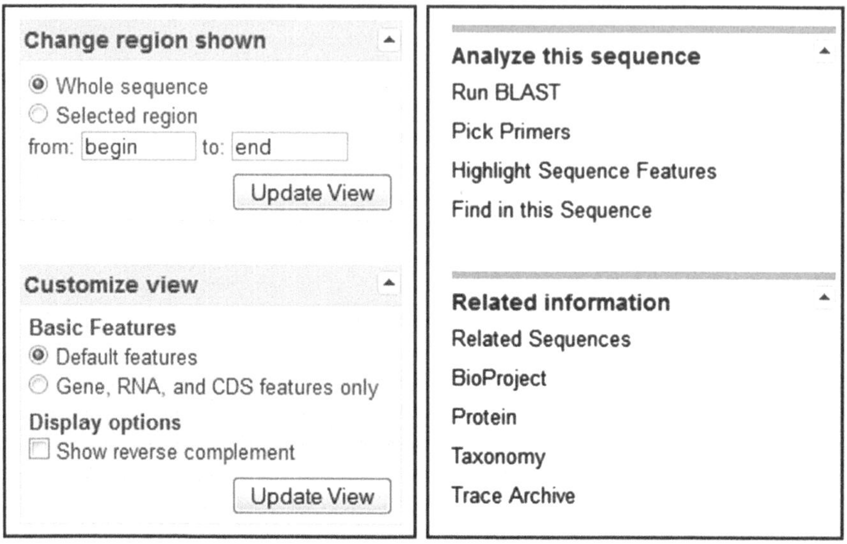


**Fig. 1.7** Options and links for easy navigation

**Exercise**

1. Define the following:
   Flatfile, Motif, Profile, Pattern, Domain, Annotation, Algorithm.
2. What do you mean by sequence retrieval?
3. What is meant by 'gi' number?
4. What is a Database and Database Management System (DBMS)?
5. Expand the acronyms EMBL, NCBI, DDBJ and PDB.
6. What are Boolean operators and how do they function?
7. Name a secondary database.
8. List the resources for literature retrieval.

## 1.2 Primer Designing

**Introduction**

A Primer is an oligonucleotide (18–27 bp) which is attached to DNA template (single stranded) and serves as a beginning point for the synthesis of complementary strand in 5′ → 3′ direction.

**Primer Selection Criteria**

- Length of the primer between 15 and 30 nucleotides.
- GC content (40–60%).
- <3 G/C nucleotide at the 3′ end to avoid nonspecific priming.

- Primers should end (3′) in a G or C, or CG or GC to increase effectiveness of priming and to prevent breathing (dissociation and re-association of primer with the DNA template).
- No self-complementary or complementary to other primers to prevent homo- and hetero-dimer and hairpin formations.
- Forward and reverse primer melting temperature (Tm) should be less than 5 °C.
- Same nucleotides present in the 3′ end of primer (3 Nos) for degenerate primers.

**Using Computer Programs**

Computer programs will be useful when many primers are to be designed and synthesized to reduce the cost as well as time. Moreover, digitized sequences of primer can be downloaded to synthesizers which can minimize the sequence error that occurs while copying. For designing of primers both free and commercial software are available.

**Applications**

- Conventional PCR
- Designing Probes for Hybridization
- Designing RAPD marker
- DNA Sequencing
- Molecular Cloning
- Designing Degenerate Primer.

**Primer3**

Before designing primers, two important things to be kept in mind are: the quality of the sequence and avoiding repetitive elements in the sequence. If the quality of bases is low it can be altered to N's or include in the **Excluded Regions** (Fig. 1.8) while use **Included Region parameter** for the presence of low quality bases at the end of sequencing read which enables the PRIMER3 program selects primers in the high-quality region of the read. This comes handy when you decide to pick primers from single-pass sequence. To combat the second problem, one can scan the Repbase database (http://www.girinst.org/repbase/), an excellent source of repeat sequences (Bao et al. 2015; Jurka et al. 2005).

Besides these, other problems include possible vector contaminants and cloning artefacts which can be solved by using BLAST, FASTA or VecScreen. PRIMER3 also having an alternative to find the mispriming in the primer sequence (Mishyb Library).

**Exploration Activity**

**Designing Primers for Conventional PCR**

In this exercise, the gene sequence encoding manganese and iron superoxide dismutase (Npun_R6491) from *Nostoc punctiforme* PCC 73102 (Accession: CP001037.1, position: 8014646-8015248) is used as the template sequence for primer designing.

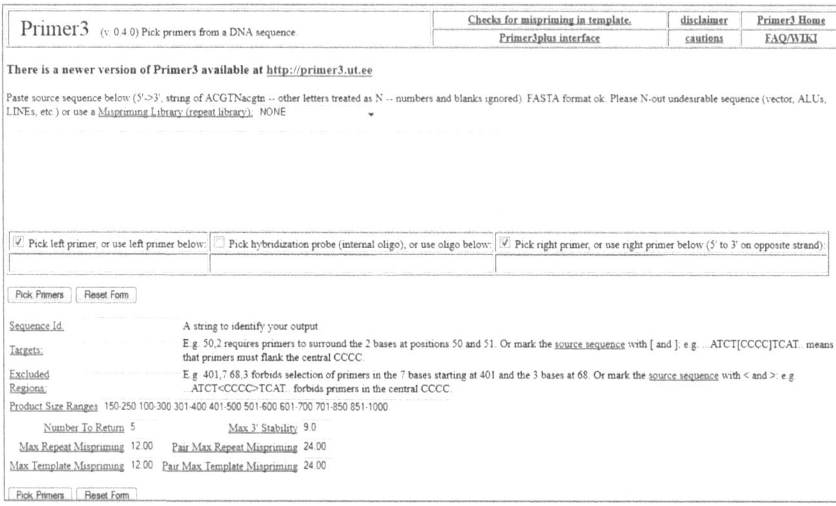


**Fig. 1.8** The Primer3 interface of Department of Bioinformatics at University of Tartu & Estonian Biocentre

1. Primer3 can be accessed from http://bioinfo.ut.ee/primer3-0.4.0/ (Fig. 1.8).
2. Collect the DNA sequence for which you want to design primers for amplification and enter by pasting the sequence in the sequence window (Npun_R6491).
3. An optional title can be provided under 'Sequence ID' that helps to identify the designed primers.
4. The '**Excluded Regions**' is particularly helpful in situations like excluding low sequence quality regions or regions having repetitive elements. The specified region will not be considered for overlapping primers. This can be specified by the format *start*, *length*. Here the *start* is the position of the first base of the region to be excluded, and *length* denotes number of bases from *start*.
5. Under '**Product Size Ranges**', **specify a particular range like 300–400**. You can also enter multiple ranges, for example 150–250, 100–300 or 301–400. If the first range is not possible, it tries the next range and move on with the other ranges until there are no more ranges. Minimum product size should be 100 base pairs.
6. You can specify the lengths like the Minimum, Optimum and Maximum (in bases) of the PCR product under '**Product Size**'. Primer3 will attempt to pick primers producing products closer to the specified 'Optimum length'.
7. The maximum number of primer pairs to be generated can be specified in **Number To Return**. The Primer pairs are ranked based on their 'quality'.

8. The size of the primers can be mentioned in '**Primer Size**' that reads Minimum, Optimum and Maximum lengths (in bases) of a primer. Primer3 will not pick primers shorter than 1 or longer than the value specified in Max (Maximum is 36). In case of default arguments, the program pick primers close with size close to the value mentioned in Opt.
9. If the user wishes to specify any particular region, like skipping a few bases from a vector, in the given sequence where primers need to be picked, it can be specified in '**Included Region**'. This can be specified by the format *start, length*. Here the *start* is the position of the first base of the region to be considered and *length* denotes number of bases in the primer-picking region.
10. Leave the other parameters as default.
11. Click the '**Pick Primers**' button.

**Interpretation of Results**

12. The results include position of the primer, its length, its predicted melting temperature and the G + C percentage (the guanine and cytosine nucleotides) for every set of oligonucleotide listed (Fig. 1.9).
13. It is followed by the mapping of the best oligonucleotide pair (both the left and right primers) based on the constraints provided at the beginning.
14. It is then followed by the four alternative primer solutions (Fig. 1.10). The descriptions of the columns are as follows:
15. '**Start**' indicates the start position of the primer's 5′ base. For a Left Primer it denotes the **leftmost** base and for a Right primer it is the **rightmost** base.
16. '**Len**' indicates the length of the designed primer.
17. '**tm**' indicates the melting temperature of the designed primer.
18. '**gc%**' indicates the percentage of G or C bases in the designed primer.
19. '**any**' refers to the self-complementary (its capacity to bind with other primers or forming secondary structure) score of the primer. This gives the inter and intra molecular interaction of the primer. When the designed primers anneal with other primers forming dimers instead of binding to target DNA, product yield is reduced. Optimally a $\Delta G$ of −5 kcal/mol at 3′ end self dimer with and an internal self dimer having a $\Delta G$ of −6 kcal/mol can be considered tolerable generally.
20. '**3′**'refers to the 3′ self-complementary (capacity to bind with itself) of the primer.
21. '**seq**' refers to the sequence of the selected primer or oligo, both left and right represented in 5′ → 3′ direction.

Once the primers have been designed, you have to verify that they will not bind anywhere other than your expected region. Also you have to make sure that selected primer sequences are not found outside your gene of interest or form a part of repeat region. This can be done checking the designed primers using a BLAST search.

**Primer3 Output**

```
No mispriming library specified
Using 1-based sequence positions
OLIGO            start  len     tm     gc%    any    3'  seq
LEFT PRIMER        182   20  60.00   55.00   4.00  0.00 ACCTTCAGCGTACCACCATC
RIGHT PRIMER       519   20  59.84   45.00   3.00  2.00 TATGGCAAGCATCACAAAGC
SEQUENCE SIZE: 603
INCLUDED REGION SIZE: 603

PRODUCT SIZE: 338, PAIR ANY COMPL: 4.00, PAIR 3' COMPL: 0.00

  1 TTAAGCTTTAGCCAAATTTTCAGCAGCAAAATCCCAGTTAACCAAGTTATCTAGGAAATT


 61 CTTGATGAACGCTGGACGAGCGTTTCTGTAGTCAATGTAGTAGGCATGTTCCCAAACATC


121 CAAGGTTAGGAGTGCCTTCTTACCATGAGCTAGAGGGTTTTCTGCATTTGGTGTTTTGAT


181 CACCTTCAGCGTACCACCATCATCAATCAGCCAAGACCACCCACTGCCAAATTGAGTTGC
     >>>>>>>>>>>>>>>>>>>>

241 AGCTGCGTTAGAGAACTCTTCCTTGAATTTGTCGAAGCTACCAAAATCTTTATCAATCCT


301 GGCTGCAAGATCGCCAGTGGGTGCGCCGCCACCATTTGGTTTTAAGGAATTCCAGAAGAA


361 GGTGTGGTTCCAAACTTGAGCAGCGTTGTTGAAGATTCCCGCCTTAGAGGAGTCTTGGAA


421 GGAAATTTTGATCACCTCTTCCAGAGGCTTATCAGCAAGTTCTGTACCTTCAGTGAGCTT


481 GTTGAGGTTGTCTACATAAGCTTTGTGATGCTTGCCATAGTGATACTCGAAAGTTTCAGC
                       <<<<<<<<<<<<<<<<<<<<

541 TTTCATGCCATTTGGCTCTAAAGCATTCTTGTCGAAGGGTAGCGGGGGCTGTACAAATAC
```

**Fig. 1.9** Parameters and a map of the best oligonucleotide pair. '>>>>>>' indicates left primer and '<<<<<<' right primer binding positions

```
ADDITIONAL OLIGOS
                    start  len     tm     gc%    any    3'  seq

 1 LEFT PRIMER        182   20   60.00   55.00  4.00   0.00 ACCTTCAGCGTACCACCATC
   RIGHT PRIMER       559   20   60.21   45.00  6.00   0.00 AGAGCCAAATGGCATGAAAG
   PRODUCT SIZE: 378, PAIR ANY COMPL: 4.00, PAIR 3' COMPL: 1.00

 2 LEFT PRIMER        155   20   59.84   45.00  4.00   0.00 GGGTTTTCTGCATTTGGTGT
   RIGHT PRIMER       519   20   59.84   45.00  3.00   2.00 TATGGCAAGCATCACAAAGC
   PRODUCT SIZE: 365, PAIR ANY COMPL: 4.00, PAIR 3' COMPL: 3.00

 3 LEFT PRIMER        200   20   60.20   45.00  2.00   0.00 TCATCAATCAGCCAAGACCA
   RIGHT PRIMER       519   20   59.84   45.00  3.00   2.00 TATGGCAAGCATCACAAAGC
   PRODUCT SIZE: 320, PAIR ANY COMPL: 4.00, PAIR 3' COMPL: 0.00

 4 LEFT PRIMER        107   20   60.21   45.00  4.00   1.00 TGTTCCCAAACATCCAAGGT
   RIGHT PRIMER       411   20   60.17   50.00  4.00   2.00 TCCTCTAAGGCGGGAATCTT
   PRODUCT SIZE: 305, PAIR ANY COMPL: 5.00, PAIR 3' COMPL: 3.00
```

**Fig. 1.10** Additional primers with their features and sequence

### Exercise

1. What are the characteristics of a primer?
2. What is the minimum product size desirable for a conventional PCR?
3. What is a melting temperature?
4. How annealing temperature for a PCR reaction can be calculated?

## 1.3 Designing Degenerate Primers

### Introduction

Degenerate primers are helpful for fishing out a part of similar genes from a variety of species or when only the protein sequence of a gene is known. For more distantly related organisms it is more difficult to design primers.

Degenerate primers can be designed by gathering sequences from a wide range of organisms and aligning the translated amino acid sequence. Based on the alignments highly conserved amino acid can be identified which can then be exploited for designing degenerate primers (Fig. 1.11).

Two important things to be kept in mind for designing degenerate primers are the product length and the degeneracy. To have a successful PCR amplification, there should be two conserved regions, preferably 6 or 7 AA long, for annealing of the forward and the reverse primers and reasonably close (around 400), though 200–600 bp should also be working. If the primers are less degenerate, they can be further apart also. The second thing is to determine which of the possible primers shows least degeneracy. Table 1.1 shows the degeneracy of all the amino acids.

While designing degenerate primers, it is better to avoid 6 fold sites (L, S and R) and increase the number of 1 or 2 fold sites in the region. This can be done by comparing the degeneracy of the possible primers. Degeneracy can be computed by multiplying the degeneracy of each of the contributing amino acids. For example, if you have a primer which matches the amino acid sequence **F H M T G**, this would correspond to a degeneracy of 2 * 2 * 1 * 4 * 4 = 64. Then weigh the factors of

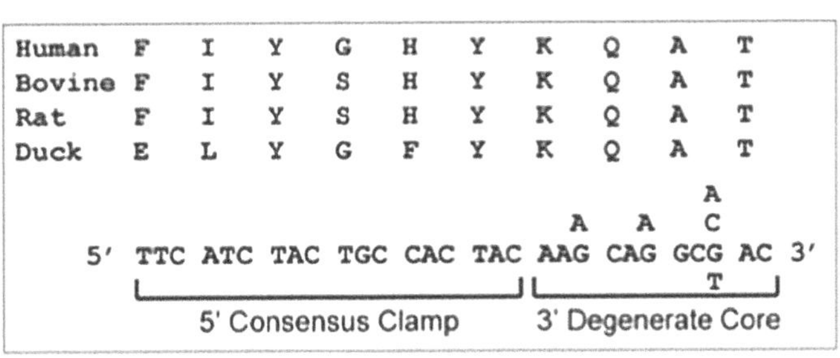


**Fig. 1.11** The sequence in degenerate primer designing

**Table 1.1** The amino acids and their corresponding number of codons

| Fold sites/number of codons | Amino acids |
|---|---|
| 1 | M W |
| 2 | F Y H Q N K D E C |
| 3 | I |
| 4 | V P T A G |
| 6 | L S R |

degeneracy and calculate the distance between the forward and reverse primers and select a pair.

Since Inosine can pair with any nucleotide, sites requiring complete degeneracy can be incorporated with Inosine. But before designing, check with the company if they can synthesize primers with it, as not all primer manufacturers use inosine. Avoid designing oligos with degeneracy at the 3′ and avoid oligos ending in inosine. Sometimes adding tails at the 5′ ends can increases the PCR efficiency as they increase the length and so the annealing temperature. Though this does not help in initial rounds of amplification, at later stages, they help in amplifying the short PCR products.

## Exploration Activity

1. D-Primer degenerate primer designing tool can be accessed from http://202.141.12.147/DGEN_tool/index.html (Fig. 1.12). The program is also available for download from http://202.141.12.147/DGEN_tool/Download.html so as to use it as standalone.

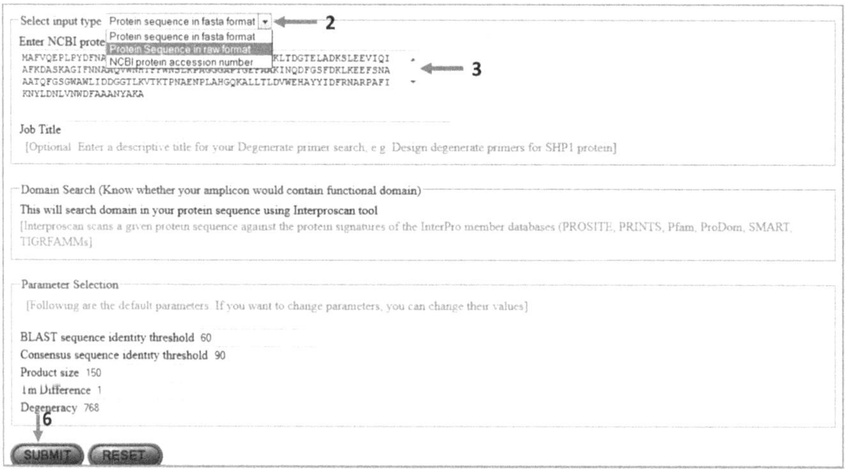


**Fig. 1.12** The D-Primer interface developed by Bioinformatics Lab, NBPGR, ICAR, New Delhi, India

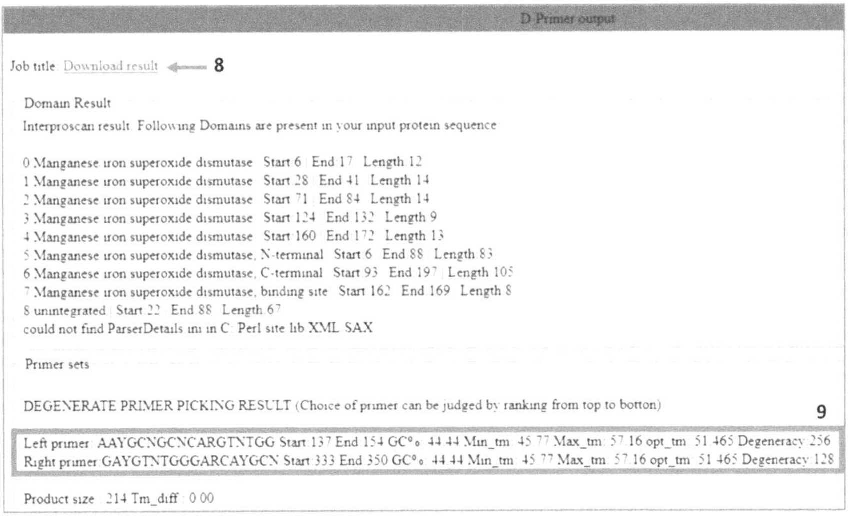


**Fig. 1.13** The main results page of D-Primer

2. The program accepts protein sequence in FASTA format, raw format or as NCBI's accession number. Select the type of your sequence you wish to give as input from the dropdown menu under 'Select input type' (Fig. 1.13).
3. Paste your sequence in the sequence window. FeSOD sequence from Nostoc sp. strain PCC 7120 (**Q8YSZ1**) has been given as input.
4. Give a title under '**Job Title**' (optional).
5. Leave the other options as default. If the user wishes to change the product size, Tm difference between primers or degeneracy, appropriate values can be entered.
6. Click the '**Submit**' button at the bottom of the page.

### Interpretation of Results

7. The program displays two results, (a) **Interproscan result**—displaying domains present in your input protein sequence, (b) **Degenerate primer list**—The choice of primer can be judged by ranking from top to bottom (Fig. 1.13).
8. The link on the top of the result allows the user to download the results (Fig. 1.13) including input sequence (seq.txt), domains (domain.txt), visualization of domain as image (domain.visual.jpg), BLAST output (BLAST_OUT.txt), primer list (Primers.txt) (Fig. 1.14) and tree (tree.eps).
9. If no primers are found with the specified parameters, the program displays message "**No Primers detected**. **Change parameters**". Change the product size, Tm difference and/or degeneracy and look if the program can give any primers or not.

```
Job title:
----------------------------------------------------------------------------------------------------

Left primer: AAYGCNGCNCARGTNTGG  Start:137  End:154  GC%: 44.44  Min_tm: 45.77  Max_tm: 57.16  opt_tm: 51.465  Degeneracy:256
Right primer:GAYGTNTGGGARCAYGCN  Start:333  End:350  GC%: 44.44  Min_tm: 45.77  Max_tm: 57.16  opt_tm: 51.465  Degeneracy:128

Product size : 214  Tm_diff : 0.00

----------------------------------------------------------------------------------------------------

Left primer: GCNGCNCARGTNTGGAAY  Start:140  End:157  GC%: 44.44  Min_tm: 45.77  Max_tm: 57.16  opt_tm: 51.465  Degeneracy:256
Right primer:GAYGTNTGGGARCAYGCN  Start:333  End:350  GC%: 44.44  Min_tm: 45.77  Max_tm: 57.16  opt_tm: 51.465  Degeneracy:128

Product size : 211  Tm_diff : 0.00

----------------------------------------------------------------------------------------------------

Left primer: GCNCARGTNTGGAAYCAY  Start:143  End:160  GC%: 38.89  Min_tm: 43.49  Max_tm: 54.88  opt_tm: 49.185  Degeneracy:128
Right primer:GTNTGGGARCAYGCNTAY  Start:336  End:353  GC%: 38.89  Min_tm: 43.49  Max_tm: 54.88  opt_tm: 49.185  Degeneracy:128

Product size : 211  Tm_diff : 0.00

----------------------------------------------------------------------------------------------------
```

**Fig. 1.14** The list of primers displayed in Primer.txt file

10. The program suggests list of degenerate primers that has been designed for the given sequence along with the information like start and end position, GC content, melting temperature and degeneracy.
11. The key to symbols that are used in primers are as follows:

| | |
|---|---|
| R = A/G | Y = C/T |
| M = A/C | K = G/T |
| S = G/C | W = A/T |
| H = A/T/C | D = G/A/T |
| B = G/T/C | V = G/A/C |
| N = A/T/G/C | |

### Exercise

1. What is a degenerate primer?
2. How a degenerate primer is different from a conventional primer?
3. What is the importance of fold sites in designing degenerate primers?
4. On what circumstances, one should design a degenerate primer?

## 1.4 Reading of Sequence Trace Files Using Finch TV

### Introduction

Sequence trace files called electropherogram contains the raw data output from automated sequencing instruments which can be visualized using variety of free DNA sequence viewers available over internet. Besides visualization, the program has features like BLAST search and reverse complement the sequences.

Geospiza's FinchTV, a electropherogram viewer, runs on Linux, Mac, Windows and Solaris. It is downloadable from the URL http://www.geospiza.com/Products/finchtv.shtml. It opens electropherogram files from Applied Biosystems, Amersham MegaBace DNA sequencers as well as SCF format electropherogram. The first part of the exercise deals with the visualization of electropherogram file using FinchTV.

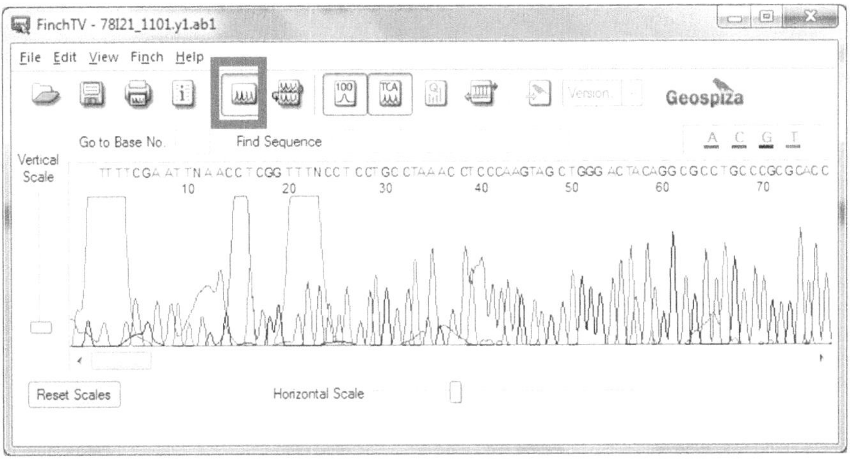


**Fig. 1.15** Standard View

## Exploration Activity

1. FinchTV can be launched by clicking on the application icon in your computer.
2. An electropherogram can be opened by selecting **File > Open** from the menu.
3. Select the electropherogram file to be viewed from your computer.
4. A electropherogram file can be viewed as **Standard View** or **Wrapped View**. The Standard View shows the trace as one continuous strip, scrollable horizontally (Fig. 1.15). The Wrapped View shows a multi-panelled electropherogram, scrollable vertically (Fig. 1.16). Select **View > Standard View** or **View > Wrapped View** from the menu bar.
5. In order to examine smaller peaks easily or eliminating the overly large peaks, the trace peaks can be scaled vertically or horizontally (see boxed), allowing better visualization of compressed areas, or in order to get an overall view of the trace all in one screen. This can be done by using the **vertical** or **horizontal scale** bars (Fig. 1.17).
6. Some electropherogram files contain quality information about the base calls known as quality value represented as **Q score**. The quality values depend on the electropherogram and it may be from the Phred, KB basecaller or from TraceTuner. If quality information is present in the electropherogram, FinchTV can display it. If this information is not available, you can see that the option is disabled. These values can be made to be displayed or hidden by selecting the **Quality Values** button in the toolbar or by choosing **View > Quality Values** from the main menu.
7. A light blue (aqua-coloured) dotted horizontal line represent the value of Q = 20 and any bar found above this line shows that that the base call have a better than 99% probability of being correct (Table 1.2). To see the score of a particular base, click on it. The score will be displayed (Q52) in the status bar in the lower left of the main FinchTV window (Fig. 1.18a).

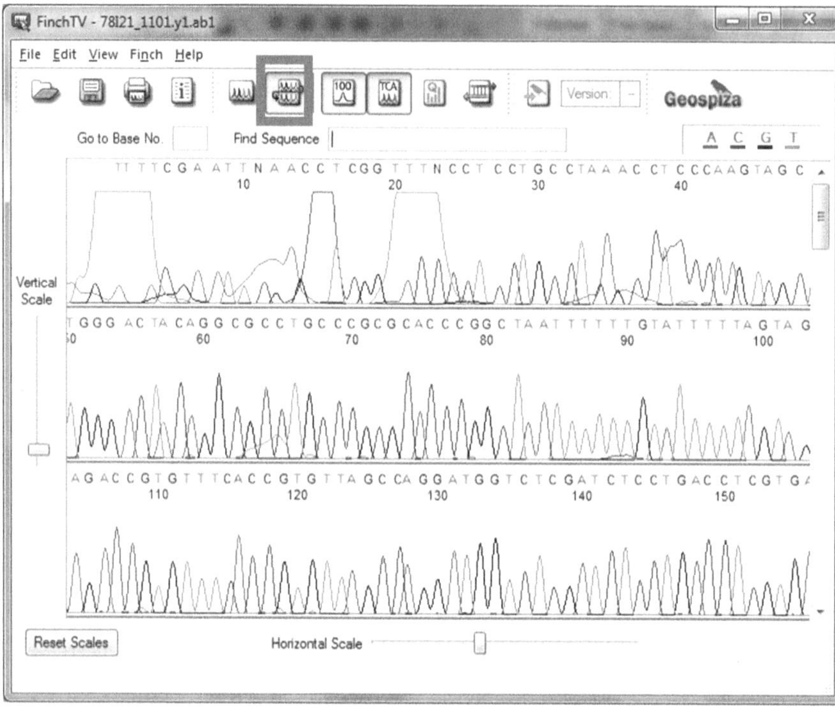


Fig. 1.16 Wrapped View

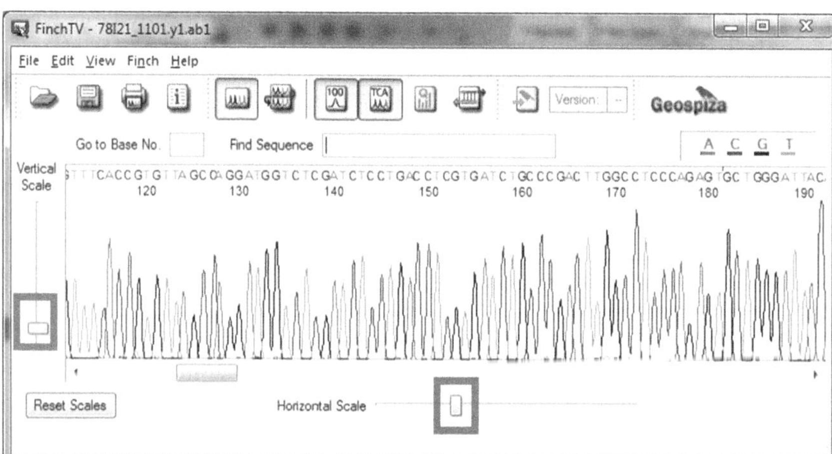


Fig. 1.17 Vertical and horizontal scale bars for better visualization

**Table 1.2** Phred quality scores are logarithmically linked to error probabilities

| Phred quality score | Probability of incorrect base call | Base call accuracy (%) |
|---|---|---|
| 10 | 1 in 10 | 90 |
| 20 | 1 in 100 | 99 |
| 30 | 1 in 1000 | 99.9 |
| 40 | 1 in 10,000 | 99.99 |
| 50 | 1 in 100,000 | 99.999 |

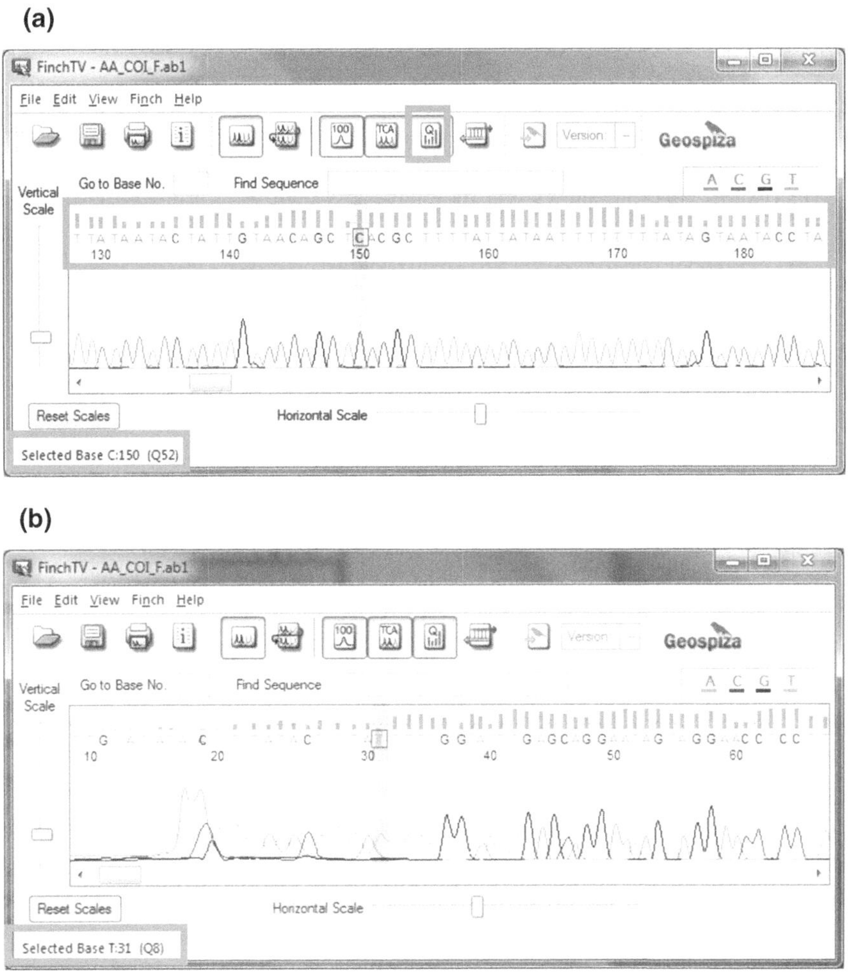


**Fig. 1.18** **a** Quality information of electropherogram file based on Q score. **b** A selected base showing a low Q score

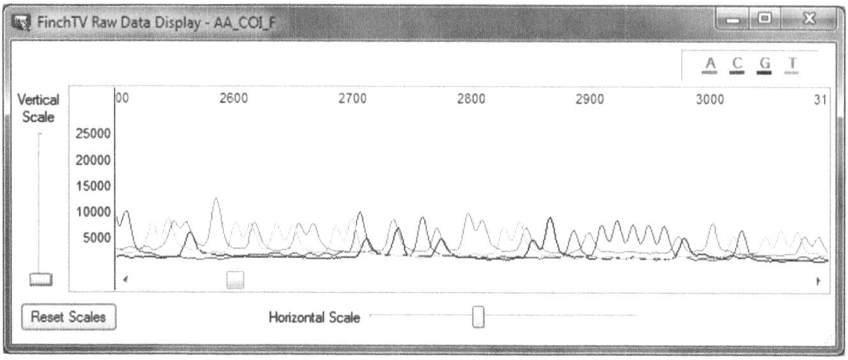


**Fig. 1.19** The raw data display

8. Bars below this line shows that they have less than 99% probability of being called correctly (Fig. 1.18b).
9. ABI formatted electropherogram usually comes with raw, pre-processed, trace data. This information is particularly useful in troubleshooting sequencing instruments. The raw trace data can be seen by clicking **View > Raw Data** from the main menu which pop up a new window showing the data (Fig. 1.19).

**Exercise**

1. What is a electropherogram?
2. What are the formats files supported by FinchTV?
3. What is meant by raw data? How it is different from the processed data?
4. What is the importance of Q score?
5. How will you interpret the phred quality score?

## 1.5 Troubleshooting DNA Sequencing Problems

**Introduction**

Sequence databases that are publically accessible, stores and distributes information in support of research. It must be accurate and reliable so as to be useful. But errors are quiet common, like in the format, or in the annotation or may be in the sequence itself. Though other errors like in the format are very less, mistakes occur in DNA or protein sequences are very common. In most nucleotide sequence databases sequence data and annotations can only be modified by the researchers who have originally submitted the sequence. Such errors have serious impact on subsequent analyses.

As the number of sequences in public sequence databases is going to increase in the future in a drastic manner, and so are the errors. Since these databases form the basis for research not only in bioinformatics and genomics, testing the strength of these foundations in extracting useful information become even more difficult.

**Troubleshooting DNA Sequencing Problems**
DNA sequencing has become one of the most common techniques done in molecular biological laboratories. It becomes more difficult to identify where it went wrong. But there are only few reasons for such failures in DNA sequencing. This helps in identifying the most common causes of various sequencing problems. It also shows how to overcome each problem type and provides tips for improving DNA sequencing quality. Though raw file show clearer picture of the problems encountered, the normalized files are taken into consideration for the purpose of following trouble shooting.

### Identifying Common Problems

#### Multiple Overlapping Peaks—Mixed Template

In case multiple colonies are picked or two or more inserts present in the cloning vector, multiple peaks with the same height or of differing heights, overlapping one another may be observed (Fig. 1.20). This may be due to:

- Clone contamination,
- Heterozygous PCR template due to indels present in a diploid (or polyploid) organism or
- Impurities in the PCR products.

#### Multiple Binding

Multiple binding results in two or more peaks with good signal strength (peaks at same height) appear at different locations (Fig. 1.21). This is observed when

- More than two templates gets mixed or
- More than two binding site for primers are present in DNA template.

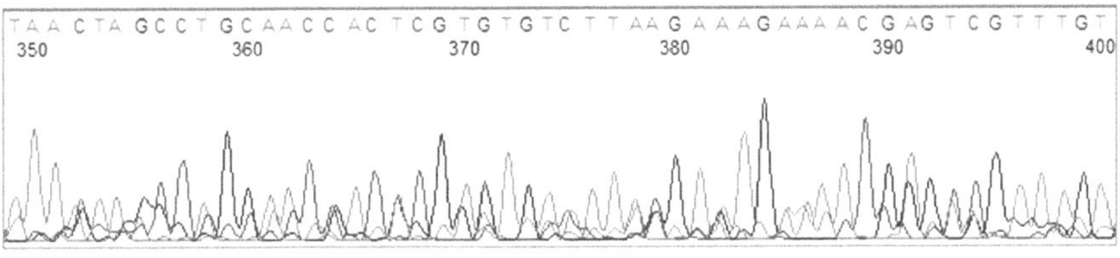


**Fig. 1.20** Multiple overlapping peaks due to mixed template

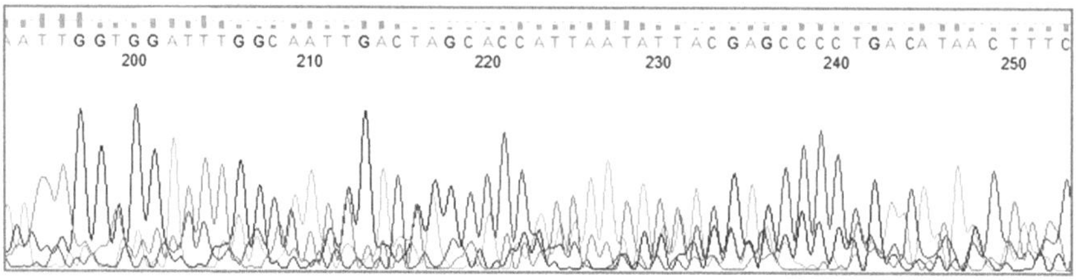


**Fig. 1.21** Contamination due to presence of two or more priming sites in DNA template

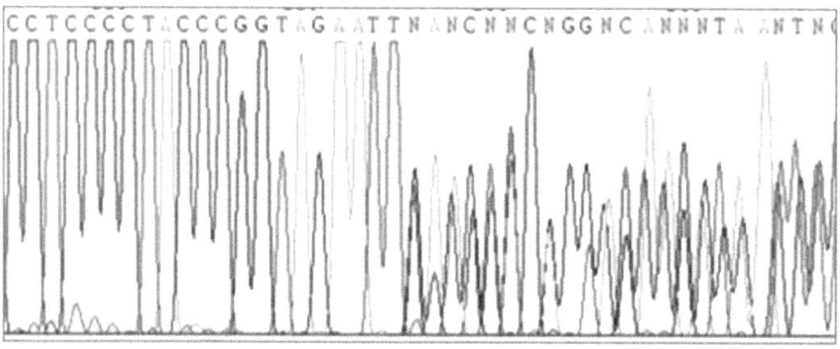


**Fig. 1.22** Compression due to secondary structure

### Compression Due to Secondary Structure

When there is a probable secondary structure formation in the fragments, fragments of different lengths show similar electrophoretic mobility. This results in the observation of multiple peaks after a particular point (Fig. 1.22). Sometimes the peaks are also compressed which may occur due to

- High G/C or A/T content regions.

### Mixed Base

Two overlapping peaks were observed but at height half to the prior peak (Fig. 1.23). This might be caused due to

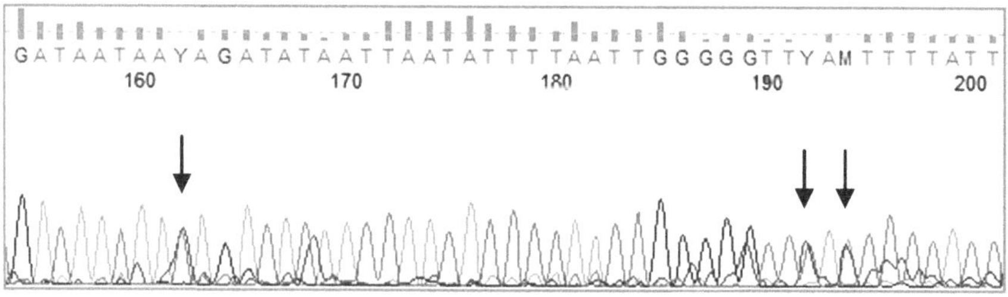


**Fig. 1.23** Mixed base

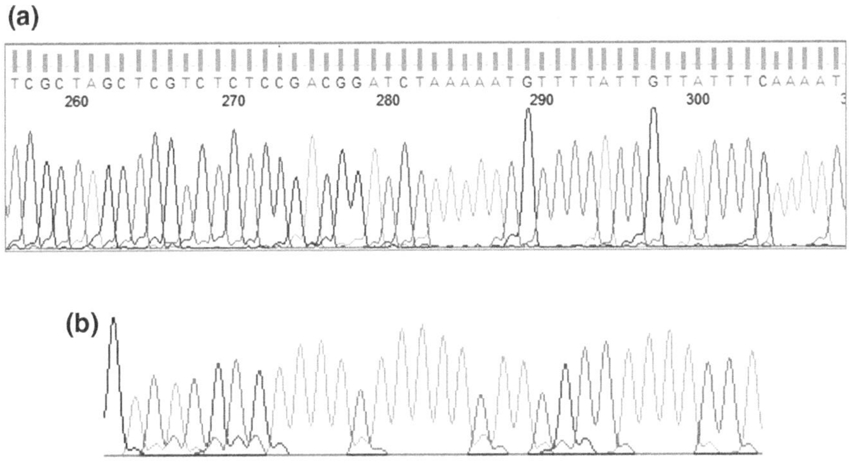


**Fig. 1.24** **a** N − 1 primer degradation. **b** N + 1 primer degradation

- A single nucleotide polymorphism (SNP) or due to
- Point mutation.

**Primer Degradation: N − 1 Primer**

From the beginning in sequencing data file, we can see low peaks appearing under the major peaks (Fig. 1.24). This is due to primer degradation. You can overcome this issue by re-synthesizing the primers. This is usually caused by,

- Presence of any unpurified byproduct during primer synthesis or
- Might be due to degradation of primer.

**N + 1 Primer**

This is possible when there is a problem in the synthesis of the primer itself. In sequencing data, low peaks will be following the major peak.

**Frameshift Mutation**

This is similar to N − 1 primer issue but minor peaks appear under the major peaks only from a specific point (Fig. 1.25). The lower peaks appear in the electropherogram only after a frame-shift has occurred.

**Abrupt Signal Loss**

Abrupt signal loss means a sudden drop in DNA sequence signals (Fig. 1.26). This may be caused by

- Presence of strong secondary structure in template DNA in that point.
- GC or AT rich regions.

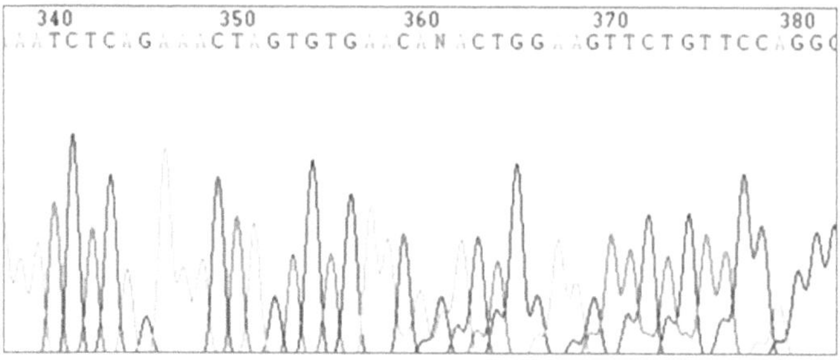

**Fig. 1.25** Frameshift mutation

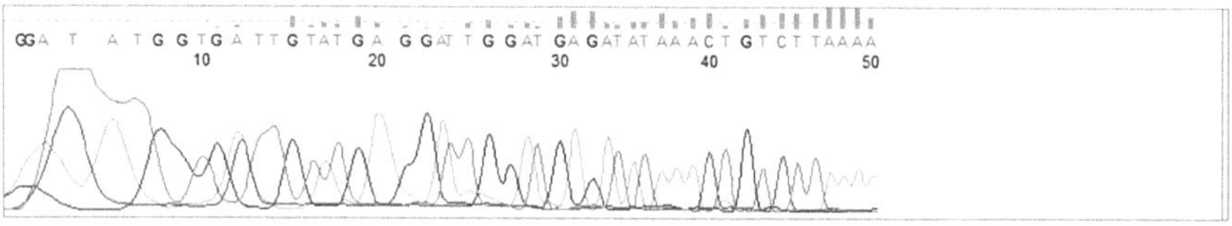

**Fig. 1.26** Abrupt signal loss

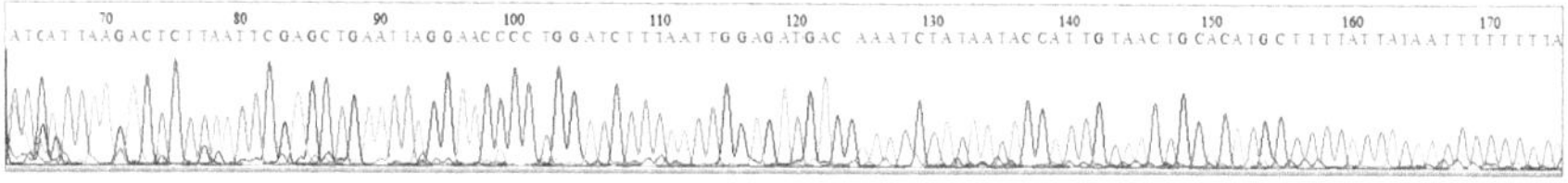

**Fig. 1.27** Gradual signal loss

### Gradual Signal Loss

Sequencing signal gradually drops off (Fig. 1.27). This may be due to:

- High concentrations of DNA template or primer used
- Template having GC or AT content

### Broad or Oddly Shaped Peaks

Broad or unusually shaped peaks on beginning of the sequence (Fig. 1.28).

- Possibility of salt or other impurities that might be carried over from DNA isolation procedure, especially when homemade protocols were followed

### Spikes or Dye blobs

A high-intensity, sharp and overlapping peaks that appear randomly in the sequence file (Fig. 1.29). These generate very strong spikes.

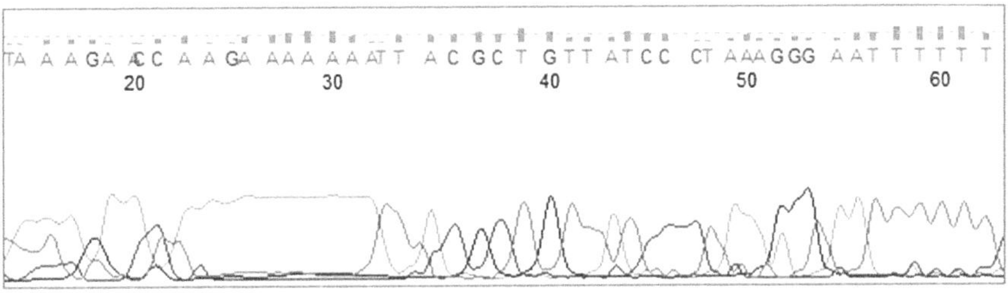


**Fig. 1.28** Broad or oddly shaped peaks

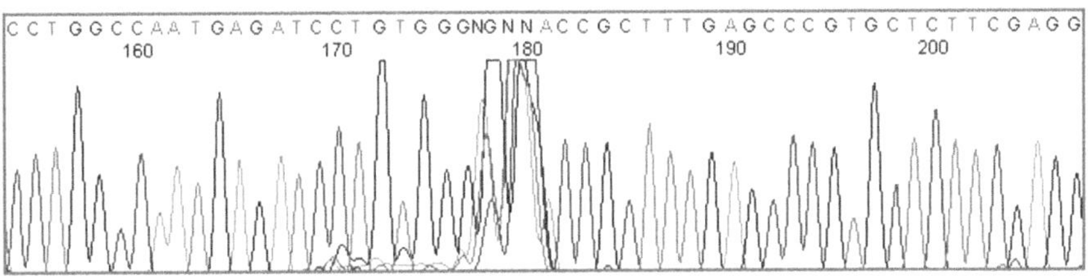


**Fig. 1.29** Dye blob resulting in spikes

One possible reason could be due to the:

- Impurity or due to excess unincorporated dyes in the reaction
- Micro bubble that passes through the detector system of the sequencer.

## An Ideal Sequence File

An ideal sequence trace file display clear peaks, with good Q score, ideally spaced without any overlap (Fig. 1.30). If these are not observed in your sequence trace file, prefer re-sequencing.

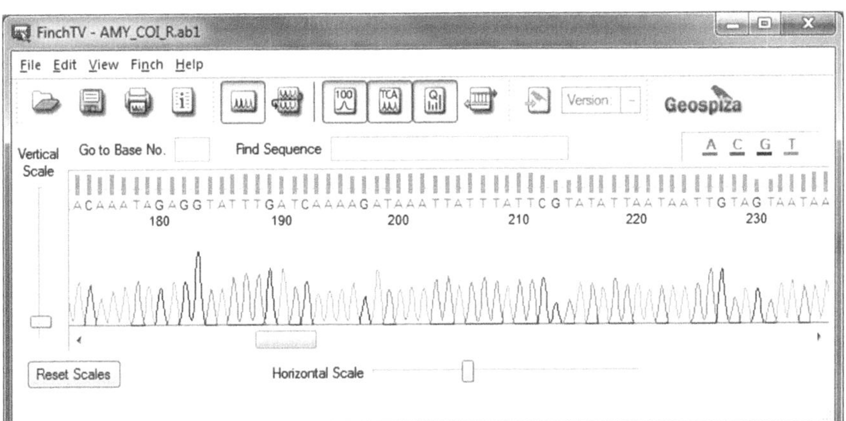


**Fig. 1.30** An ideal sequence file

**Exercise**

1. How contamination affects sequencing?
2. What are the common errors encountered during sequencing and how will you prevent that?
3. In what cases, one would prefer re-sequencing of the sample?

## 1.6 Editing Sequence Data

**Introduction**

Ambiguity at certain sites along the DNA sequence can be when viewing electropherogram. This happens when the sequencing machine cannot precisely determine what the nucleotide is. This is usually represented by either a broad peak or a set of overlaying peaks and the bases are commonly represented as N. In such cases the sequencing requires careful proofreading and careful evaluation of the quality of the sequence as it can give an idea whether it needs re-sequencing or have both a forward and reverse sequencing reaction to be done.

**Exploration Activity**

1. Editing options offers us with replacing **X**'s or **N**'s, inserting or deleting bases single bases with any IUPAC nucleotide designation or even changing regions of bases to uppercase or lowercase (Fig. 1.31).
2. For editing single bases, click on the base requires editing and type the new base. The new base entered appears in **Bold** font, **underlined** and is **coloured** appropriately. Bases that are non-ACGT appear black.

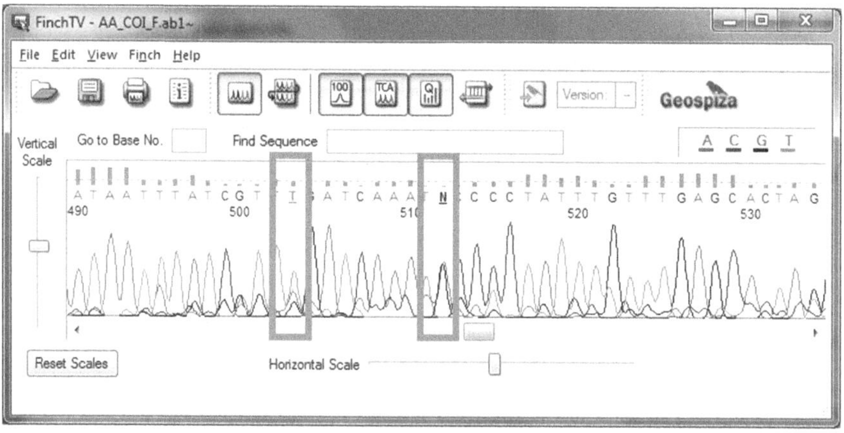


**Fig. 1.31** Visualization of edited and non-ACGT bases

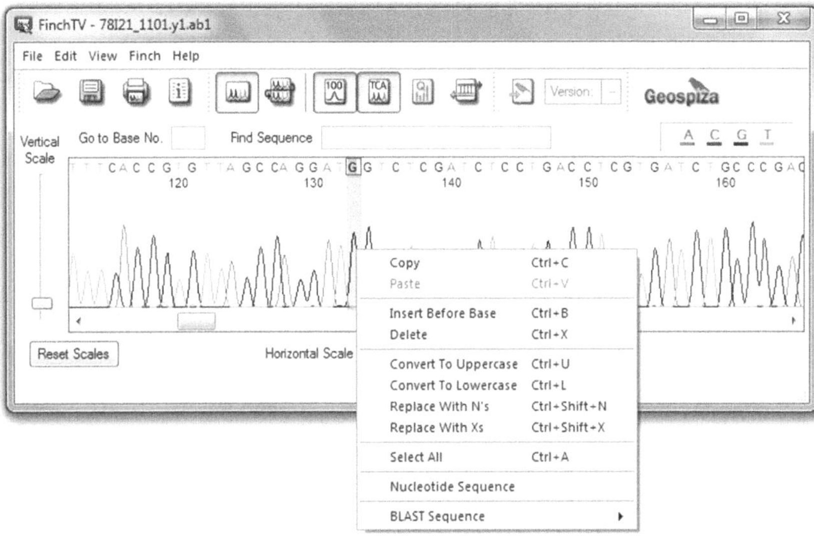


**Fig. 1.32** Contextual menu of a selected base

3. For editing a range of bases, select the group of bases to be edited, right click your mouse button and select appropriate options like **Convert to Uppercase**, **Convert to Lowercase**, **Replace with X's**, **Replace with N's** from **Contextual** menu (Fig. 1.32).
4. Base can be inserted by clicking on any single base and then right click your mouse button. It shows option like **Insert Before Base** and type the base. Newly inserted bases appear in bold font, underlined and also include a dot above it (Fig. 1.33). If trace file displays the quality, then a quality value (Q10) appears above the inserted base position.
5. Bases can also be deleted by selecting and clicking the **backspace** or **delete** keys from the keyboard or by selecting **Delete Base** from either the **Edit** or **Contextual** menu. A deleted base is usually indicated by "_" (underscore) character (Fig. 1.34). Though immediately quality value (if present) disappears, subsequent saving of the file removes it.
6. Undoing and redoing the edits can be done by selecting **Edit > Undo** (or pressing the key **ctrl Z)** and **Edit > Redo** (or pressing the key **ctrl Y)** from the menu, respectively.

**Navigating to a Desired Base Position**

7. The positions of a base are shown as numbers just above the trace with a default spacing of 10 peaks like 10, 20, 30, etc., (Fig. 1.35).
8. Users can move on to specific location of the base set by typing the position into the **Go to Base No**. box and pressing **return** or **enter** key (Fig. 1.35).

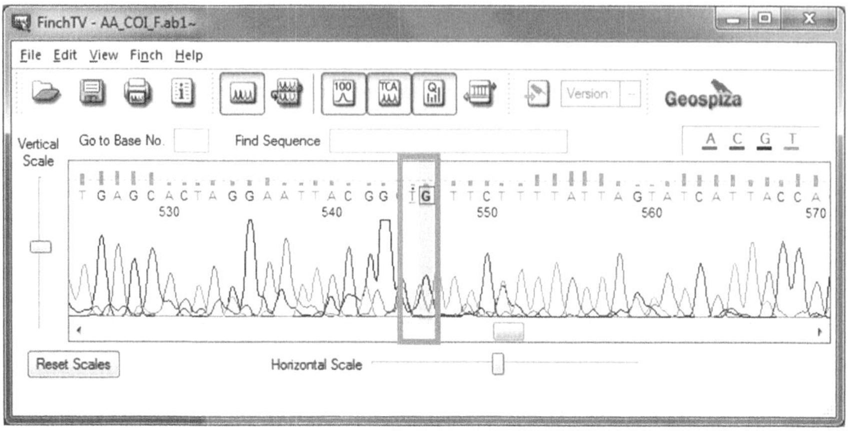


**Fig. 1.33** Insertion of base at a desired position

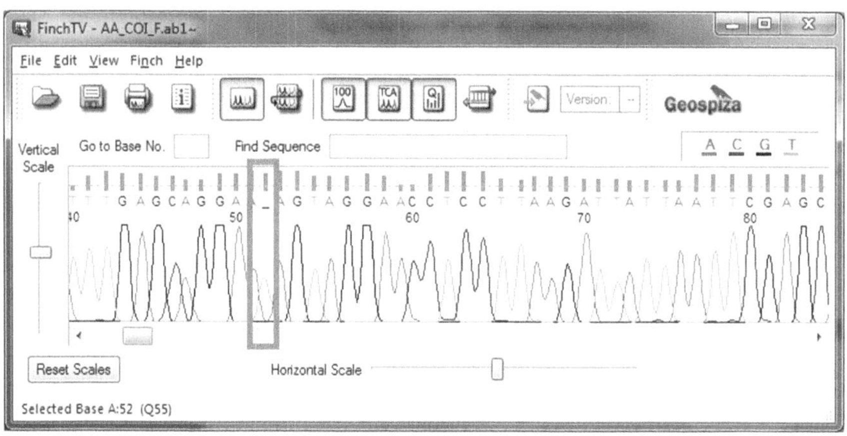


**Fig. 1.34** Deletion of base at a desired position indicated by '_' (underscore)

## Search for Sequence Pattern

9. A specific sequence pattern can also be searched by typing a sequence string into the **Find Sequence** box (Fig. 1.35) and pressing **return** or **enter** key in the keyboard. Subsequuent pressing of **return** or **enter** key, displays the next matching sequence pattern.
10. Beyond simple text matching, the program also supports regular expression also called regexes, that uses special characters like ".", "*", "|" for expanding the search terms and to match complicated queries. For example, if the tradce has any open reading frame it can be searched using the expression "ATG ((TAG|TAA|TGA) ....) + (TAG|TAA|TGA)".

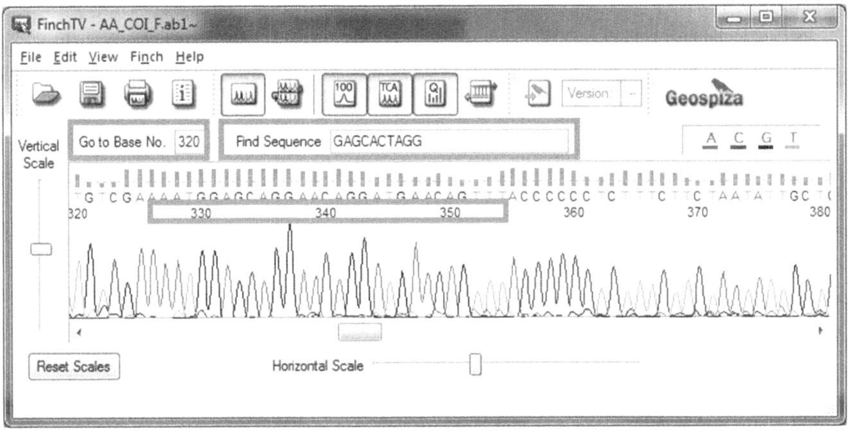


**Fig. 1.35** Navigating to desired position and searching sequence pattern

## BLAST Searches

BLAST searches from NCBI can be initiated by selecting a range of bases in the trace window and by selecting **Edit > BLAST Sequence** from the main menu (Fig. 1.36). If no region is selected complete sequence is considered for the search.

## Exporting Sequence Read in FASTA Sequence

1. FinchTV can export both a FASTA sequence and the reverse complement (In FinchTV 1.3, and above) of a sequence based on the electropherogram file.

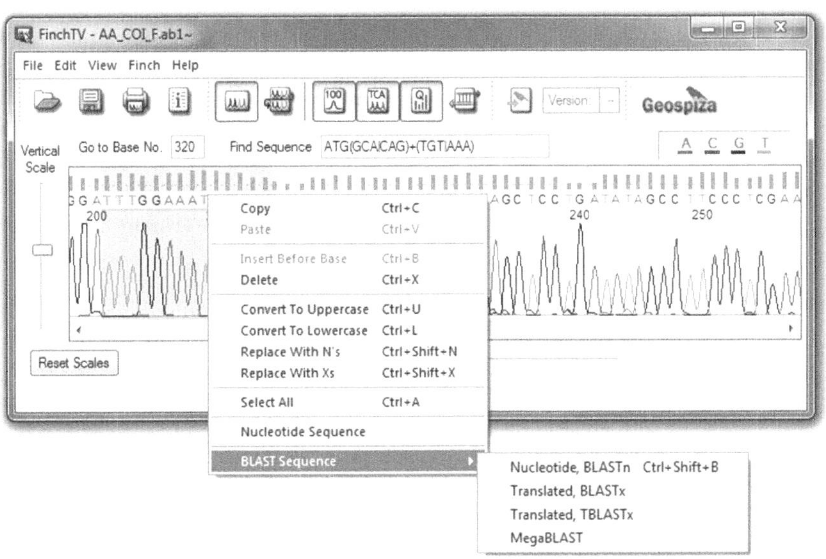


**Fig. 1.36** BLAST searching a sequence segment

2. To export a FASTA file, select **File > Export > DNA sequence**: **FASTA** from the main menu and specify a location for saving the file.
3. If you wish to export the sequence having bases of *vector* or *low quality* regions replaced with either X's or N's (version 1.4 with data from a Finch Server), you do so by selecting **File > Export > Export Options** … and selecting the appropriate options. The next time when a sequence is exported as a FASTA file, these settings will be used by the program.
4. If the reverse complement of the sequence is displayed, the FASTA sequence exported will be as the reverse compliment.

**Exercise**

1. How a Contextual Menu can be accessed and what are its components?
2. How will you export the sequence into FASTA format from FinchTV?
3. How will you go to a desired nucleotide position in the sequence trace file during editing?
4. What is SCF format?
5. What will happen to the Q score when a base is deleted?

## 1.7 Sequence Assembly—CAP3 Program

**Introduction**

Most of the DNA sequencing reactions produces approximately 800–900 base pairs of high-quality sequence data. But genes over 900 bp requires multiple sequencing reactions in order to obtain the complete sequence. In case of 16S rRNA gene, which is around 1550 bp, requires two to three sequencing reactions to cover the entire length of the gene. In such cases, the fragment of sequences requires careful assembly. The assembly process order the sequence and then aligns. This is then extended to a step called 'finishing' that involves checking and editing the assembled data. In case there are gaps due to poor coverage, new sequencing has to be done.

The (global) sequence assembly process involves three steps (Huang and Madan 1999):

Step 1 Find all possible overlaps between readings by comparing each one, in both orientations, to all the others;

Step 2 From the list of overlaps, produce the best layout of the readings;

Step 3 From the alignment of the readings in the final layout, derive a consensus sequence.

The first step is achieved in two stages which involves finding all the pairs of reading that shares exact match by 'rapid comparison' and then uses **dynamic**

**programming** methods for aligning the matches. Kececioglu and Myers (1995) have described the sequence assembly problems and algorithms.

**Sequence Assembly Using CAP3**
Contig assembly program ver. 3 from Pôle Bioinformatique Lyonnais (PBIL) is a web-based program used for assembling sequences.

**Exploration Activity**

**Prepare the Sequences**
The set of sequences that are to be assembled are derived in FASTA format in a .txt file or a Word document. Each individual sequence must be in FASTA format and have its own header.

1. CAP3 program can be accessed from http://pbil.univ-lyon1.fr/cap3.php.
   The form appears containing one sequence input window, a Submit button and a Clear button. The Lyon server allows only a total of 50 kb of nucleotide sequences to be processed (Fig. 1.37).

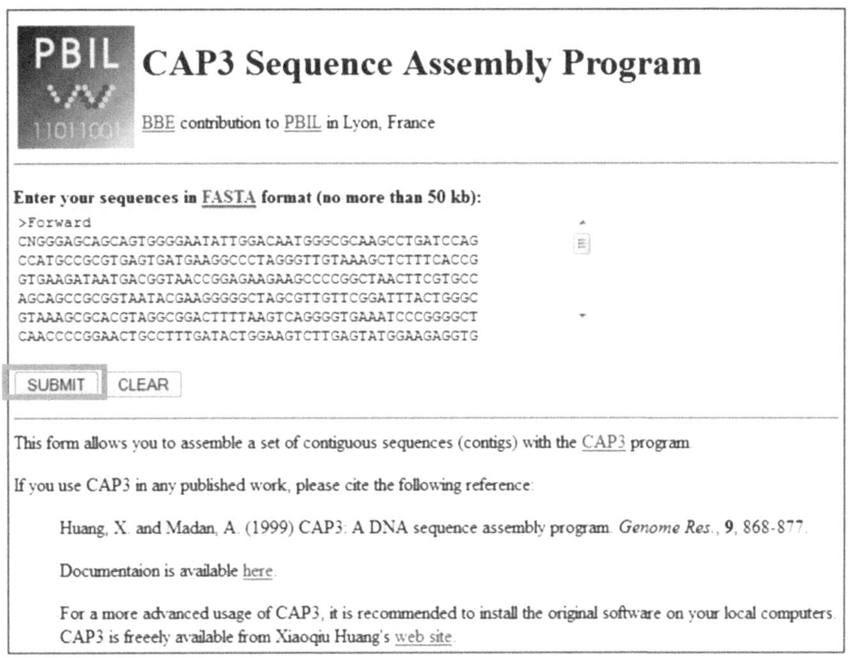


**Fig. 1.37** CAP3 interface at Pôle Bioinformatique Lyonnais (PBIL)

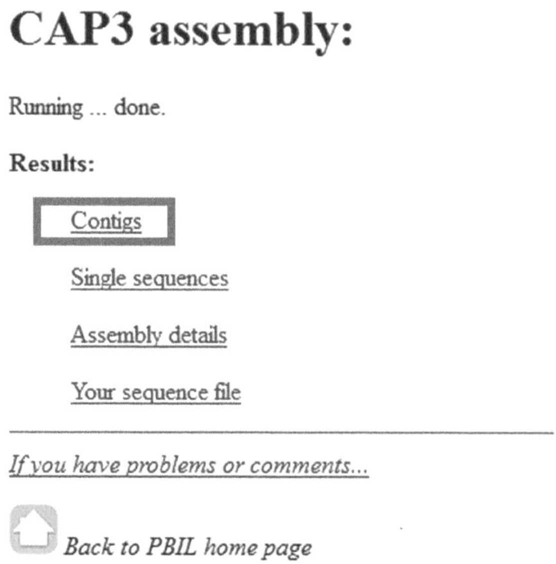


**Fig. 1.38** The output of CAP3

2. Paste the set of sequence into the Sequence input box.
   Even if the sequences are in different orientations, the program tries them in both orientations automatically (Fig. 1.37).
   1. Click the SUBMIT button to run the assembly.

**The Output**

The output will display four links (Fig. 1.38):

**Contigs**: Contains the final assembled sequence(s)

**Single Sequences**: Contains the input sequence fragment not incorporated in the assembly.

**Assembly Details**: The precise relationship between the fragments within the contigs.

**Your Sequence File**: A summary of the input sequence fragments.

These links can be clicked to access the contents.

**Interpreting the Results**

Clicking the 'Contigs' link displays contig consensus sequence (Fig. 1.39). If there is any problem in assembly, then the page will not display any result and will be blank. This happens when there is no significant overlap between the files used.

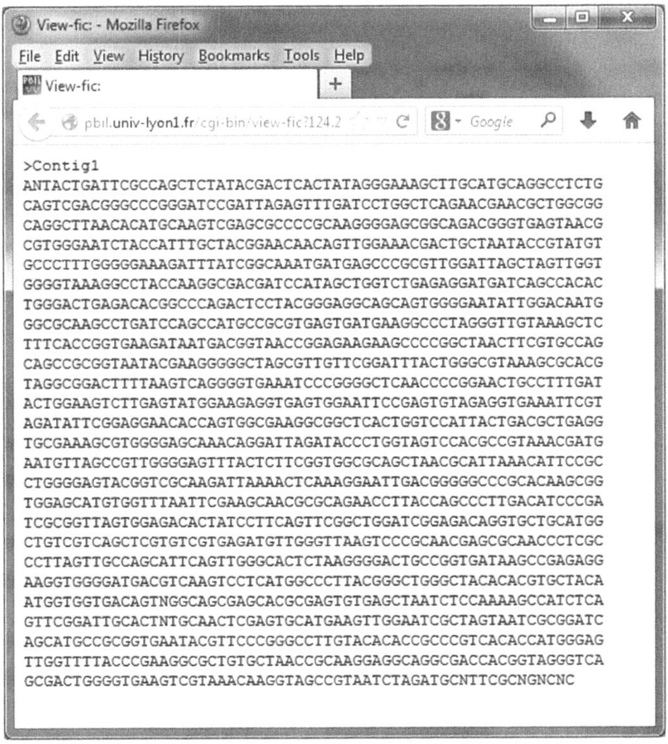


**Fig. 1.39** The contig generated by CAP3

Clicking the Assembly Details link displays the overlaps and contig structure (Fig. 1.40).

The '+' and '−' indicate whether the sequences are used as such or reverse complemented, respectively. The consensus sequence displayed is final assembled sequence. In case of any gaps found, sequence file has to be referred because the program always displays the consensus sequence alone.

In case of 'N's the program replaces with the corresponding nucleotide that is found overlapping (Fig. 1.41). If there is no overlapping region, then it retains the N as such.

## Exercise

1. What is a contig?
2. What is 'finishing'? Name the steps involved in it.
3. What do you mean by '+' and '−' strands?
4. What is the importance of 'Assembly details'?
5. What happens in a consensus sequence when one of the sequences has a gap?

```
Number of segment pairs = 6; number of pairwise comparisons = 3
'+' means given segment; '-' means reverse complement

Overlaps            Containments  No. of Constraints Supporting Overlap

******************* Contig 1 ********************
Reverse+
Middle+
Forward-

DETAILED DISPLAY OF CONTIGS
******************* Contig 1 ********************
                      .    :    .    :    .    :    .    :    .    :    .    :
Reverse+             ANTACTGATTCGCCAGCTCTATACGACTCACTATAGGGAAAGCTTGCATGCAGGCCTCTG
                     ____________________________________________________________
consensus            ANTACTGATTCGCCAGCTCTATACGACTCACTATAGGGAAAGCTTGCATGCAGGCCTCTG

                      .    :    .    :    .    :    .    :    .    :    .    :
Reverse+             CAGTCGACGGGCCCGGGATCCGATTAGAGTTTGATCCTGGCTCAGAACGAACGCTGGCGG
                     ____________________________________________________________
consensus            CAGTCGACGGGCCCGGGATCCGATTAGAGTTTGATCCTGGCTCAGAACGAACGCTGGCGG

                      .    :    .    :    .    :    .    :    .    :    .    :
Reverse+             CAGGCTTAACACATGCAAGTCGAGCGCCCCGCAAGGGGAGCGGCAGACGGGTGAGTAACG
                     ____________________________________________________________
consensus            CAGGCTTAACACATGCAAGTCGAGCGCCCCGCAAGGGGAGCGGCAGACGGGTGAGTAACG

                      .    :    .    :    .    :    .    :    .    :    .    :
Reverse+             CGTGGGAATCTACCATTTGCTACGGAACAACAGTTGGAAACGACTGCTAATACCGTATGT
                     ____________________________________________________________
consensus            CGTGGGAATCTACCATTTGCTACGGAACAACAGTTGGAAACGACTGCTAATACCGTATGT

                      .    :    .    :    .    :    .    :    .    :    .    :
Reverse+             GCCCTTTGGGGGAAAGATTTATCGGCAAATGATGAGCCCGCGTTGGATTAGCTAGTTGGT
                     ____________________________________________________________
consensus            GCCCTTTGGGGGAAAGATTTATCGGCAAATGATGAGCCCGCGTTGGATTAGCTAGTTGGT
```

Fig. 1.40 Assembly details of the contig

```
      .    :    .    :    .    :    .    :    .    :    .    :
GGCGCAAGCCTGATCCAGCCATGCCGCGTGAGTGATGAAGGCCCTAGGGTTGTAAAGCTC
GGCGCA-GCCTGANCCAGCCATGCCGCGTGAGTGATGAAGGCCCTAGGGTTGTAAAGCTC
____________________________________________________________
GGCGCAAGCCTGATCCAGCCATGCCGCGTGAGTGATGAAGGCCCTAGGGTTGTAAAGCTC
```

Fig. 1.41 The effect gap and 'N' in the consensus sequence

## 1.8 Checking for Vector Contamination

**Introduction**

When DNA/RNA inserted into a cloning vector for sequencing purposes, the raw sequences include segments derived from those vectors, or transposable elements from the cloning host (generally bacteria or yeast) contaminating the finished sequence. Theses pose serious problems unless they are identified and removed.

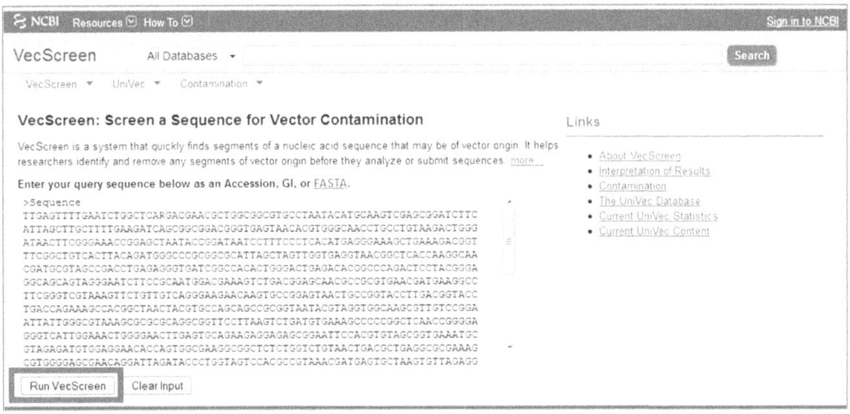


Fig. 1.42 VecScreen interface at NCBI

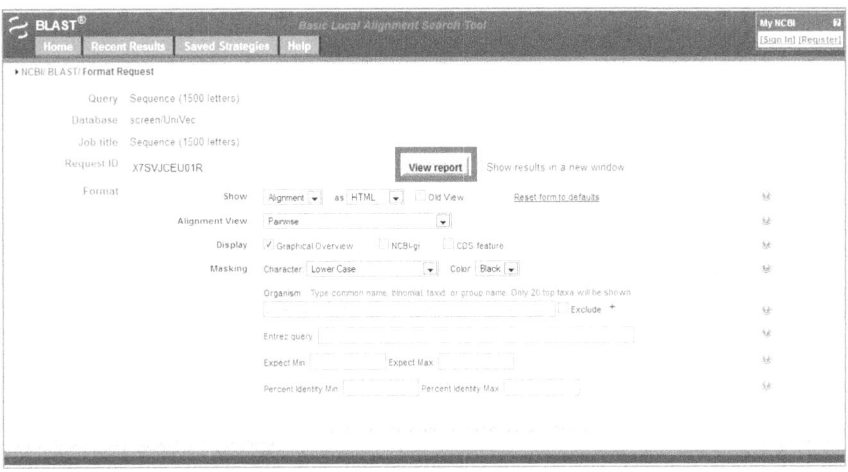


Fig. 1.43 BLAST output formatting options

## Exploration Activity

- Go to NCBI's VecScreen server page (http://www.ncbi.nlm.nih.gov/tools/vecscreen/)
- Copy and paste the sequence of the contig you generated earlier in the box.
- Click the 'Run VecScreen' button below the box (Fig. 1.42).
- You are provided with an intermediate page called 'Format Request'. Click 'View Report' button (Fig. 1.43).
- You can see red colour patch on the sequence if your sequence is contaminated (Fig. 1.44).

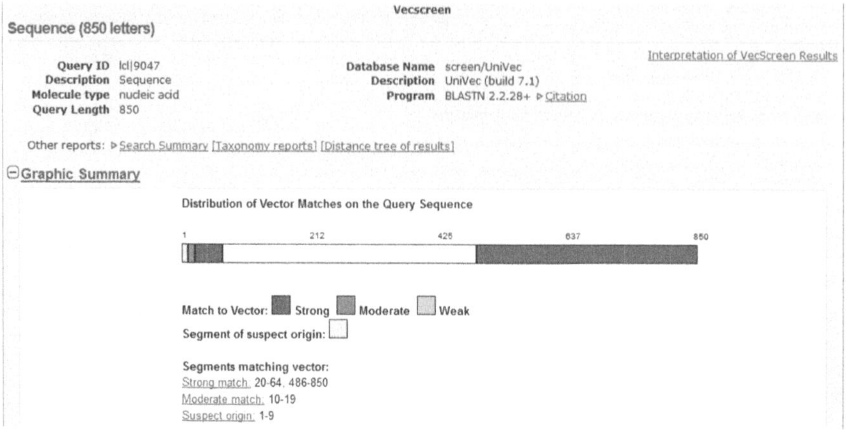


**Fig. 1.44** Vector match distribution on the query sequence

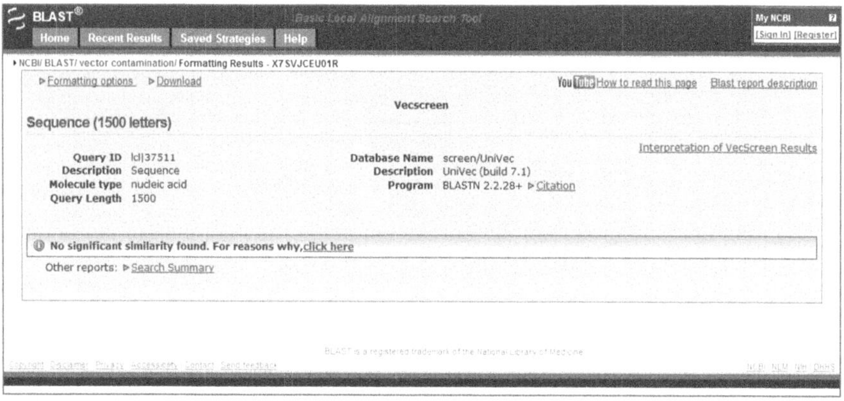


**Fig. 1.45** Query sequence without vector contamination

- If you get any such result look for the information 'Strong match: 20–64, 486–850' (Fig. 1.44) under 'Segments matching vector' for the location of the contamination.
- Once the location is identified then the segment can be removed manually.
- If there is no contamination, you might get message 'No significant similarity found' (Fig. 1.45).

## Exercise

1. What is the purpose of using VecScreen?
2. What will you do when your sequence trace file is contaminated with vector sequence?

3. What is the basis to remove vector sequence from the sequence trace file after performing VecScreen?
4. Name the organization hosting VecScreen.

## 1.9 Restriction Mapping Using NEBcutter

**Introduction**
A restriction map describes the location of cleavage sites of restriction endonuclease within a DNA fragment. Restriction enzymes are an enzyme that recognize a specific sequence in a DNA molecule and cuts the DNA strand. It is of two types:

| | |
|---|---|
| Type I | Recognizes specific sequence, but cuts elsewhere |
| Type II | Cuts only within the recognition site |
| Type III | Recognizes two separate non-palindromic sequences that are inversely oriented and they cut DNA about 20–30 base pairs after the recognition site |

Based on arrangement of breaks, it is further classified as blunt end enzymes (E.g. HaeIII) and cohesive end enzymes (E.g. EcoR1). A particular RE generates a unique family of fragments from a particular DNA molecule. The very first step in characterizing an unknown DNA is generating a restriction map with 6 base cutter. The uncharacterized DNA is usually within the multiple cleavage sites (MCS) to facilitate mapping.

**Applications**

- Identifying the size of the insert
- Check the orientation of gene in cloning
- Mutation and Polymorphism studies
- Removing part of gene
- **To check whether the sequence you get back from sequencing lab/company corresponds to the piece of DNA you sent out**
- To check whether the sequence assembly is correct in case of shotgun sequencing of genomes.

**Using a Computer to Generate Restriction Maps**
To generate restriction maps using computer you need to have the DNA sequence of yours. Just feeding into any number of computer programs can generate restriction map at ease.

**NEBcutter**
NEBcutter is a free, online program that identifies all restriction sites in your query sequence. This program supports sequences of most formats and can generate restriction maps for the enzymes sold by NEB and other commercially available enzymes. The restriction maps can be saved as high-quality raster or vector images.

Apart from generating the digest, there is a provision for selecting the enzymes (on the enzyme selection page) and generate a virtual gel. Options to add DNA marker to the virtual gel has also been provided.

**Exploration Activity**

The objective of this activity is to check whether the sequence you got back from sequencing lab/company corresponds to the piece of DNA you sent for sequencing.

1. NEBcutter program can accessed from http://tools.neb.com/NEBcutter2/
2. **Paste** a nucleotide sequence for analysis. Nucleotide sequence of Mn and Fe superoxide dismutase from *Nostoc punctiforme* is given as input for restriction digestion analysis. You can cut and paste either plain DNA sequence or a FASTA formatted sequence from your local computer into the box or alternatively you can use the "Browse" button to select the files in your computer in '**Local sequence file**'. If you know the Genbank accession number for the sequence, you can directly enter under '**GenBank number**' If not, you can use the "Browse GenBank" button to search GenBank (Fig. 1.46).
3. Indicate whether the sequence is **linear or circular**. In case of plasmids, they are circular. If a GenBank file is used containing a circular sequence, NEBcutter will automatically recognize that.
4. Choose the **type of enzyme** to be used like 'which are available in NEB' or 'user defined'.

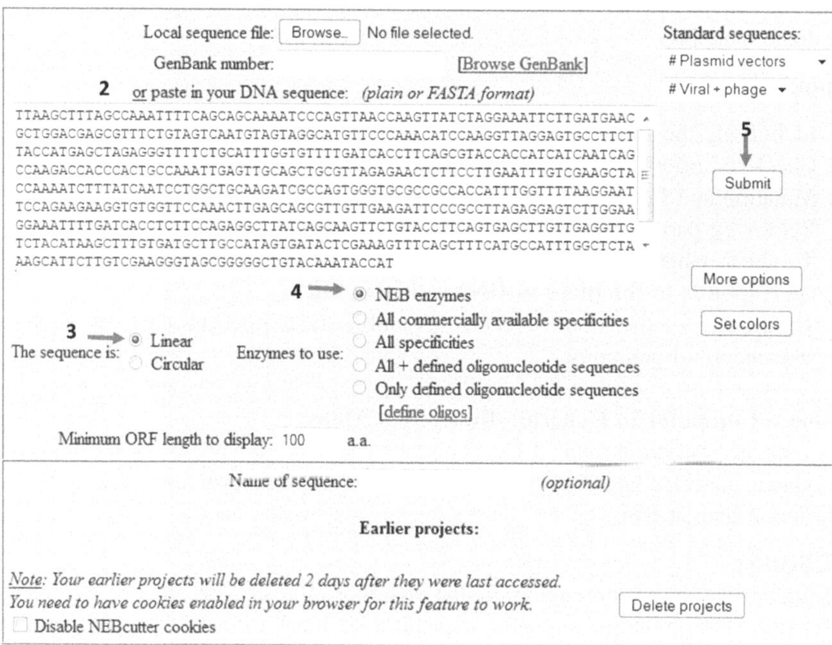


**Fig. 1.46** NEBcutter interface

5. Click on '**Submit**' button to analyze your sequence by NEBcutter.
6. Other options (Optional)

**Minimum ORF Length to Display**
The minimum length of ORF (open reading frame) that is to be shown can be entered. The program by default displays all ORFs with 100 or more amino acids.

**More Options**
Click on the '**More options**' button to see further options like ignoring the methylation sensitivity or using different genetic code for translation or even to specify a particular region within the input (Fig. 1.47).

**Name of Sequence**
You can identify the input sequence by the name when a name is entered here. By default it displays as "unnamed sequence". You can see this name in the "Earlier projects" list for accessing this sequence at any time. But it should be noted that the sequences will deleted from this list if no activity within after 2 days is seen.

**Delete Projects**
This option removes your entire project files saved in the server.

**Interpretation of Results**

7. The result for the Mn and Fe superoxide dismutase from *Nostoc punctiforme* is displayed with information like the AT and GC content, a graphical output representing the location of the enzymes' sites on the sequence etc. (Fig. 1.48).
8. Under 'Main options' click 'Custom digest' to view the enzymes that have restriction sites on the sequence provided in a table form (Fig. 1.49).

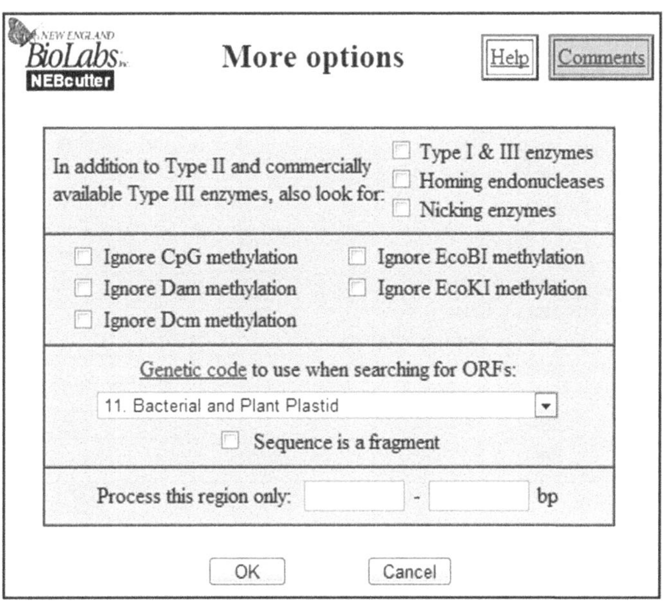


**Fig. 1.47** Parameters under 'More options' in NEBcutter

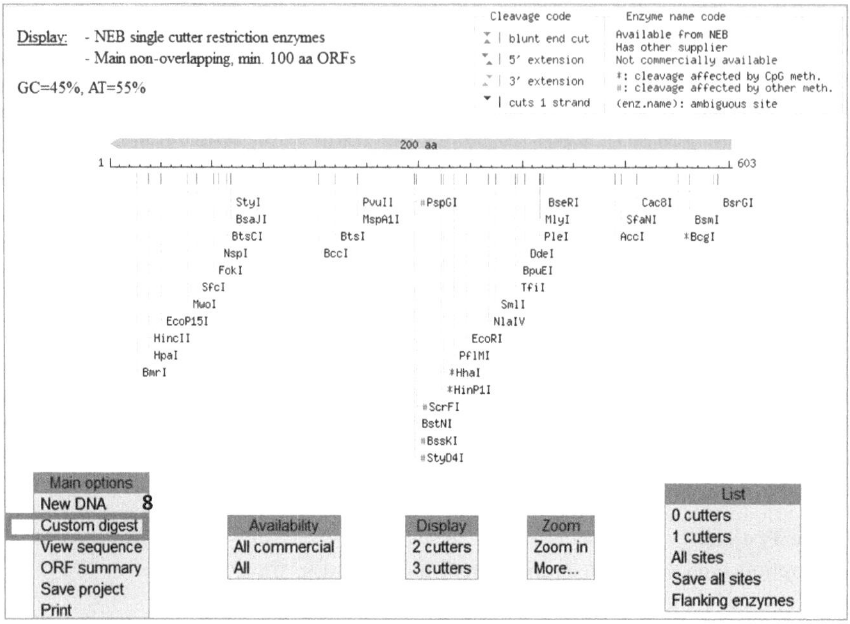


**Fig. 1.48** Graphical output of the restriction map

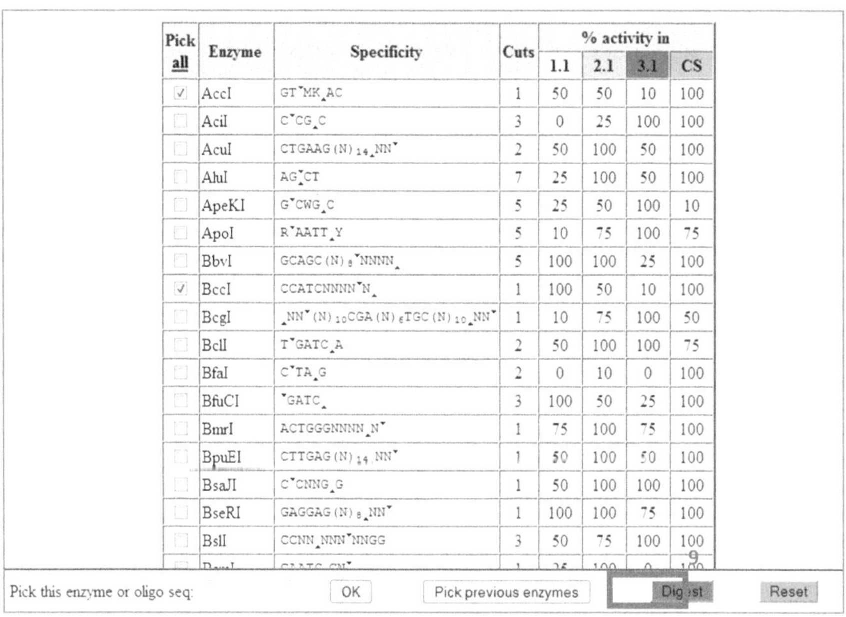

| Pick all | Enzyme | Specificity | Cuts | % activity in | | | |
|---|---|---|---|---|---|---|---|
| | | | | 1.1 | 2.1 | 3.1 | CS |
| ✓ | AccI | GT^MK_AC | 1 | 50 | 50 | 10 | 100 |
| | AciI | C^CG_C | 3 | 0 | 25 | 100 | 100 |
| | AcuI | CTGAAG(N)14_NN^ | 2 | 50 | 100 | 50 | 100 |
| | AluI | AG^CT | 7 | 25 | 100 | 50 | 100 |
| | ApeKI | G^CWG_C | 5 | 25 | 50 | 100 | 10 |
| | ApoI | R^AATT_Y | 5 | 10 | 75 | 100 | 75 |
| | BbvI | GCAGC(N)8^NNNN_ | 5 | 100 | 100 | 25 | 100 |
| ✓ | BccI | CCATCNNNN^N_ | 1 | 100 | 50 | 10 | 100 |
| | BcgI | _NN^(N)10CGA(N)6TGC(N)10_NN^ | 1 | 10 | 75 | 100 | 50 |
| | BclI | T^GATC_A | 2 | 50 | 100 | 100 | 75 |
| | BfaI | C^TA_G | 2 | 0 | 10 | 0 | 100 |
| | BfuCI | ^GATC_ | 3 | 100 | 50 | 25 | 100 |
| | BmrI | ACTGGGNNNN_N^ | 1 | 75 | 100 | 75 | 100 |
| | BpuEI | CTTGAG(N)14_NN^ | 1 | 50 | 100 | 50 | 100 |
| | BsaJI | C^CNNG_G | 1 | 50 | 100 | 100 | 100 |
| | BseRI | GAGGAG(N)8_NN^ | 1 | 100 | 100 | 75 | 100 |
| | BslI | CCNN_NNN^NNGG | 3 | 50 | 75 | 100 | 100 |



**Fig. 1.49** The enzyme selection list

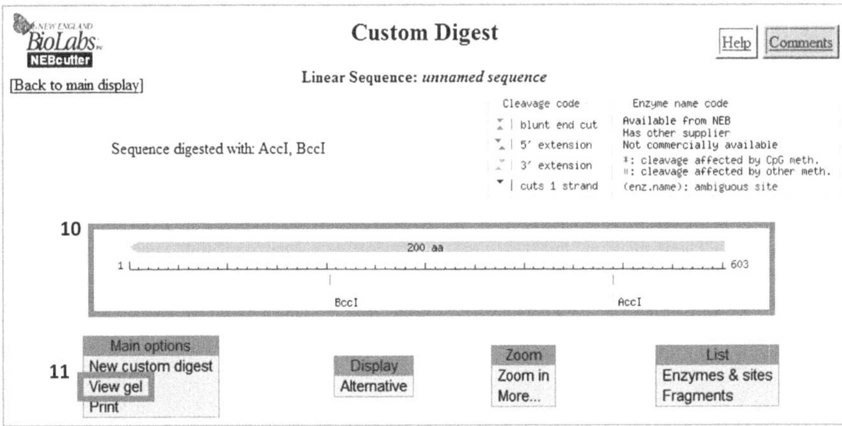


**Fig. 1.50** The restriction map of the selected enzymes

9. Chose the enzymes of interest (AccI and BccI) preferably a single cutter one and click 'Digest' button (Fig. 1.49).
10. This displays the sites of the chosen enzyme(s) alone (Fig. 1.50).
11. To view the restriction pattern on gel, click '**View gel**' found on the '**Main options**' menu.
12. You can also have provisions to choose the gel percentage and ladders of your choice. Here 1% agarose and 100 bp DNA ladder were selected. Three bands of sizes 112, 205 and 286 bp were generated. The total should be equal to the size of the product before digestion (Fig. 1.51).

While sending the PCR products for sequencing, it is mandatory to generate a wet lab restriction digestion. Once the sequence is received from the sequencing company, process (sequence assembly and vector contamination) and check for in silico digestion with NEBcutter. This tells whether the sequence you got back from sequencing lab/company corresponds to the piece of DNA you sent out and also to check whether the sequence assembly is correct.

## Exercise

1. What is restriction mapping?
2. What is a restriction enzyme? Name its types.
3. What are the applications of restriction mapping?
4. What are the advantages of doing a restriction map using computer programs?
5. What is double digestion?

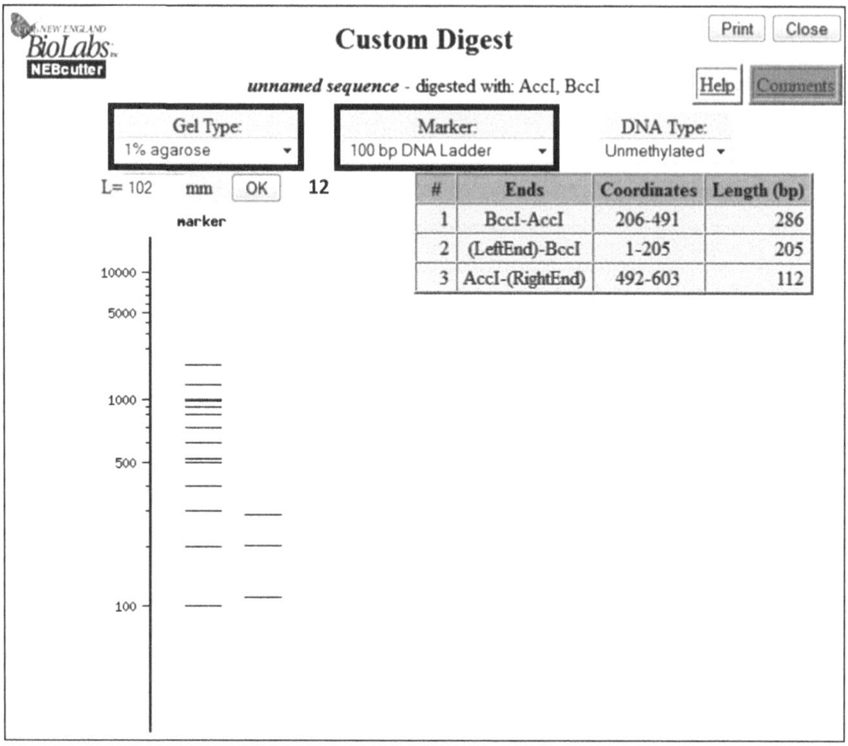


Fig. 1.51 The gel picture generated for the selected enzymes

## 1.10 Gene Prediction Using ORF Finder (Open Reading Frame Finder)

### Introduction

Once the genome is sequenced and assembled, the first ensuring task is to locate all the protein-coding genes hidden within the genome. This helps in understanding the functional content of the genome. Computational gene prediction aims to predict all the genes with near 100% accuracy and thereby could reduce the amount of experimental work required.

### Differences Between Prokaryotic and Eukaryotic Gene Organization

| Prokaryote | Eukaryote |
|---|---|
| Have relatively small genome sizes | Nuclear genomes are much larger than prokaryotes |
| Have high gene density (>90% of coding) | Have a very low gene density. The space between genes is often very large |
| Few repeats | Rich in repetitive sequences and transposable elements |

(continued)

(continued)

| Prokaryote | Eukaryote |
|---|---|
| Single ORF, found adjacent to one another | Eukaryotic genomes are characterized by a mosaic organization in which a gene is split into pieces (called exons) by intervening non-coding sequences (called introns) |
| No post transcriptional modifications | The nascent transcript from a eukaryotic gene is modified in three different ways (Capping, Splicing and Polyadenylation) before becoming a mature mRNA for protein translation |
| Less complex | More complex |

**Gene Prediction Methods**

Location of genes is achieved through one or combination of the following methods:

- Intrinsic Method or Template Method
  - Searching for **Signal**
  - Searching by **Content**
- Extrinsic Method or Lookup Method
  - **Homology** based prediction
  - **Comparative** gene prediction

**Basic signals** in gene prediction include the start/stop codon, the 5′/3′ splice site, ribosomal binding site, transcription factor binding site and polyadenylation (poly-A) sites (Fig. 1.52). Searching for gene **content** refers to coding statistics, which includes non-random nucleotide distribution, amino acid distribution, synonymous codon usage and hexamer frequencies. The **homology-based method** makes predictions based on significant matches of the query sequence with sequences of known genes. **Comparative gene prediction** involves comparisons made with large number of genomes available now-a-days. It is based on the rational that functional regions tend to be more conserved than non-protein coding regions.

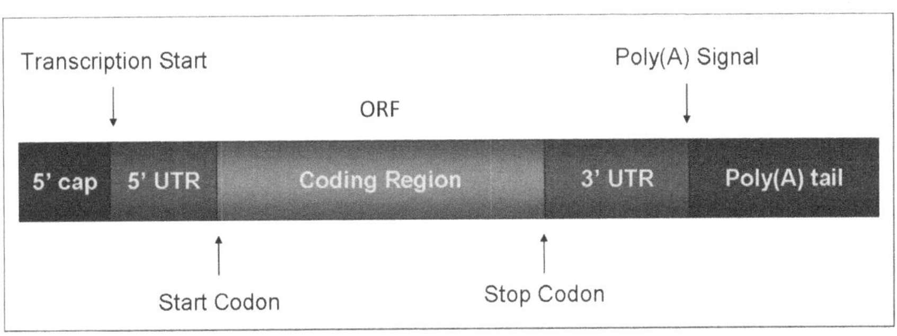


**Fig. 1.52** The structure of a gene

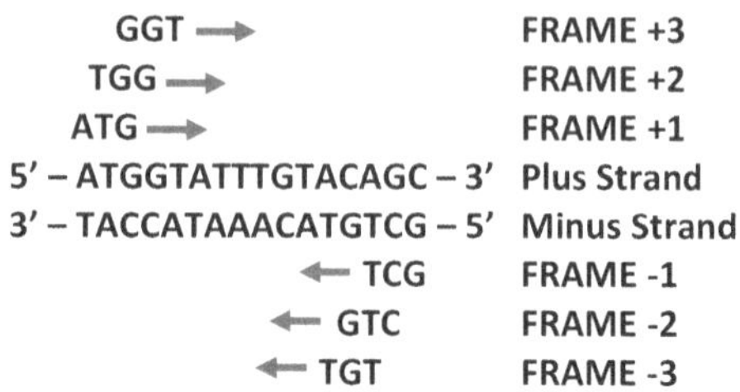


**Fig. 1.53** The six reading frames of translation

In silico prediction of gene involves determination of correct **Open Reading Frame** (ORF). Generally an open reading frame starts with the codon ATG, (that codes for methionine) and ends with codons TAA, TAG or TGA in most species. This can be determined by translating to six possible reading frames, that is, three in each direction (Fig. 1.53). It has to be noted that even with six possible translations only one, probably the longest, will be used in translating a gene. When that is identified, then the amino acid sequence can be determined by translating the appropriate frame.

The ORF Finder which is available at NCBI is a graphical analysis tool which finds all open reading frames (six frames) for user provided sequence or in a sequence already found in the database using the standard or alternative genetic codes.

The translated amino acid sequence can be then searched against the sequence database using BLAST server. This program also helps in processing the protein-coding sequences for submissions. Thus this program uses a **combination of intrinsic and extrinsic methods** for gene prediction.

**Genetic Codes**

The following genetic codes can be used in ORF Finder (Griffiths et al. 2000):

- The Standard Code
- The Vertebrate Mitochondrial Code
- The Yeast Mitochondrial Code
- The Mold, Protozoan, and Coelenterate Mitochondrial Code and the Mycoplasma/Spiroplasma Code
- The Invertebrate Mitochondrial Code
- The Ciliate, Dasycladacean and Hexamita Nuclear Code
- The Echinoderm and Flatworm Mitochondrial Code
- The Euplotid Nuclear Code
- The Bacterial, Archaeal and Plant Plastid Code
- The Alternative Yeast Nuclear Code
- The Ascidian Mitochondrial Code
- The Alternative Flatworm Mitochondrial Code
- Blepharisma Nuclear Code
- Chlorophycean Mitochondrial Code
- Trematode Mitochondrial Code

- Scenedesmus Obliquus Mitochondrial Code
- Thraustochytrium Mitochondrial Code
- Pterobranchia Mitochondrial Code.

**Exploration Activity**

Nucleotide sequence of *Actias maenas* voucher MZUWSM2 (Accession: **KC759169**) containing partial 16S ribosomal RNA, complete tRNA-Leu gene sequence and partial coding sequence (cds) of NADH dehydrogenase subunit 1 gene from mitochondrial origin is used as an example here.

1. ORF Finder can be accessed from http://www.ncbi.nlm.nih.gov/projects/gorf/
2. Copy and paste the sequence of interest (Accession: **KC759169**) in the textbox. The sequence can be in FASTA format or in raw format. Alternatively you can also provide the accession number of the sequence if you wish to use the sequence from database (Fig. 1.54).

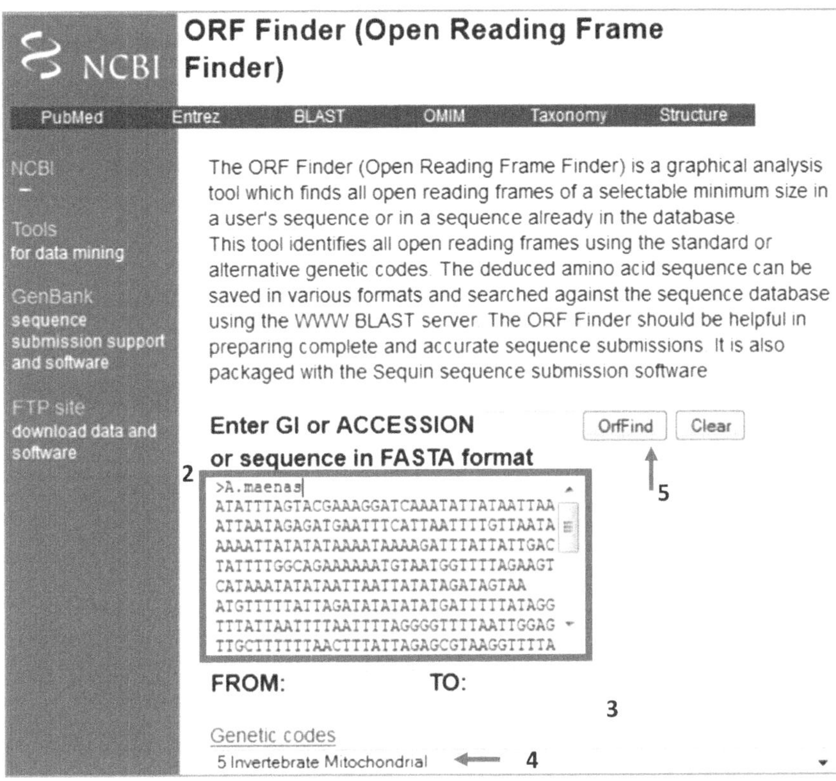


Fig. 1.54 The ORF Finder interface at NCBI

3. Specify the start and end nucleotide numbers if you want to translate only a portion of the sequence or leave blank to translate the complete sequence.
4. Choose appropriate '**Genetic code**' for translation. This can be selected from the dropdown menu. Since the organism is an invertebrate and the source of the gene is mitochondrial, '**Invertebrate Mitochondrial**' should be selected.
5. Click button '**OrfFind**' to run the program.

**Interpretation of Results**

6. The name provided in the FASTA format description line will be displayed first. If no name was provided, it will display '**Anonymous**' (Fig. 1.55).
7. Graphical representation of the translations in all the six frames will be presented.
8. Details like which frame (+ or −), start/stop positions and the length (nucleotides) will be presented on the right side of the graphical output. The results are sorted based on the length of amino acids obtained after translation.
9. Click on the shaded regions (preferably the longest) on the graphical output or on the small square found near the frame number to view the amino acids coded.
10. The nucleotide and the corresponding amino acids will be displayed below the graphical output and the chosen portion will be highlighted (Fig. 1.56).
11. Presence of stop codons, **if any**, are represented as '*'.
12. Make a note of the start position of the translation start. Here it is '173'.
13. To check if the translated protein is found in the database, click on the BLAST button on the top. This will perform a BLAST protein (blastp) search against the 'nr' database.

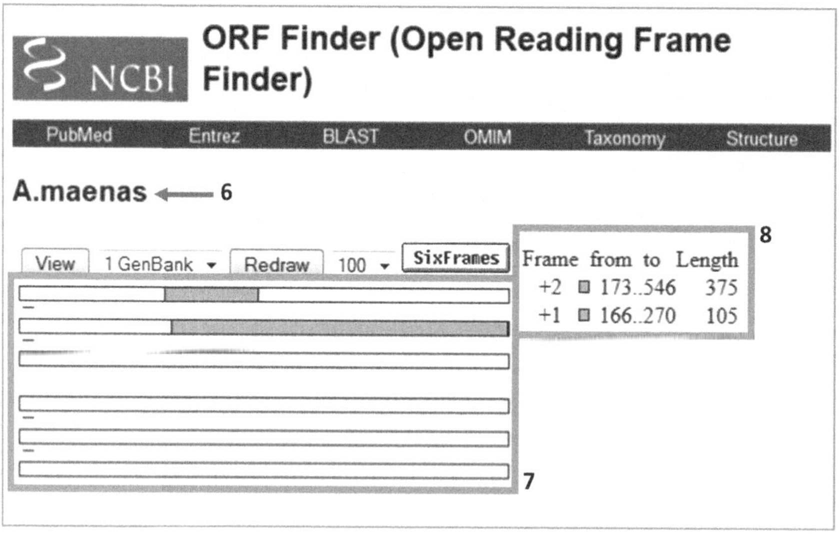


**Fig. 1.55** ORF Finder showing six frame translations for the user provided sequence

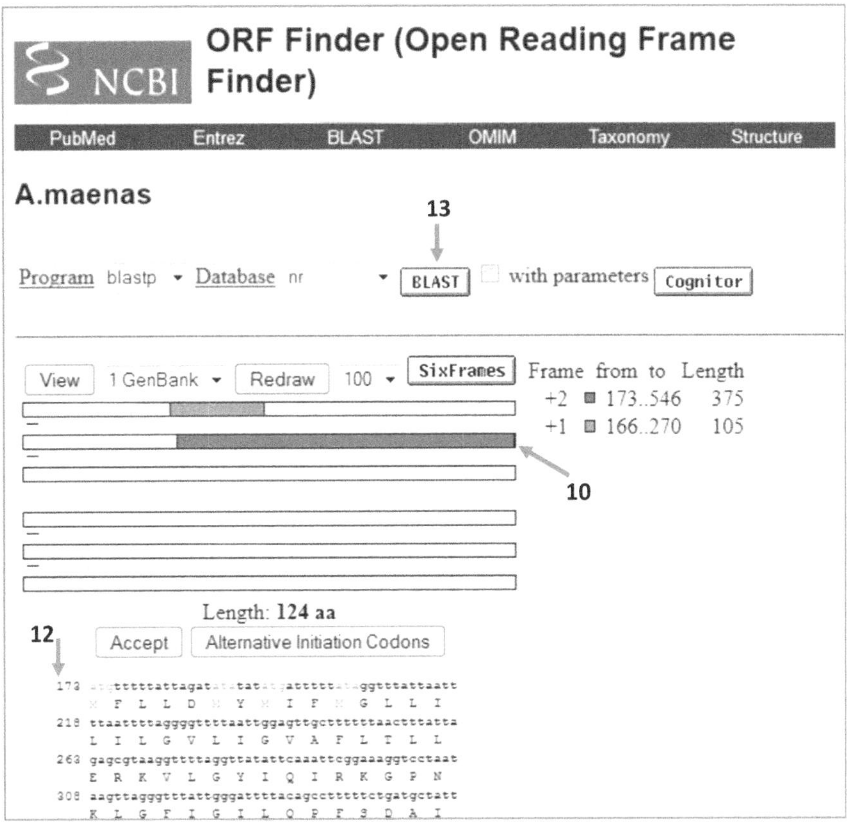


Fig. 1.56 Translated amino acids for a selected ORF

14. The user can format the output of the BLAST results with the '**format request**' page. Click on the '**View report**' button to view the BLAST result (Fig. 1.57).
15. If the translated protein does not find a match on the database, it displays '**No significant similarity found**' [like if you click the other ORF (size 105) predicted]. This may be due to many reasons, e.g. the query may be too short or might have been filtered out if it is largely of low complexity.
16. If a match is found, then on the BLAST result page, scroll down to the alignment session (Fig. 1.58).
17. Look for the start of the translation in the Sbjct (subject). If the number starts from one (amino acid), then the first amino acid (from where translation starts) is the start codon of our sequence. This indicates 5′ end of the predicted gene is complete. In our case it starts from 173rd position of nucleotide sequence.
18. If translation starts from the beginning (first, second or third) nucleotide in the user provided sequence (not like in our case where it starts from 173rd position

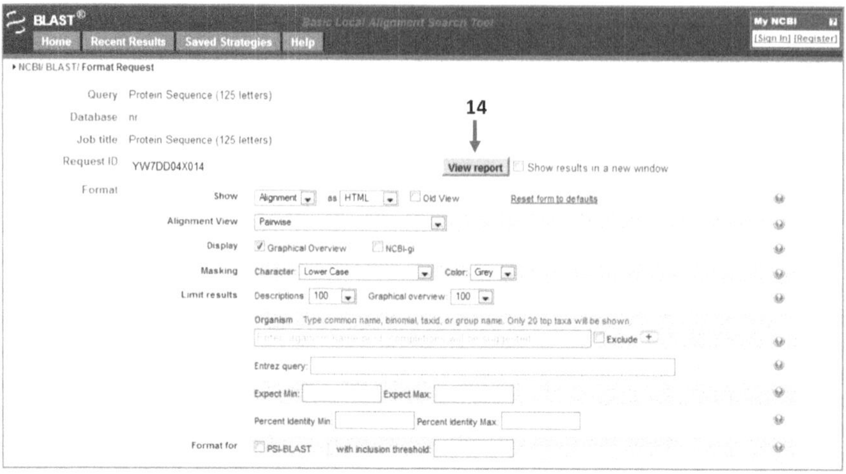

Fig. 1.57 Protein BLAST format request page

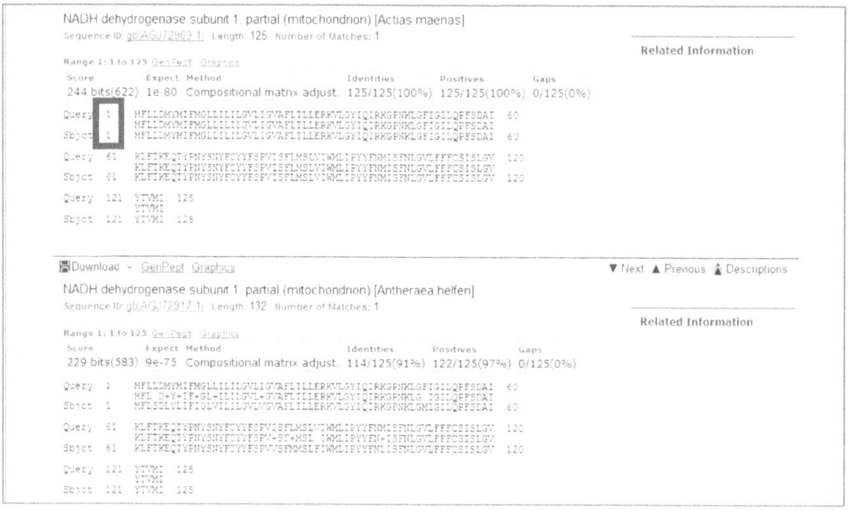

Fig. 1.58 Alignment in the BLAST result

of nucleotide sequence) and Sbjct (subject) number does not start from one, then it indicates that the gene is partial at the 5′ end and may have an N terminal extension.

Hence, from the results it is clear that the longest translation (length 375) codes for a valid protein, here NADH dehydrogenase. Its translation starts from 173rd position and has no stop codon, which indicates that the protein is partial at the 3′ end.

### Exercise

1. What is ORF and its significance?
2. What do you mean by 'no significant similarity found'?
3. What is the significance of choosing the genetic code for translation?
4. What are alternative start codons?
5. What do you infer when there is no start and stop codon?

## 1.11 Gene Prediction Using FGENESB

### Introduction

**FGENESB** is a suite of bacterial operon and gene prediction programs which is pattern/Markov chain-based and is reported to be the fastest and the most accurate ab initio bacterial gene prediction program available. This program has been trained with **FgenesB-train script that** takes only the DNA sequence as input and creates gene prediction parameters automatically. The program use ABsplit program for community sequence annotation that could separate archebacterial and eubacterial sequences.

### Exploration Activity

In this example, *Serratia marcescens* strain ACE2 plasmid pRK10 (Accession: **EU697813**), complete sequence is used as input.

1. FGENESB Bacterial Operon and Gene Prediction page can be accessed from the 'Softberry Inc' webpage at http://linux1.softberry.com/berry.phtml (Fig. 1.59)

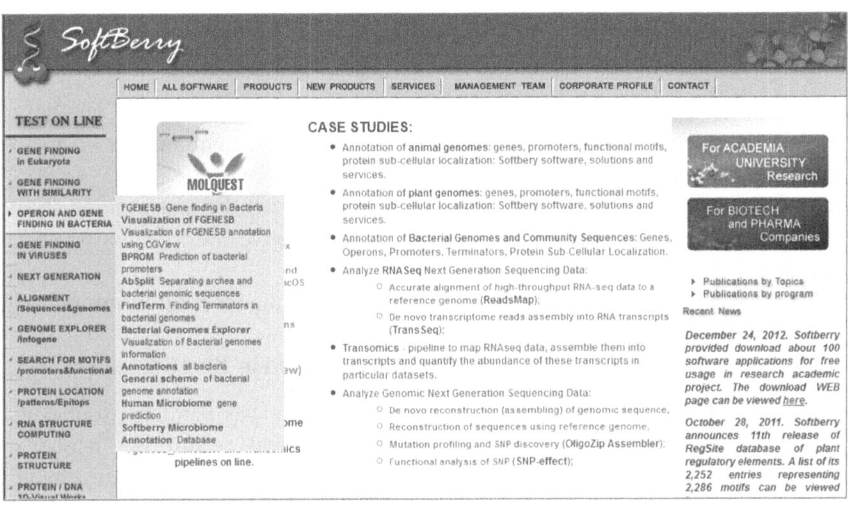


**Fig. 1.59** List of programs in Softberry

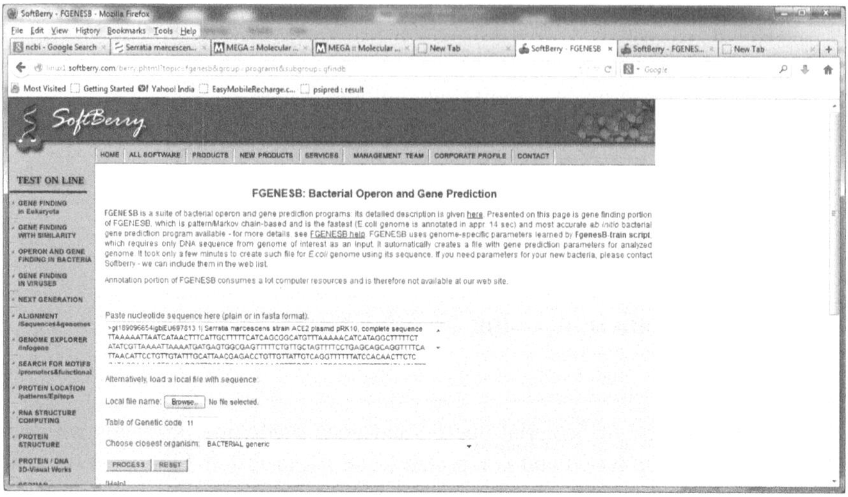

Fig. 1.60 FGENESB interface at Softberry Inc

and selecting '**FGENESB**' from '**Operon and Gene finding in Bacteria**' from the menu.

2. Input your sequence **(EU697813)** by pasting in the text box provided. This tool reads sequence in plain or in FASTA format (Fig. 1.60).
3. Alternatively, file containing valid nucleotide sequence in FASTA format can be uploaded using the '**Browse** …' button (provided below the text box) as input.
4. From the '**Choose closest organism**:' dropdown menu, select the option 'BACTERIAL generic' since *Serratia* is not listed.
5. Click on the '**Process**' button.

**Interpretation of Results**

The result is divided into three sections:

6. The **header** part displays information like sequence name, its length, number of genes predicted, number of transcription units predicted and number of operon units predicted.
7. It is then followed by a **table** displaying the start and the end position in the sequence with score for the predicted genes. It also gives information on which strand the protein is coded.
8. It is then followed by the third section displaying the **translated amino acid** sequence of predicted genes in FASTA format.
9. The user can validate the translations by **doing protein BLAST** search.

From the results it can be inferred that the program has predicted four genes (Fig. 1.61). But when compared to the original GenBank record, the number of predicted genes is less. Also one has to note that some of the genes predicted by this program are not being annotated in the GenBank record.

```
 Prediction of potential genes in microbial  genomes
 Time:   Tue Jan  1 00:00:00 2005
 Seq name: pRK10
 Length of sequence - 4241 bp
 Number of predicted genes - 4
 Number of transcription units - 3, operons - 1
     N      Tu/Op   Conserved  S                   Start          End     Score
                   pairs(N/Pv)
     1      1 Tu   1      .      -     CDS             1 -       1071      481
     2      2 Tu   1      .      -     CDS          1440 -       1673       76
     3      3 Op   1      .      +     CDS          2202 -       2516      160
     4      3 Op   2      .      +     CDS          2506 -       4005      699
Predicted protein(s):
>GENE     1          1  -       1071     481     356 aa, chain -
MKAMVFYGENDVRFEERKKPEIINPGDAIIRMIKTTICGTDLGILKGKNPEIEKNAKEKT
GSFNGRILGHEGVGIIEEIGSAVKNFKIGDKVIISCVSRCGTCDNCQKQLYAHCRNDGGW
IMGYMIDGTQAEFVRTPYADNSLYLFPDELNDDAGVLLSDALPTAHEIGIQNGMVKPGDT
VAIVGAGPVGMSCLLTGMLYSPSQLIMIDMDENRLAMAKDMGATHTINSAKENAINRVME
LTNQRGVDCAIEAVGIEPTWNICQNIIKEGGHLANVGVHGNPVSFAIEKLWIKNLTITTG
LVNANTTGMLMKTCCSGKLATEKLATHHFNFNDIEKAYDVFKHAADEKAMKVMINF
>GENE     2       1440   -       1673      76      77 aa, chain -
VRSPVPAIRFAGVIPLLWPRFLIPRLTLGSGKAVRSKLDCMHEPPVQSDCCALSGNYRLE
SNPVRHAKAPLAAATGN
>GENE     3       2202   -       2516     160     104 aa, chain +
MWVTEDEHRRLLERCDGKQLAAWMRQTCLDEKPARACKLPSISPALLRQLAGMGNNLNQI
ARQVNAGGGSAHDRVQIVSALMAINAGLERLRHAVLEKGADDDR
>GENE     4       2506   -       4005     699     499 aa, chain +
```

**Fig. 1.61** Results from the FGENESB

Remember, most of the gene predicting programs do not handle overlapping or nested genes. They also cannot find non-protein coding genes (tRNAs). Gene prediction should always be checked by other means like BLAST search, cDNA sequencing, etc.

**Exercise**

1. How prokaryotic gene prediction is different from eukaryotic gene prediction?
2. What are the methods available for gene prediction?
3. What are the signals that are used in gene prediction?
4. What do you mean by searching for content in gene prediction?
5. What is the need for predicting genes?

## 1.12 Dot-Plot

**Introduction**

A Dot plot is basically a data analysis tool that usually does a comparison of two sequences, say of lengths $l$ and $m$. In order to compare those two sequences, a table of size $l \times m$ is constructed. The two sequences are placed on the two axis of the table. Whenever the match is observed between symbols, a dot is marked in a grid. If there is no match, then the grid is left blank, thus representing all possible comparisons of characters in either sequences. This graphical representation is called a dot plot (Fig. 1.62). Thus, when two sequences are identical, they produce a continuous diagonal line. When a dot is placed outside the diagonal line, it is called noise. Noise increases when one or more nucleotides occur more in number like AT rich or GC rich regions. It is also related to the length of the sequences that are being compared.

The dot plot technique was developed in the field of genetics during 1960s and early 1970s. Even today, there are domains that require application for dot plots.

Dot plot applications are very much useful in the identifying interspersed repeats such as transposons and tandem-repeat motifs such as microsatellites or even loss or gain of whole motifs. Besides that these are also used in predicting secondary structures in protein-coding sequences and stem-loop formation or structural domains in case of RNAs. Thus, the method allows investigating patterns in sequences.

Dotlet is a Java based program that can run in a Web browser for comparing sequences using diagonal plot method (Junier and Pagni 2000).

**Exploration Activity**

1. Dotlet program can be accessed from the site http://myhits.isb-sib.ch/cgi-bin/dotlet
2. Click on the '**Input**' button to open the Sequence Input Dialog box (Fig. 1.62).
3. In the dialogue box text area, type or paste a sequence (Fig. 1.63).
4. You can give your sequence a name. If not provided, Dotlet program will create one.

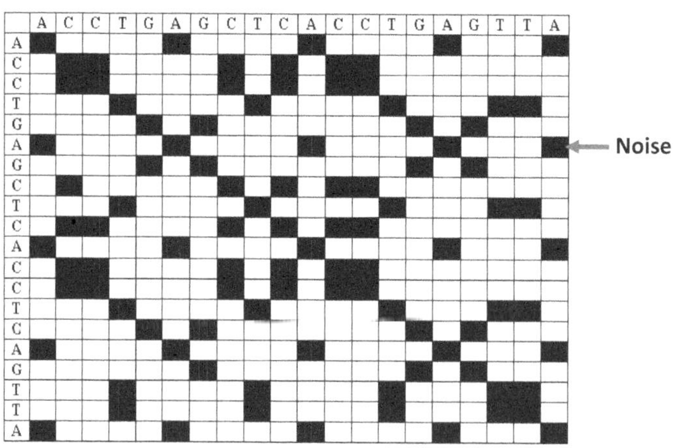


**Fig. 1.62** The matrix showing comparison of two sequences of length 20. The resulting matrix is of size ($l \times m$) 400

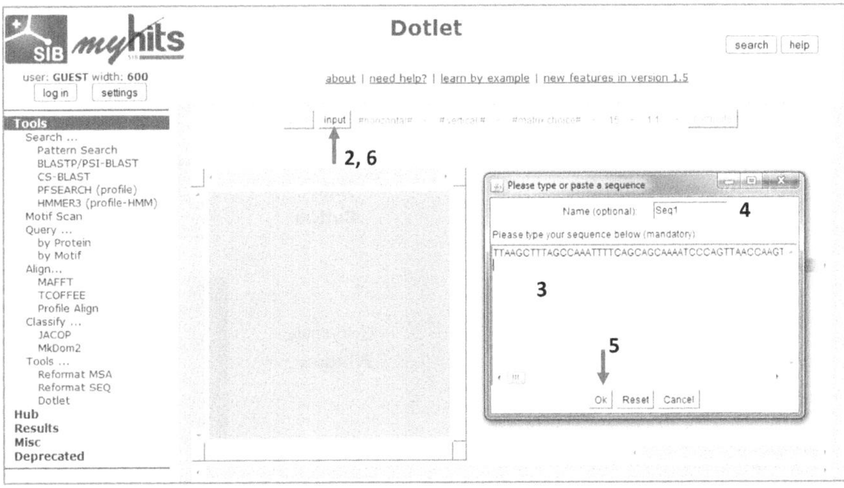


**Fig. 1.63** The Dotlet interface at SIB

5. When you're done, click the '**Ok**' button. *Note: Characters other than letters are ignored, so it is better to cut-and-paste a sequence. Even if the sequence has spaces or position numbers in it, it will be ignored and so the case of characters.*
6. Repeat this step, if you need to enter a second sequence.
7. Once the second sequence is also loaded, select the sequence 2 from the dropdown menu (Fig. 1.64).

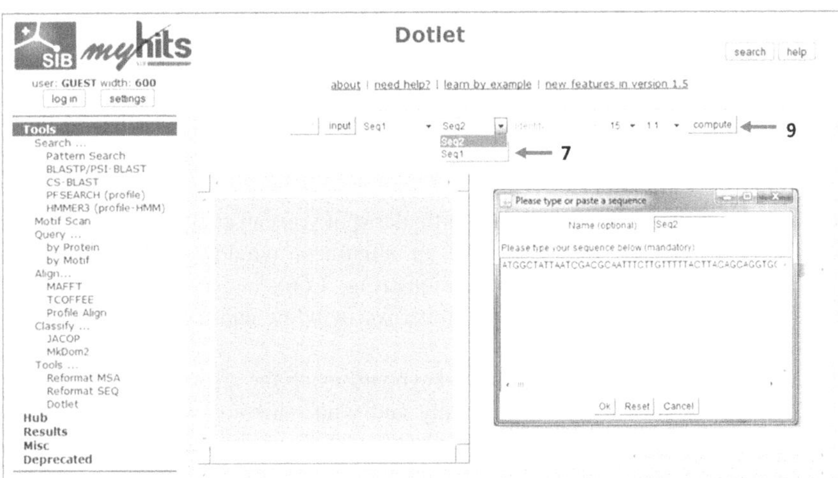


**Fig. 1.64** Input of second sequence and selecting the second sequence for computation

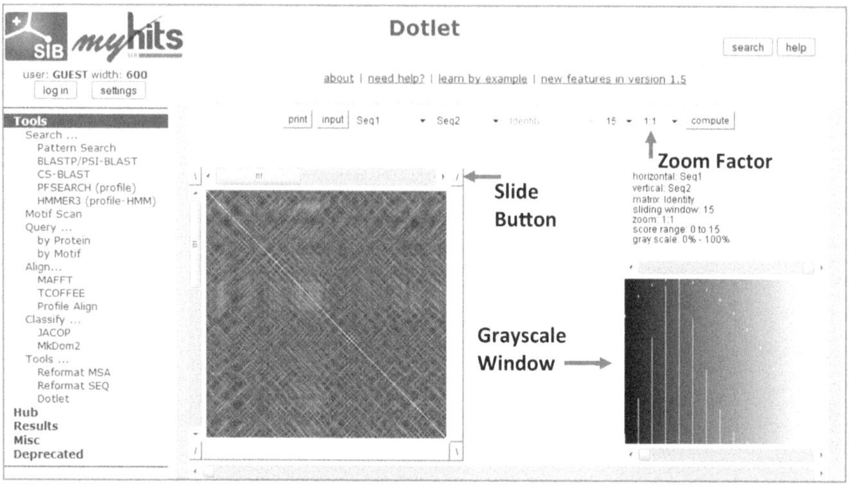


**Fig. 1.65** The dotplot result

8. When entered the programs detects automatically whether the sequence is nucleic acid or protein and selects the matrices accordingly. The user can also choose the matrix from the matrices menu. Likewise sliding window's size and zoom factor can also be selected from the corresponding menus.
9. Click the '**Compute**' button.
10. After a while, the **dots window** will display the diagonal plot (Fig. 1.65).
11. Every pixel represents a residue (and also a score) in the horizontal and vertical sequence and its colour reflects the similarity. Higher scores means good matches and lighter the pixel intensity.
12. You can reduce the background noise and bring on the similar regions distinctly by adjusting the greyscale factor. This comes handy where there is situation with higher number of darker pixels (low scores) and very few lighter pixels (high scores) (Fig. 1.65).
13. For lengthier sequences choose a smaller zoom factor using zoom factor menu and re-compute. You can also use vertical and horizontal scrollbars to move around.
14. For inspecting the alignment at a particular spot you can click on the spot which will show a blue cross bar and an alignment window that displays the sequences around the cursor's positions (Fig. 1.66).
15. Matching Residues shown in a comparison window and are coloured in blue (Fig. 1.66).
16. The can be moved around with the keyboard using the arrow keys, for moving the cursor vertically and horizontally and with 'shift+ <' (up left), 'shift+>' (down right), '[' (up right), and ']' (down left) for moving the cursor diagonally (Table 1.3).

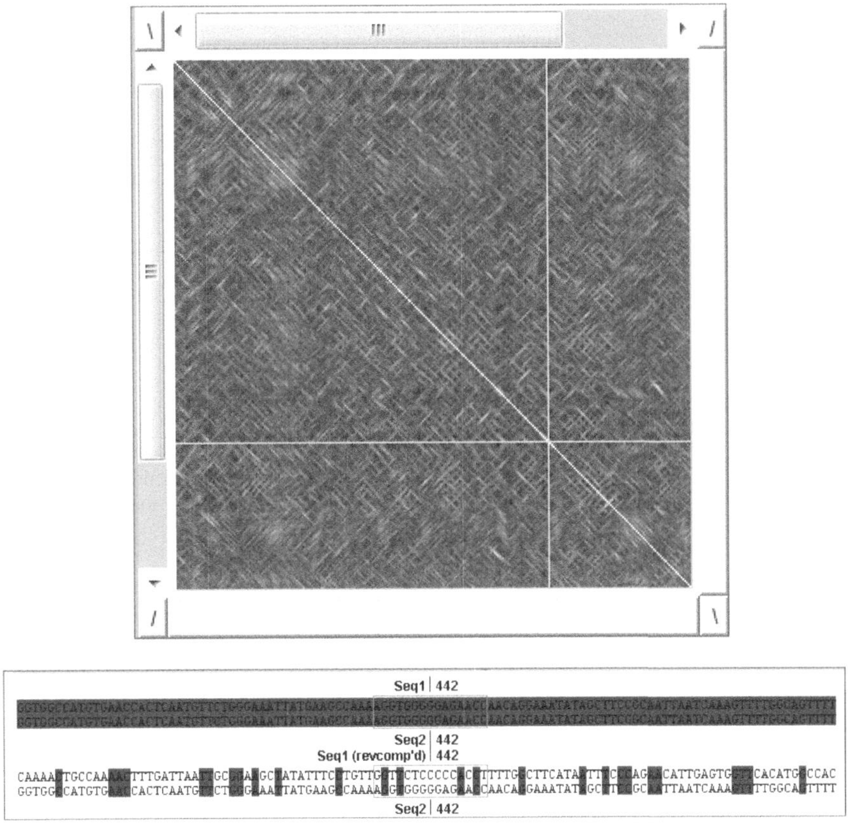


**Fig. 1.66** The *blue cursor* showing a selected pixel in the dot window and the alignment window and *magenta box* indicating the window size (15)

## Interpreting Dot Plots

## Exercise

1. What are the applications of dot plot analysis?
2. What information you gain from the following dot plot analysis:

   a. Continuous diagonal line
   b. Discontinuous diagonal line
   c. Bold block
   d. Parallel line
   e. Continuous Perpendicular line
   f. Discontinuous Perpendicular line

**Table 1.3** Overview of characteristic patterns appearing in dot plots and their probable inferences

| Pattern | Description | Inference |
|---|---|---|
| 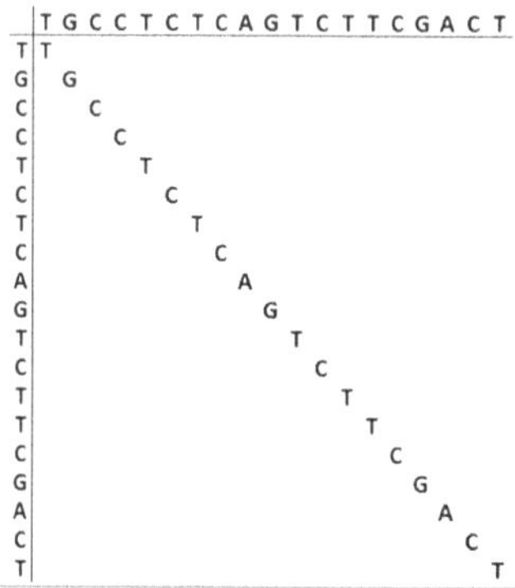 | Continuous main diagonal | Indicates a perfect similarity |
| 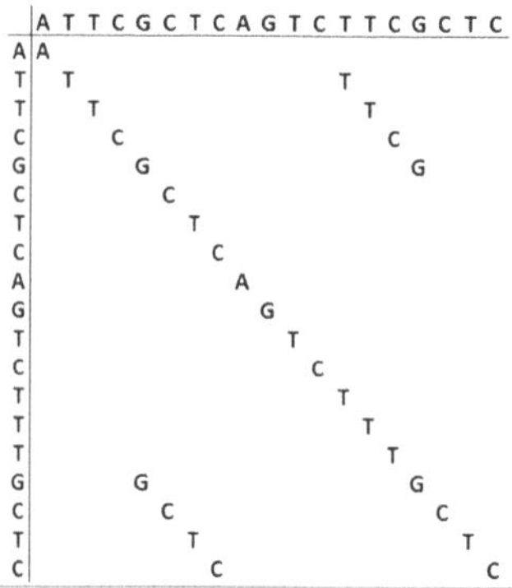 | Parallel lines on either sides to the main diagonal | Repeated regions are present in the same reading direction on different parts of the sequences |
| 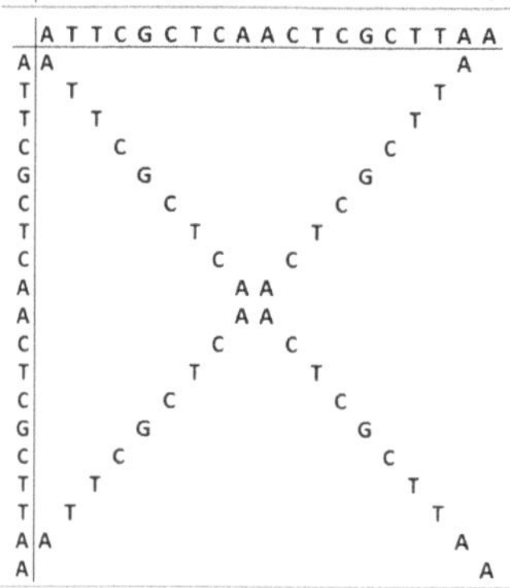 | Perfect perpendicular lines to the main diagonal | Indicate palindromic regions in the sequences |

(continued)

**Table 1.3** (continued)

| Pattern | Description | Inference |
|---|---|---|
| 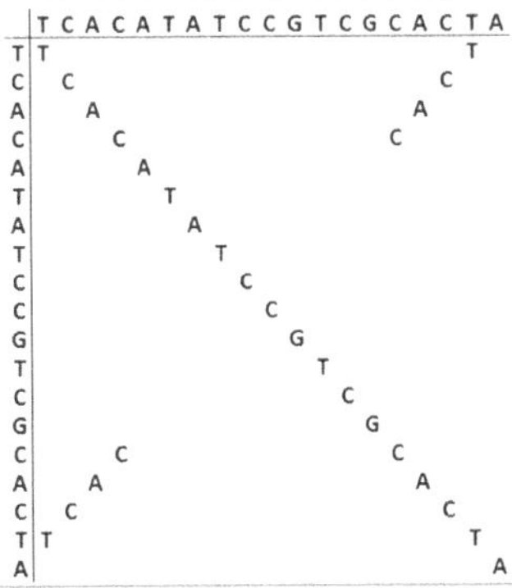 | Partial perpendicular lines to the main diagonal | Indicates that a portion of the sequence matching with the reverse complement of the other. This is observed in case of many transposable elements |
| 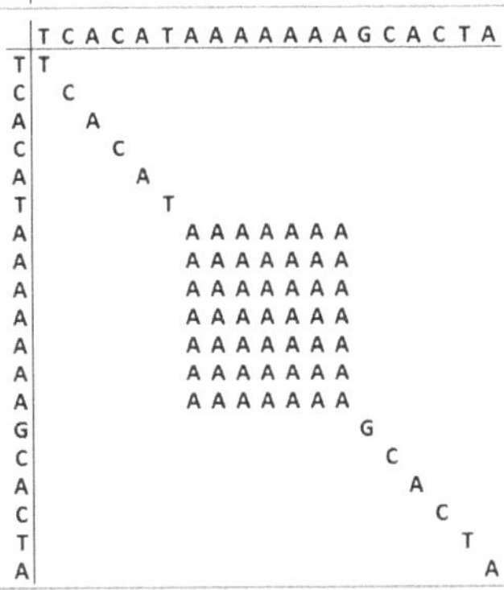 | Bold blocks on the main diagonal | Indicate that same nucleotide is repeated in both the sequences. This is observed in case of microsatellite repeats |
| 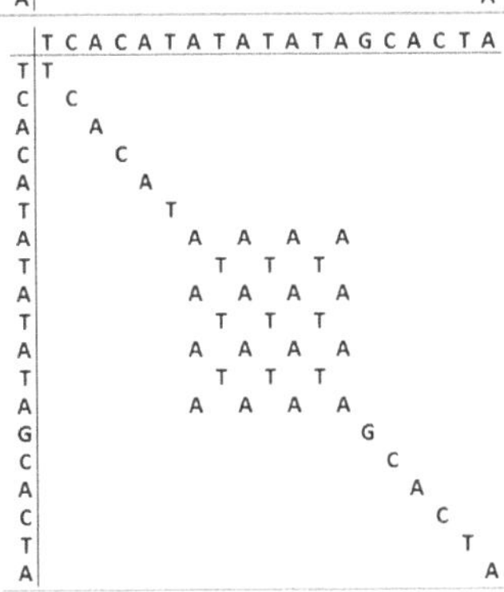 | Parallel lines to the main diagonal | Indicate tandem repeats like a motif in both sequences. The distance between the diagonals reflects the distance of the motif in the sequences |

(continued)

**Table 1.3** (continued)

| Pattern | Description | Inference |
|---|---|---|
| 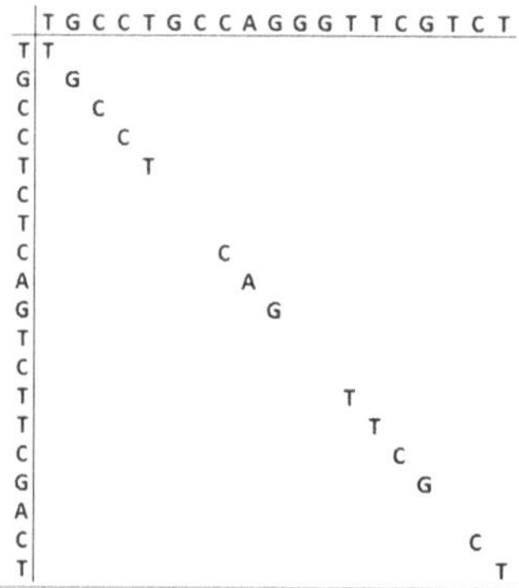 | A discontinuous diagonal | Indicates that both the sequences share a common source. The interruptions reflect the changes (mutations) accumulated over a period of time |
| 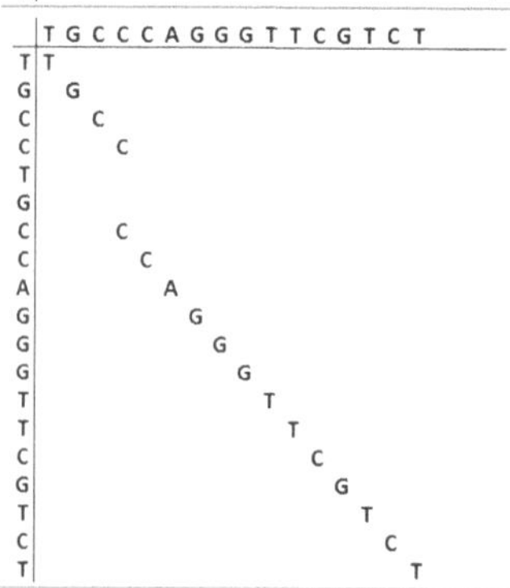 | Break in the main diagonal along with a shift | Indicates the presence of 'indels', which may be due to an artial deletion in one sequence or insertion in the other sequence. This case is observed in proteins losing or substituted with motifs (Beaussart et al. 2007). Even comparing mRNA sequences without introns (cDNAs) against the DNA sequence which is unspliced produces this kind of pattern |

3. What is zoom factor?
4. What is window size?
5. What is the use of grayscale window?
6. What do you mean by noise?

## 1.13 Global Sequence Alignment

### Introduction

Sequence alignment a process of arranging the sequences of DNA, RNA or protein so as to predict regions of similarity. These similarities may be due to functional conservation, structural and or may be due to evolutionary relationships. The alignment explains the how likely they are derived from a common ancestor. Difference in sequences is not only due to substitution, but also due to insertion and deletion. Hence, they should be optimally aligned to maximize similarity. Moreover

```
GTTCGTTTAGCACAGCCTTTTTCTATGCGCTACATGTTGGTTGACGGTCAGGGTAACTTC
          ||||||||||..||||||||||||||||||...||||||||
----------CACAGCCTTTAACTATGCGCTACATGTTGGAACACGGTCAG---------
```

**Fig. 1.67** Global alignment occurs across the complete lengths of both sequences

it should be kept in mind that optimal alignment is not always unique as there could be more than one alignment with the same best score. Making slight changes in the scoring system can adversely affect the ranking of alignments making a different one to appear as the best. Using two sequences for alignment is referred as pairwise alignment and using multiple (more than two) sequences is referred as multiple sequence alignment.

As global alignments produce end to end alignment of the two sequences for identifying conserved regions and differences, there are expected to be potentially related.

Global alignment tools uses Needleman–Wunsch (1970) dynamic programming algorithm for alignment and can be used to compare two genes with same or similar function (Fig. 1.67).

**Global Alignment Using EMBOSS Needle @ EBI**

**Introduction**

EMBOSS Needle takes two input sequences and presents an optimal global sequence alignment of two sequences along their entire length using Needleman–Wunsch alignment algorithm.

**Exploration Activity**

1. Go to the 'Pairwise sequence alignment' tools page by typing http://www.ebi.ac.uk/Tools/psa in your browser.
2. Under Global Alignment choose appropriate program (protein or nucleotide) (Fig. 1.68).
3. Input your first and second sequence by pasting in the text boxes provided in **Step 1** (Fig. 1.69). The input sequence can be in GCG, FASTA, EMBL, GenBank, PIR, NBRF, Phylip or UniProtKB/Swiss-Prot format. ***Note***: *Avoid using data directly from word processors as it may yield unpredictable results due to the presence of hidden/control characters.*

   Alternatively, files containing valid sequences in any format (GCG, FASTA, EMBL, GenBank, PIR, NBRF, Phylip or UniProtKB/Swiss-Prot) can be uploaded using the 'Browse' options (provided below the text box) as input. ***Note***: *Avoid using word processors files. Preferably generate a file with sequence in FASTA format and save the file with extensions .fas or .fasta.*
4. Set the alignment options in **Step 2**. Generally you can work with default parameters. The default parameters are given in Table 1.4. You can also change these parameters and look how it affects the result.

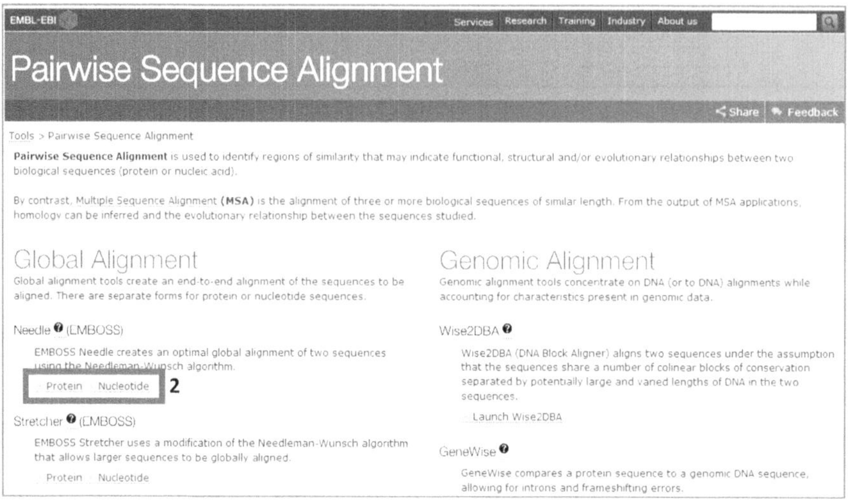


**Fig. 1.68** Pairwise sequence alignment tools at EBI

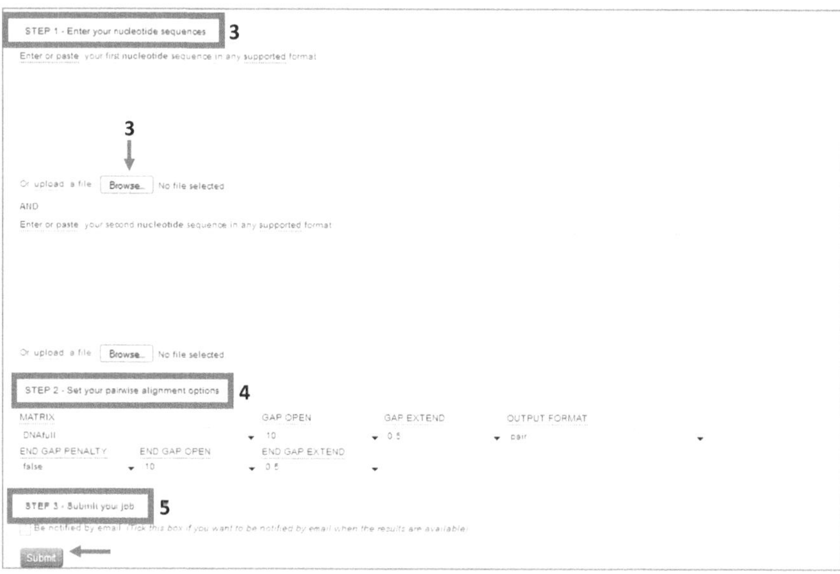


**Fig. 1.69** EMBOSS Needle pairwise alignment—nucleotide interface at EBI

**Table 1.4** Default of EMBOSS needle alignment parameters for protein and nucleotide sequences

| Parameters | Default value for protein | Default value for nucleotide |
|---|---|---|
| Matrix (substitution scoring matrices) | BLOSUM62 [EBLOSUM62] | Default value is: DNAfull [EDNAFULL] |
| Gap open penalty (pairwise alignment score for the first residue in a gap) | 10 | Default value is: 10 |
| Gap extend penalty (pairwise alignment score for each additional residue in a gap) | Default value is: 0.5 | Default value is: 0.5 |
| Output formats (pairwise sequences format) | Default value is: pair | Default value is: pair |
| End gap penalty (apply end gap penalty) | Default value is: false | Default value is: false |
| End gap open penalty (score taken away when an end gap is created) | Default value is: 10 | Default value is: 10 |
| End gap extend penalty (penalty is added to the end gap penalty for each base or residue in the end gap. This is how long end gaps are penalized) | Default value is: 0.5 | Default value is: 0.5 |

5. Click '**Submit**' button for getting the results interactively in **Step 3**. If you wish to receive the results as link via. e-mail, check the box before 'Be notified by email'.

### Interpreting Results of Alignments

6. The results are presented under different tabs.
7. The '**Alignment**' tab shows the alignment of the two sequences.
8. In case of **amino acids**, matching amino acids are connected with a "|" symbol whereas mismatches are connected with a **space**. A gap in the alignment is represented with a "-" symbol. Similar conservative (that share same physicochemical properties) amino acids (Table 1.5) are connected via. a ":" symbol and non-conservative amino acids are connected via. a "." symbol (Fig. 1.70).
9. In case of **nucleotides**, matching nucleotides are connected with a "|" symbol whereas mismatches are connected via. a "." symbol. A gap in the alignment is represented with a "-" symbol.
10. The page also displays information like number of sequences, their length, identity, similarity gaps and score. The identity % shows the percentage of identical matches between the two sequences and similarity % reports aligned region having scoring matrix value is more than or equal to 0.
11. The second tab '**Submission Details**' displays the information regarding the program used, its version, input parameters, etc.
12. Clicking the final tab '**Submit another Job**' takes to the EMBOSS Needle—Alignment submission form.

**Table 1.5** Classification of amino acids based on physicochemical properties

| Nature | Amino acid | Three letter code | Single letter code |
|---|---|---|---|
| Basic amino acids (positively charged) | Arginine | Arg | R |
| | Histidine | His | H |
| | Lysine | Lys | K |
| Acidic amino acids (negatively charged) | Aspartic acid | Asp | D |
| | Glutamic acid | Glu | E |
| Polar amino acids (*hydrophilic*) | Asparagine | Asn | N |
| | Cysteine | Cys | C |
| | Glutamine | Gln | Q |
| | Glycine | Gly | G |
| | Serine | Ser | S |
| | Threonine | Thr | T |
| | Tyrosine | Tyr | Y |
| Nonpolar amino acids (*hydrophobic*) | Alanine | Ala | A |
| | Isoleucine | Ile | I |
| | Leucine | Leu | L |
| | Methionine | Met | M |
| | Phenylalanine | Phe | F |
| | Proline | Pro | P |
| | Tryptophan | Trp | W |
| | Valine | Val | V |

```
# Length: 258
# Identity:      88/258 (34.1%)
# Similarity:   123/258 (47.7%)
# Gaps:          61/258 (23.6%)
# Score: 443.0
#
#
#=======================================

O._sativa_FeS      1 MAAFASALRVLPSPPAAVPRRLRSREQRQGCRSRRYSKVVAYYALTTPPY     50
                                                              .....||.
N._punctiform      1 ------------------------------------------MVFVQPPL      8

O._sativa_FeS     51 KLD--ALEPY-ISKRTVELHWGKHQQDYVDSLNKQLATSMFYGYTLEELI     97
                     ..|  ||||. :...|.|.|:|||.:.|||:|||...:.....|||:|
N._punctiform      9 PFDKNALEPNGMKAETFEYHYGKHHKAYVDNLNKLTEGTELADKPLEEVI     58

O._sativa_FeS     98 KEAYNNGNPLPEYNNAAQVWNHHFFWESMQPEGGGSPGRGVLQQIEKDFG    147
                     |.::.:.:....:|||||||||.|||.|::|.|||:|...:..:|:||||
N._punctiform     59 KISFQDSSKAGIFNNAAQVWNHTFFWNSLKPNGGGAPTGDLAARIDKDFG    108

O._sativa_FeS    148 SFTNFREEFIRSALSLLGSGWVLLVLKRKERKLSVVHTQNAISPLALGDI    197
                     ||..|:|||..:|.:..||||..|:  .....|.|:.|.||.:|||.|..
N._punctiform    109 SFDKFKEEFSNAAATQFGSGWSWLI--DDGGTLKVIKTPNAENPLAHGKK    156

O._sativa_FeS    198 PLINLDLWEHAYYLDYKDDRRMYVTNFIDHLVSWDTVTLRMMRAEAFVNL    247
                     .|:.||:||||||:||::.|..::.||:|:||:||.....:.:|
N._punctiform    157 ALLTLDVWEHAYYIDYRNARPAFIKNFLDNLVNWDFAAENLAKA------    200

O._sativa_FeS    248 GEPNIPVA    255

N._punctiform    201 --------    200
```

**Fig. 1.70** Global alignment of protein sequences

**Exercise**

1. What is the significance of sequence alignment?
2. What is global alignment?
3. When a global alignment can be done?
4. What is meant by Gap Penalty?

## 1.14 Local Sequence Alignment

**Introduction**

Local alignment tools determine the identical or similar region or regions within an alignment. In this type of alignment, the sequences need not be related to each other as it looks whether a substring aligns with the other or not (Fig. 1.71). Local alignment tools uses Smith–Waterman (1981) dynamic programming algorithm for alignment. This can be helpful in detecting local similarities in large newly sequenced genomes or predicting conserved domains or motifs between two proteins.

### Local Alignment Using EMBOSS Water @ EBI

**Introduction**

EMBOSS Water program uses the modified Smith–Waterman algorithm for calculating the local alignment between two protein or nucleotide sequences.

**Exploration Activity**

1. Go to the 'Pairwise sequence alignment' tools page by typing http://www.ebi.ac.uk/Tools/psa in your browser (Fig. 1.72).
2. Under Local Alignment choose appropriate program (protein or nucleotide).
3. Input your first and second sequences by pasting in the text boxes provided in **Step 1**. Input sequence can be in GCG, FASTA, EMBL, GenBank, PIR, NBRF, Phylip or UniProtKB/Swiss-Prot format. ***Note***: *Avoid using data directly from word processors as it may yield unpredictable results due to the presence of hidden/control characters.*

```
CACAGCCTTTTTC------CTACATGTTGGTTGACGGTCAG
||||||||||..|      |||||||||||...||||||||
CACAGCCTTTAACTATGCGCTACATGTTGGAACACGGTCAG
```

**Fig. 1.71** Local alignment—occurs at a region of high-quality alignment within an overall larger alignment

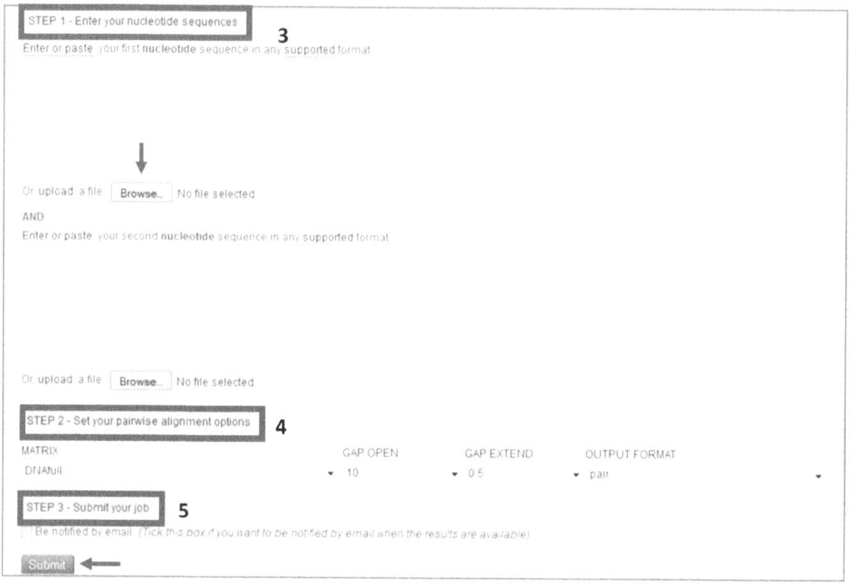


**Fig. 1.72** EMBOSS water pairwise alignment—nucleotide interface at EBI

**Table 1.6** Default of EMBOSS water alignment parameters for protein and nucleotide sequences

| Parameters | Protein | Nucleotide |
|---|---|---|
| Matrix (default substitution scoring matrices) | Default value is: BLOSUM62 [EBLOSUM62] | Default value is: DNAfull [EDNAFULL] |
| Gap open penalty (pairwise alignment score for the first residue in a gap) | Default value is: 10 | Default value is: 10 |
| Gap extend penalty (pairwise alignment score for each additional residue in a gap) | Default value is: 0.5 | Default value is: 0.5 |
| Output formats (pairwise sequences format) | Default value is: pair | Default value is: pair |

Alternatively, files containing valid sequences in any format (GCG, FASTA, EMBL, GenBank, PIR, NBRF, Phylip or UniProtKB/Swiss-Prot) can be uploaded using the 'Browse' options (provided below the text box) as input. ***Note***: *Avoid using word processors files. Preferably generate a file with sequence in FASTA format and save the file with extensions .fas or .fasta.*

4. Set the alignment options in **Step 2**. Generally you can work with default parameters. The default parameters are given in Table 1.6. You can also change these parameters and look how it affects the result.
5. Click '**Submit**' button for getting the results interactively in **Step 3**. If you wish to receive the results as link via. e-mail, check the box before 'Be notified by email'.

### Interpreting Results of Alignments

6. The results are presented under different tabs.
7. The '**Alignment**' tab shows the alignment of the two sequences.
8. In case of **amino acids**, matching amino acids are connected with a "|" symbol whereas mismatches are connected with a **space**. A gap in the alignment is represented with a "**-**" symbol. Similar conservative (that share same physicochemical properties) amino acids (Table 1.5) are connected via. a ":" symbol and non-conservative amino acids are connected via. a "." symbol (Fig. 1.70).
9. In case of **nucleotides**, matching nucleotides are connected with a "|" symbol whereas mismatches are connected via. a "." symbol. A gap in the alignment is represented with a "**-**" symbol.
10. The page also displays information like number of sequences, their length, identity, similarity gaps and score. The identity % shows the percentage of identical matches between the two sequences and similarity % reports aligned region having scoring matrix value is more than or equal to 0.

```
#=======================================
#
# Aligned_sequences: 2
# 1: O._sativa_FeSOD
# 2: N._punctiforme_FeSOD
# Matrix: EBLOSUM62
# Gap_penalty: 10.0
# Extend_penalty: 0.5
#
# Length: 194
# Identity:      86/194 (44.3%)
# Similarity:   123/194 (63.4%)
# Gaps:           3/194 ( 1.5%)
# Score: 445.5
#
#
#=======================================

O._sativa_FeS      49 PYKLDALEPY-ISKRTVELHWGKHQQDYVDSLNKQLATSMFYGYTLEELI     97
                      |:..:||||. :...|.|.|:|||.:.|||:|||...:......|||:|
N._punctiform       9 PFDKNALEPNGMKAETFEYHYGKHHKAYVDNLNKLTEGTELADKPLEEVI     58

O._sativa_FeS      98 KEAYNNGNPLPEYNNAAQVWNHHFFWESMQPEGGGSPGRGVLQQIEKDFG    147
                      |.::.:.:....:|||||||||.|||.|::|.|||:|...:..:|:||||
N._punctiform      59 KISFQDSSKAGIFNNAAQVWNHTFFWNSLKPNGGGAPTGDLAARIDKDFG    108

O._sativa_FeS     148 SFTNFREEFIRSALSLLGSGWVLLVLKRKERKLSVVHTQNAISPLALGDI    197
                      ||..|:|||..:|.:..||||..|:  .....|.|:.|.||.:|||.|..
N._punctiform     109 SFDKFKEEFSNAAATQFGSGWSWLI--DDGGTLKVIKTPNAENPLAHGKK    156

O._sativa_FeS     198 PLINLDLWEHAYYLDYKDDRRMYVTNFIDHLVSWDTVTLRMMRA    241
                      .|:.||:||||||:||::.|..::.||:|:||:||.....:.:|
N._punctiform     157 ALLTLDVWEHAYYIDYRNARPAFIKNFLDNLVNWDFAAENLAKA    200
```

Fig. 1.73 Local alignment of protein sequences

11. The second tab '**Submission Details**' displays the information regarding the program used, its version, input parameters, etc.
12. Clicking the final tab '**Submit another Job**' takes to the EMBOSS Needle—Alignment submission form.

**Exercise**

1. How local alignment is different from global alignment?
2. What is the application of local alignment?
3. What is the difference between 'identity' and 'similarity'?

## 1.15 Basic Local Alignment Search Tool (BLAST)—Nucleotide BLAST

**Introduction**

**BLAST Algorithm**

BLAST algorithm finds similar sequences in the database by splitting the query sequence into short subsequences called 'words' and tries to find the exact matches of those words in the database and comes out with 'hits'. The program then extend these word hits bi-directionally in multiple steps to produce high scoring pairs (HSPs) and generate the final gapped alignments.

**BLAST Interface**

BLAST search has four components: query, database, program, and search purpose/goal. The links in the **Basic BLAST** section of the 'Home' tab lead to BLAST forms that share a common design (Johnson et al. 2008). Each form has four sections:

**Section 1: Enter Query Sequence**

This section allows the user to input the query sequence by copy paste or by browsing the file from your computer. There is also a provision to specify a sub-range to be searched. An optional title for the job can also be provided. BLAST by default provides title automatically for each search. If the entered sequence is in FASTA format, the description provided in the sequence will be taken as the title for the job (Fig. 1.74).

**Section 2: Choose Search Set**

For nucleotide (**blastn**), you can select a database for searching and can also limit your search by an organism, optionally (Fig. 1.75).

The default database is '**Others (nr etc.)**', which search the non-redundant database. If your search is restricted to Human or Mouse, you can choose, respective database.

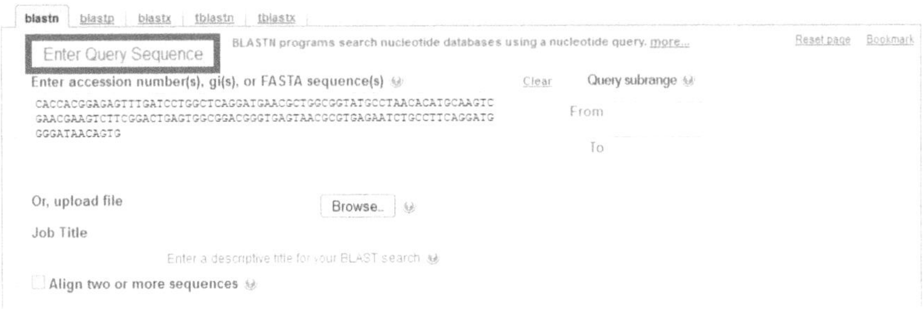


**Fig. 1.74** Enter query sequence

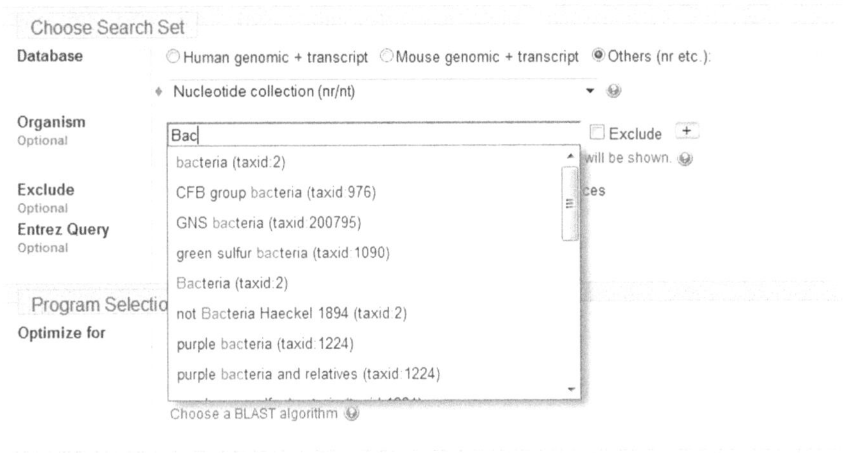


**Fig. 1.75** Choose search set

If you want to target BLAST against a particular organism, then typing a text in the **Organism** box, pops up an auto-complete text box where you can select the (scientific or common) name or taxid of an organism. For protein (blastp), the default database is **nr** (non-redundant).

**Section 3: Program Selection**

This allows you to using different scenarios (say, intra- vs. inter-species searches). The choices include megablast, discontiguous megablast, and blastn in case of nucleotide (Fig. 1.76).

- **Megablast** works very fast and works best if the target percent identity is 95% or more
- **Discontiguous megablast** can be used for cross-species comparisons as it allows some mismatch in the initial rounds of searching
- **BlastN** though slow, it can allows a word size as lower as seven bases.

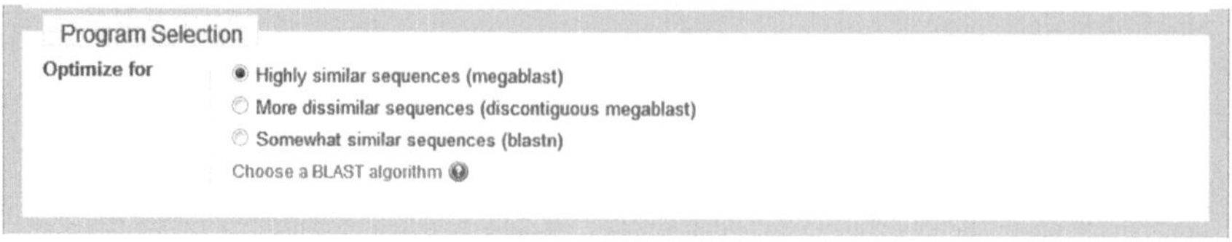


**Fig. 1.76** Program selection

**MEGABLAST is the Tool to Identify Identical Nucleotide Sequence**.
Though programs like MEGABLAST, discontiguous megablast, and blastn can identify an unknown sequence by checking if that sequence already exists in a public database, for accessing biological information, MEGABLAST efficiently find **identical match** to the query sequence by providing an **adjustable percent identity cut-off** for the alignment, in addition to the significance cut-off threshold set which is set by Expect value.

Discontiguous MEGABLAST on the other hand can find similar nucleotide sequences, but not identical, to the nucleotide query. In this program, third base wobbling is considered for finding matches at the first and second codon positions while the mismatches in the third position is ignored. Discontiguous MEGABLAST surpasses blastn when same word size is used for searching, but the word size is adjustable in blastn which can be reduced from the default value of 11 to a minimum of 7 so as to increase the search sensitivity.

**Searching Primer or Short Nucleotide Searches**
Due to stringent parameter settings searching short sequences like less than 20 bases, will not returns any significant matches. But BLAST page at NCBI have provisions to adjust both the word size and expect value with preset values to give optimal results with short sequences (Table 1.7).

This finds application in checking the specificity of PCR or hybridization primers. This doesn't require reverse complementing the reverse primer, as BLAST automatically searches both strands using local alignments.

**Table 1.7** Parameter settings for searching short and nearly exact matches

| Program | Word size | DUST filter setting | Expect value |
|---|---|---|---|
| Standard blastn | 11 | On | 10 |
| Search for short nearly exact matches | 7 | Off | 1000 |

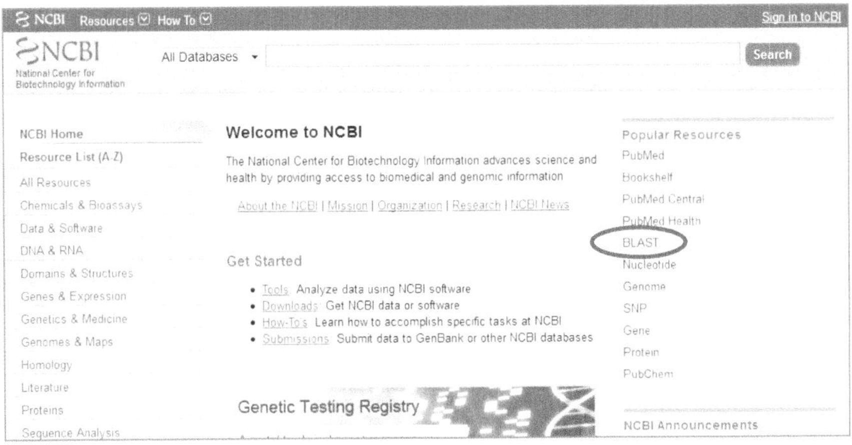


Fig. 1.77 NCBI home page

## Exploration Activity

## Doing Nucleotide BLAST Search

1. Prepare the **query nucleotide sequence** (sequence to be BLAST) in raw format or in FASTA format.
2. Go to NCBI home page available at http://www.ncbi.nlm.nih.gov/ and choose BLAST program (Fig. 1.77).
3. Choose the '**nucleotide BLAST**' program (Fig. 1.78).
4. Copy and paste the sequence in the sequence box or alternatively you can also browse and upload the sequence in FASTA format from your computer (Fig. 1.79).

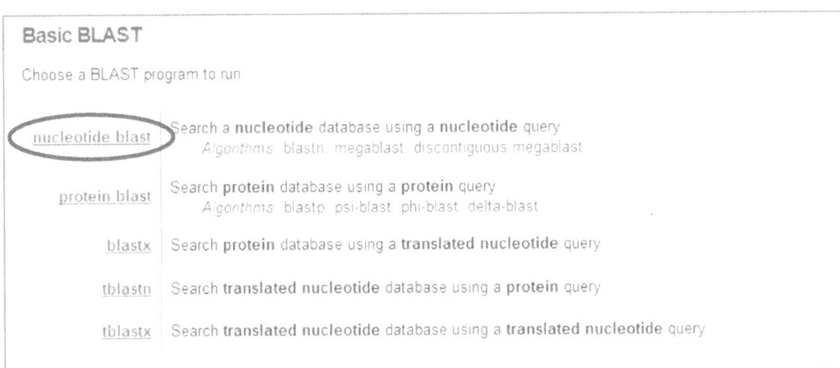


Fig. 1.78 Basic BLAST programs in NCBI

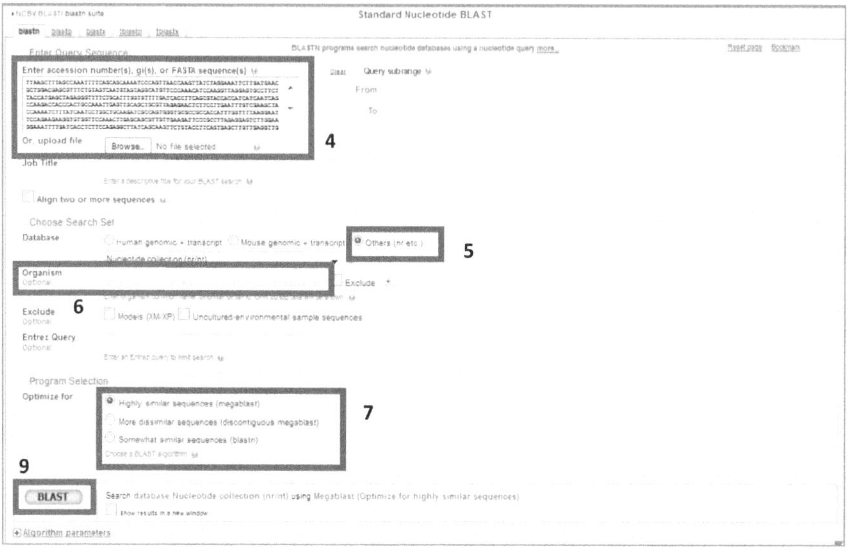


**Fig. 1.79** Nucleotide BLAST interface in NCBI

5. Select the **database** you want to search against (preferably 'Others' (nr—non-redundant).
6. If you wish to search against any specific organism, under 'Organism', enter first few characters of the organism's name and select the organism of interest from the dropdown list.
7. Choose the appropriate program under 'Program Selection'.
   a. MEGABLAST—to identify identical nucleotide sequence (Fig. 1.80).
   b. Discontiguous MEGABLAST—to identify similar nucleotide sequences, but not identical (Fig. 1.81).
   c. Blastn—to identify related nucleotide sequences from other organisms (distantly related) (Fig. 1.82).
8. Adjust any parameter, if required. Any parameter changed is highlighted in yellow.
9. Click 'BLAST' button at the end of the page.

## Exercise

1. What is meant by Identical, Similar and Homologous sequences?
2. Explain BLAST algorithm.
3. Define threshold value, hit and HSP?
4. Differentiate Megablast from Blastn?
5. What is meant by non-redundant (nr) database?

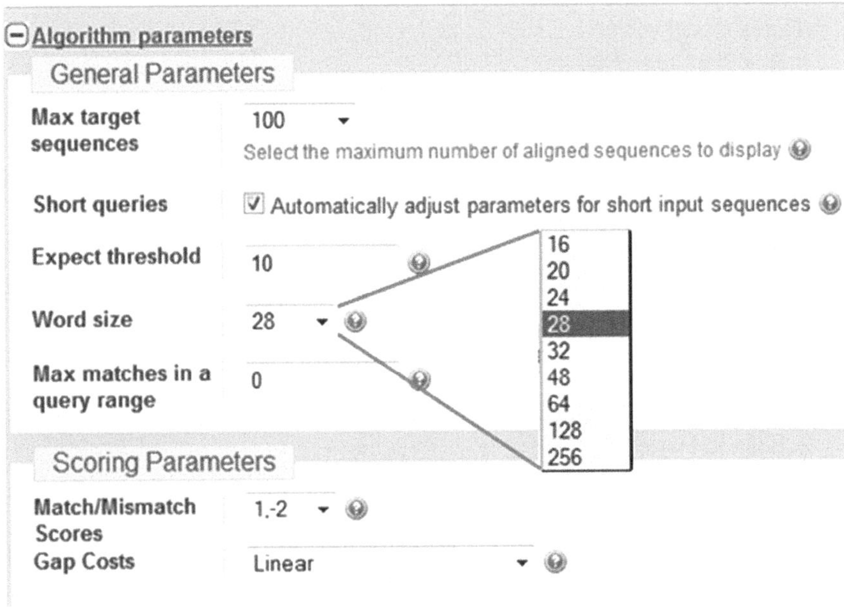


Fig. 1.80 Megablast algorithm parameters

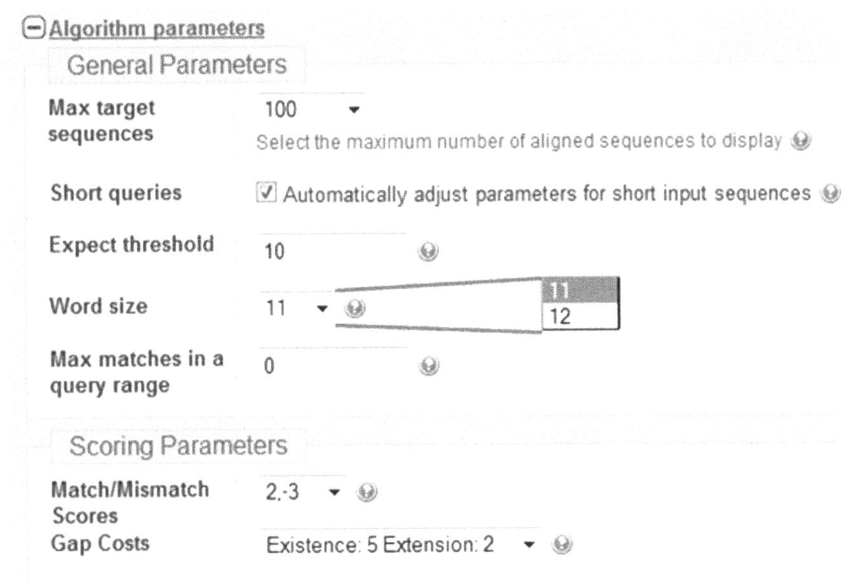


**Fig. 1.81** Discontiguous megablast algorithm parameters

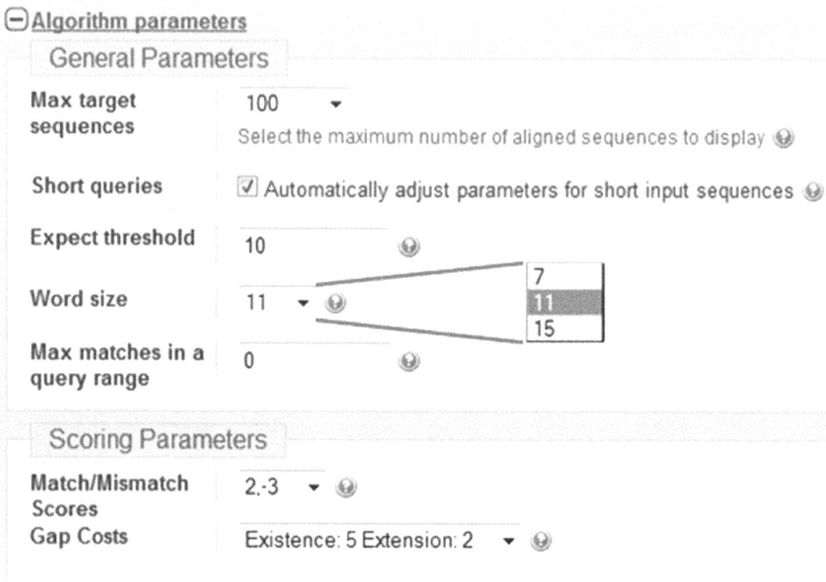


**Fig. 1.82** Blastn algorithm parameters

## 1.16 Interpreting BLAST Result

**Introduction**
Interpreting BLAST result requires prior knowledge on the following terminologies (Fassler and Cooper 2008):

**Description**: Information about the sequence record for a particular hit.
**Hit**: Matched sequence in the database.
**Score**: A score is numerical value that describes the overall quality of the alignment.
**Bit score**: It is a normalized score expressed in bits. It is log scaled version of the score. For example, a bit score of 30 indicates $2^{30}$ = 1 billion independent segment pairs to find a score this core by chance.
**Max score**: Highest alignment score (bit score) between query and database sequence segment.
**Total score**: Sum of alignment scores of all segments from the same database sequence that match the query sequence (calculated over all segments). This score is different from the max score, if several parts of the database sequence match different parts of the query sequence.
**Query coverage**: Percent of query length that is included in the aligned segments. This coverage is calculated over all segments.

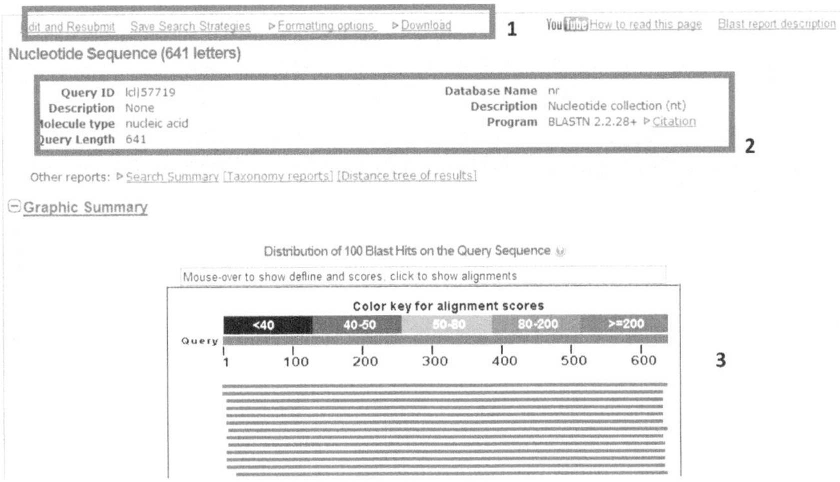


Fig. 1.83 The header section of the BLAST result

**E value (Expect value)**: Number of alignments expected by chance with a particular score or better. It is derived from P value. The E-value below $10^{-4}$ usually indicates the evidence for homology.
**Max identity**: BLAST calculates the percentage identity between the query and the hit in a nucleotide-to-nucleotide alignment. If there are multiple alignments with a single hit, then only the highest percent identity is shown.
**Accession**: A unique identifier number of a specific (sequence) GenBank record.

The BLAST result consists of three major sections:

(1) **The Header**—containing information about the query sequence, the database searched, BLAST version and its release date. It also provides a graphical overview of the query coverage;
(2) **The Descriptions**—shows the descriptions of each sequence found to match in the database in one line. It also provides a quick overview.
(3) **The Alignments**—displaying the pairwise alignments of the query sequence against each database sequence matched.

**The Header Section**

1. The top line provide links (Fig. 1.83) to:
   **'Edit and Resubmit'**—for editing and doing again the blast search,
   **'Save Search Strategies'**—to save search parameters as ASN.1 (Abstract Syntax Notation ver. 1—file format),
   **'Formatting options'**—to modify the formatting options and

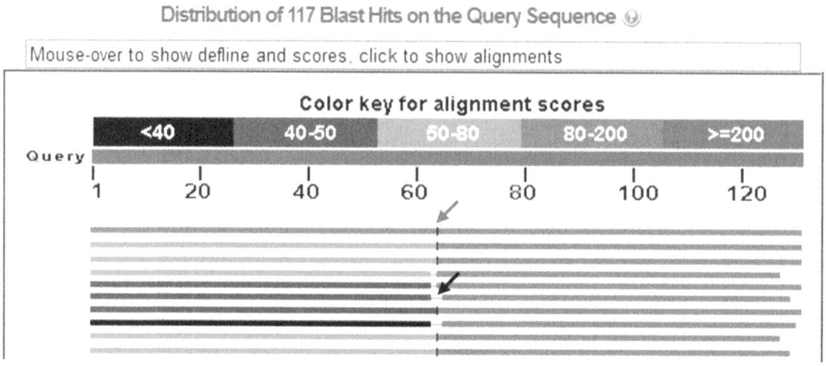


**Fig. 1.84** HSPs are from the same sequence

'**Download**'—to download the result in Text, XML, ASN.1, Hit Table (as text), and Hit Table (as csv).

2. It is followed by the information about molecule type, it length, the database selected, the type of program and the version.
3. It is then followed by the **graphical output**. The query sequence is represented by the scaled red bar above of the figure.
4. The hits from the database are shown below the red bar; the most similar sequences closest to the query are listed. Sometimes a thin **black vertical line (|)** is observed in the bar (hit). This indicates two HSPs are from the same sequence. Here a perpendicular line (arrow) indicates that the distance between the two HSPs is less whereas, the horizontal line (arrow) indicates more distance between the two HSPs (Fig. 1.84).
5. The figure also provides a colour key to represent the match of the hits. The length of the colour bar indicates the query coverage. Moving the mouse cursor over the bars displays the definition line above the graphic for that sequence (Fig. 1.84).

### The Descriptions

6. The **One-line descriptions** (Fig. 1.85) is composed of (a) a brief textual description, (b) the max score, (c) the total score, (d) the query coverage, (e) the E value, (f) the maximum identity and the accession number. Clicking the hyperlinks will open corresponding GenBank record.
7. The one-line descriptions are useful to have a quick overview of the search results.

### The Alignments

8. Above the pairwise sequence alignment is the sequence identifier followed by a full definition line and the length of the target (matched) sequence.

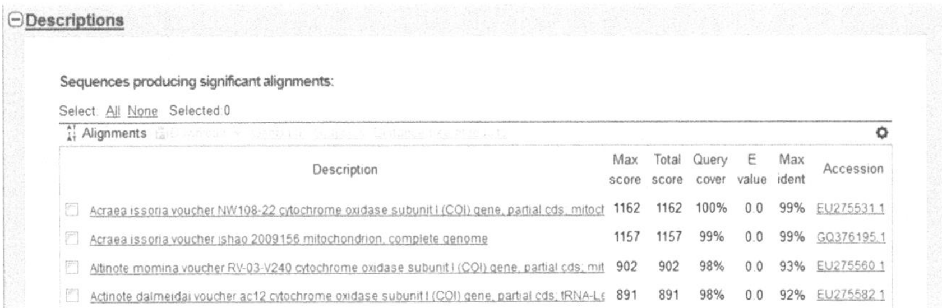


**Fig. 1.85** Description section in BLAST result

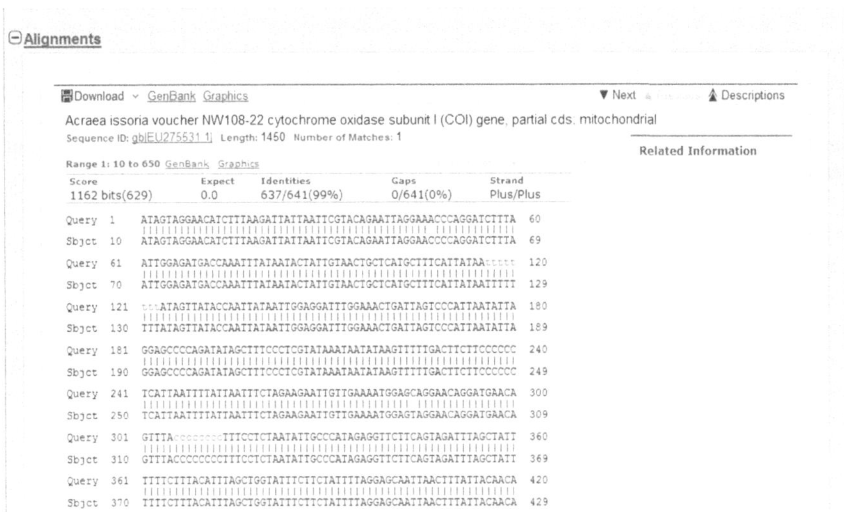


**Fig. 1.86** Alignments section in BLAST result

The 'Range 1:' indicates the first segment that is matched and the following values tell the range that is being matched with our query sequence (Fig. 1.86).

9. It then followed by the bit score (the raw score represented in parentheses) and then the E value, the percentage of identities, gaps and finally the strand. A plus/plus indicates that both the query and subject are in forward strand. A plus/minus indicates that the query is in reverse strand. This is because all the sequence in the database is stored only as plus strand.
10. The '**Query**' refers to our query sequence and the '**Sbjct**' refers to the subject sequence (the match) from the database.
11. By default, BLAST displays 100 sequence matches. This can be changed on the BLAST page under '**Algorithm parameters**' option. Components in the

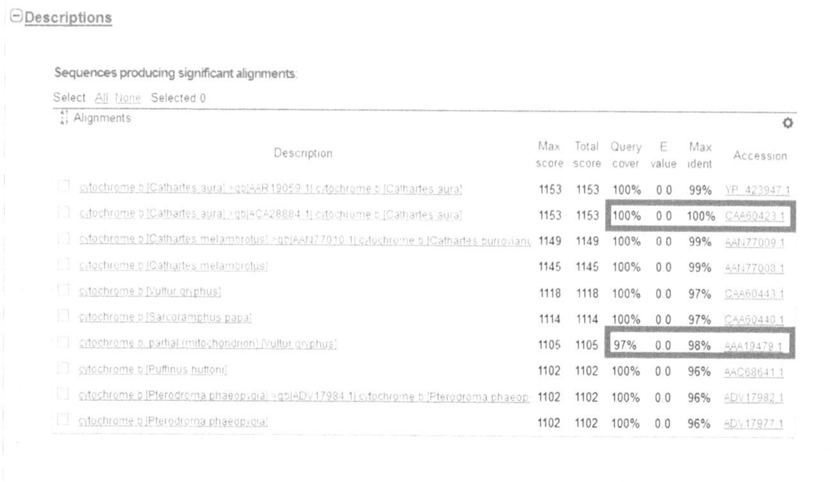


**Fig. 1.87** Blast results for cytochrome b from *Cathartes aura*. Protein Blast with default parameters was used

BLAST results are hyperlinked so as to provide more information about the sequence an also includes links to abstracts in PubMed.

In the above Fig. 1.87, our query sequence for BLAST search is cytochrome b from *Cathartes aura* (Sequence ID: emb|CAA60423.1|Length: 342). The second hit shows 100% Max identity (This hit is our query sequence) with E value = 0. This is because of both query and subject having same length (1–342) (Fig. 1.88).

**Fig. 1.88** Alignment for cytochrome b from *Cathartes aura*

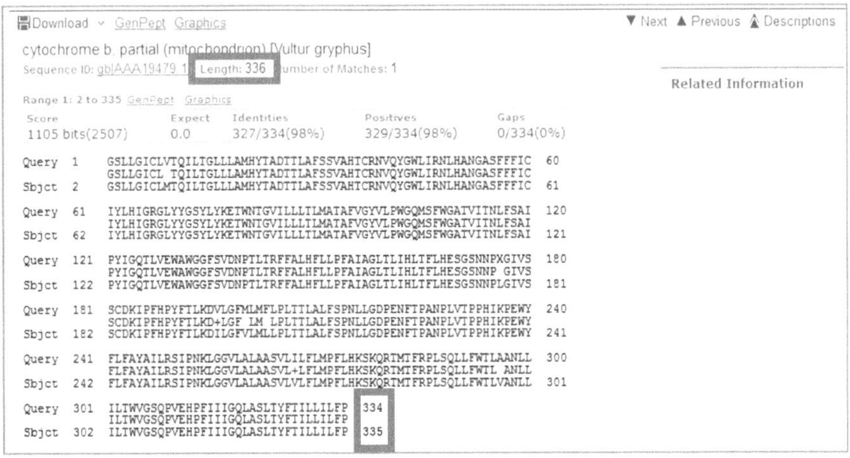


**Fig. 1.89** Alignment of cytochrome b protein sequence of *Vultur gryphus* and *Cathartes aura*

But with *Vultur gryphus*, the query coverage is only 98%. This is because only 334 amino acids of our sequence is matched to the query sequence (hit of length 336) with E value = 0 (Fig. 1.89).

The Query coverage, Max identity, Total score and Max score are different between *V. gryphus* and *C. aura* due to alignment scoring system (Fig. 1.87). Though the query coverage is 100% and E value is 0, still difference is observed in Max identity and Max score. This is reflected in bit score which was calculated from the Max identity and Max score.

Since BLAST has provisions for using different parameters like substitution matrices, it report results in a manner that can compare alignments made with different scoring matrices or gap penalties. This is done by reporting $S'$ values (bit scores) which is calculated:

$$S' = (\lambda S - \ln K)/\ln 2$$

where which $\lambda$ and $K$ reflect the matrices and penalties used. Stopping at this level will produce a list of sequences arranged by bit scores that would show the degree of similarity to the query sequence.

But whether the sequences were homologs to the query sequence cannot be addressed as larger databases reports sequences with matches to the query that are merely due to chance and not due to homology.

To solve this issue, E-values are calculated from bit scores using the expression

$$E = (n \times m)/2^{S'}$$

where $n$ is the total number of residues (amino acids or nucleic acids) in the database, and $m$ is the length of the query sequence.

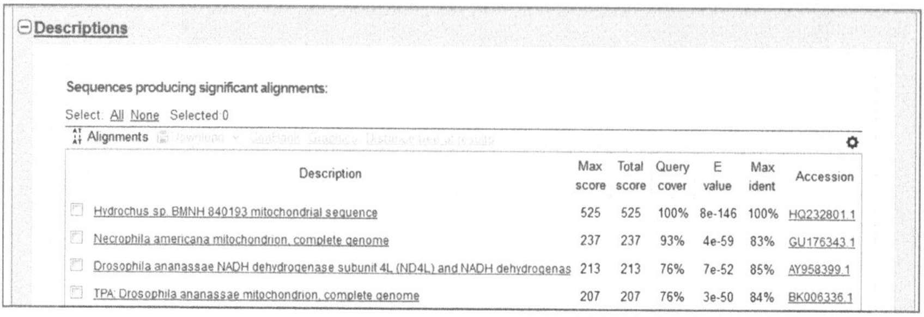
Descriptions

Sequences producing significant alignments:

Select: All None Selected 0

Alignments

| Description | Max score | Total score | Query cover | E value | Max ident | Accession |
| --- | --- | --- | --- | --- | --- | --- |
| Hydrochus sp. BMNH 840193 mitochondrial sequence | 525 | 525 | 100% | 8e-146 | 100% | HQ232801.1 |
| Necrophila americana mitochondrion, complete genome | 237 | 237 | 93% | 4e-59 | 83% | GU176343.1 |
| Drosophila ananassae NADH dehydrogenase subunit 4L (ND4L) and NADH dehydrogenas | 213 | 213 | 76% | 7e-52 | 85% | AY958399.1 |
| TPA: Drosophila ananassae mitochondrion, complete genome | 207 | 207 | 76% | 3e-50 | 84% | BK006336.1 |

**Fig. 1.90** The arrangement of results based on E-values

E-values are the number of hits that can be expected to be taken from the database having a bit score equal to or greater than the one calculated from the alignment of the query and hit based on chance alone. E-values for hits that are very similar to the query sequence will be small, i.e., nearer to zero and are widely used as a measure to assess the confidence with which one should claim the subject sequence(s) and the query sequence as homologs. E-value ranging from 0 to 0.0001 ($10^{-4}$) normally indicates the evidence for homology (Fig. 1.90). In blast result, it is based on the E-values, the hits are ranked.

**Exercise**

1. Differentiate E-value from bit score.
2. What is the difference between Total score and Max score?
3. What do you mean by Max identity?
4. What do mean by plus/plus and plus/minus strand?
5. Arrange the E-value in ascending order: 8e−146, 7e−52, 0.0, 3e−45, 4e−44.

## 1.17 Multiple Sequence Alignment: T-Coffee for Small Alignments

**Introduction**

**Multiple Sequence Alignment (MSA)** aligns of three or more protein or nucleic acid sequences of similar length. This is not only useful in inferring homology by identifying conserved regions but also inferring evolutionary relationships between the sequences. This is designing experiments to test and modify the function of specific proteins and in identifying new members of protein families.

Like pairwise alignments, sequences can be aligned considering their entire length as in the case of global alignment or only in particular regions, where it is called as local alignment in multiple alignments. Global alignments use gaps (representing insertions or deletions), while local alignments align the regions between gaps.

## Conserved Substitutions

Substitution of one amino acid residue with another residue having similar physiochemical properties is termed as conserved or conservative substitution. Apart from the evolutionary pressures favouring the conservation of secondary structure, from the evolutionary point of view, it was observed that the hydrophobicity and the molecular bulk of the side chain tend to be conserved (French and Robson 1983).

## Progressive Algorithm

For the sequences to be aligned, the possible number of pairs is calculated first (Fig. 1.91). Then between the pairs, pairwise alignments are done and from the scores, distance is calculated (Higher the lesser the distance). From the distance, a guide tree is then generated. Based on the guide tree, again alignments are done in a progressive manner.

**T-Coffee** (**T**ree-based **C**onsistency **O**bjective **F**unction **F**or alignment **E**valuation) is a multiple sequence alignment program that allows you to merge results obtained from several alignment methods. It is suitable for small alignments and generates consistent alignments by overcoming the drawbacks of the progressive alignment methods. By default, T-Coffee produces a global alignment by comparing all the sequences two by two and then a series of local alignments using the lalign algorithm and finally produces a multiple alignment by combining all these alignments. Though T-Coffee uses progressive approach like ClustalW, it is incorporated advanced features like identifying the presence of motifs and evaluating the quality of the alignments (Notredame et al. 2000).

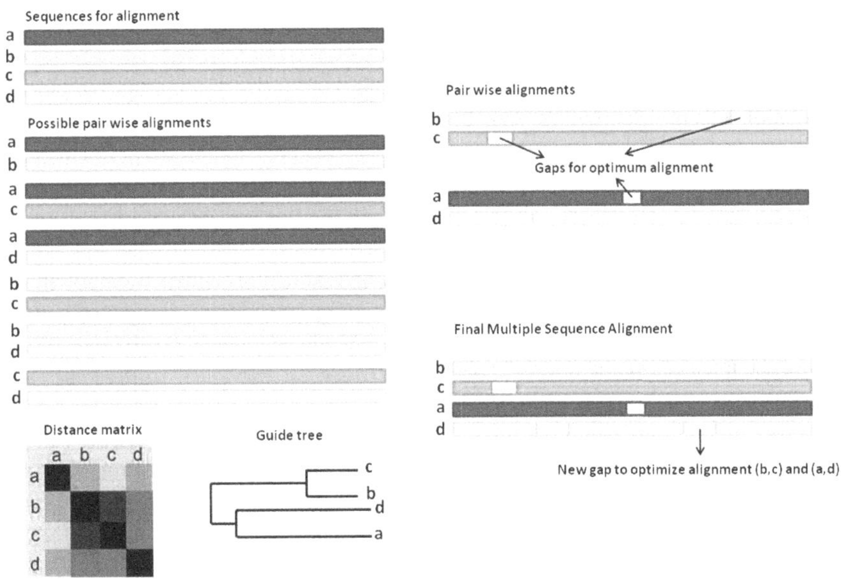


**Fig. 1.91** Progressive algorithm

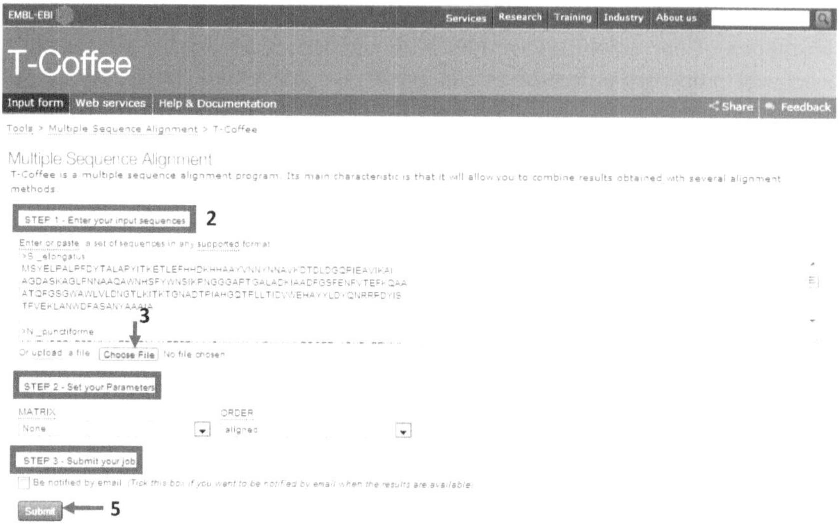


Fig. 1.92 T-Coffee multiple sequence alignment interface at EBI

**Exploration Activity**

To do multiple sequence alignment of iron superoxide dismutase (FeSODs) of *Oryza sativa* subsp. *indica* (B8B2C9), *Arabidopsis thaliana* (P21276), *Escherichia coli* (P0AGD3), *Nostoc punctiforme* (B2IZB2) and *Synechococcus elongatus* strain PCC 7942 (P18655). The sequences are to be prepared in FASTA format as a single file (Notepad).

1. Go to the T-Coffee multiple sequence alignment page by typing http://www.ebi.ac.uk/Tools/msa/tcoffee/ in your browser (Fig. 1.92).
2. Input your sequences by pasting in the text boxes provided in **Step 1**. This tool reads three or more (protein or nucleotide) sequences. Input sequence can be in GCG, FASTA, EMBL, GenBank, PIR, NBRF or UniProtKB/Swiss-Prot format. But they do not accept partially formatted sequences. Generally, adding a return (pressing ENTER button on keyboard) to the end of the sequence may help, in most cases, understand the input by these programs. ***Note***: *Avoid using data directly from word processors as it may yield unpredictable results due to the presence of hidden/control characters. There is currently a limit of 500 sequences and 1 MB of data.*
3. Alternatively, file containing three or more valid sequences in any format (GCG, FASTA, EMBL, GenBank, PIR, NBRF or UniProtKB/Swiss-Prot) can be uploaded using the '**Choose File**' options (provided below the text box) as input. ***Note***: *Avoid using word processor files. Preferably generate a file with sequence*

*in FASTA format and save the file with extensions .fas or .fasta. There is currently a limit of 500 sequences and 1 MB of data.*

4. Click on '**more options**' button to set the alignment options in **Step 2**. Generally you can work with default parameters. You can also change these parameters and look how it affects the result.
   **Matrix**
   Matrix series to be used when generating the multiple sequence alignment. The program goes through the chosen matrix series, spanning the full range of amino acid distances.
   Default value is: '**None**'
   Other options: BLOSUM and PAM
   **Order**
   The order in which the sequences appear in the final alignment
   Default value is: '**aligned**'
   Other option: 'input'
5. Click '**Submit**' button for getting the results interactively in **Step 3**. If you wish to receive the results as link via. e-mail, check the box before '**Be notified by email**' (Fig. 1.92).

### Interpreting T-Coffee Results

6. The results are presented under different tabs. The '**Alignment**' tab (default) shows the alignment in aln format (Fig. 1.93).

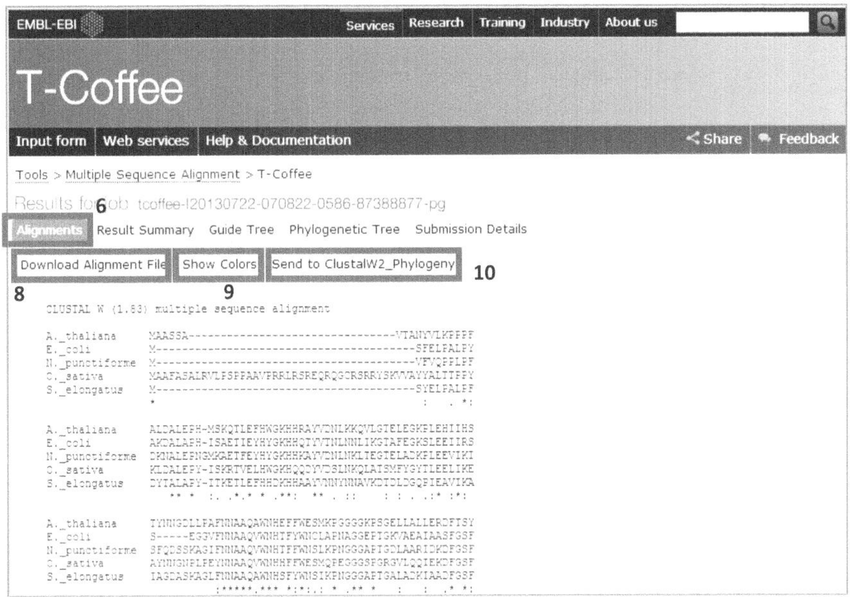


Fig. 1.93 The multiple sequence alignment in aln format

## Consensus Symbols

7. By default an alignment will display the following symbols that denote the degree of conservation observed in each column:

   "*" refers to the residues or nucleotides in that column are **identical** in all sequences in the alignment.
   ":" indicates that **conserved substitutions** have been observed.
   "." indicates that **semi-conserved substitutions** are observed.

8. Clicking the '**Download Alignment File**' will allow you to download the alignment in .aln format.
9. Also you can change the colour option in the output results. By clicking the button '**Show Colors**', the alignment will be shown in colour (Fig. 1.94).
10. The multiple sequence alignment can be directly parsed to ClustalW2 Phylogeny program by clicking the button '**ClustalW2_Phylogeny**'. This allows the user to control the method of tree construction.
11. The second tab '**Result Summary**' displays the result files (Fig. 1.95) comprising the input sequences for the alignment (.input), tool output (.output), which is a log file created during the alignment, alignment in HTML format (.html), alignment in **PHYLIP format** (.phylip), alignment in CLUSTAL format (.clustalw), alignment in **MSF format** (.msf) and guide tree (.dnd) that contains the information for building the cladogram or phylogram.
12. The '**Start JalView**' button under JalView triggers JalView, a Java based editor in new window (Fig. 1.96). **This requires Java program to be preinstalled**.

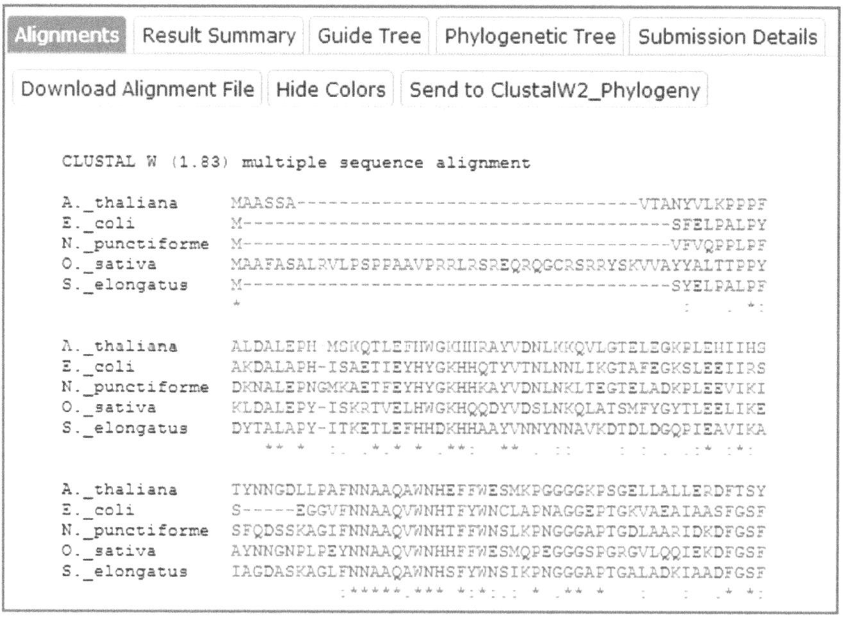


**Fig. 1.94** The alignment displayed in colour based on their physicochemical properties

**Fig. 1.95** The result summary along with the JalView trigger button

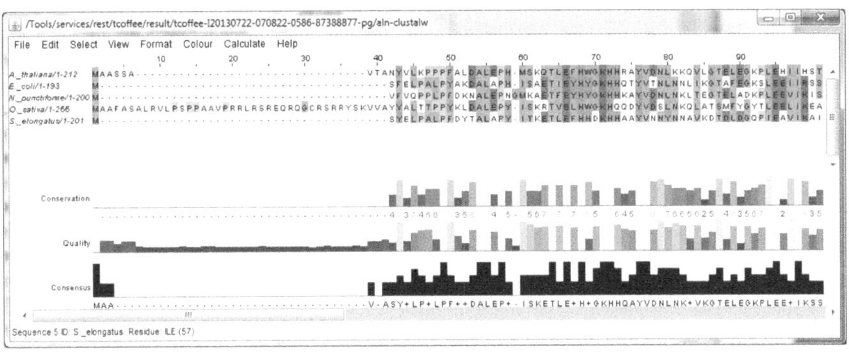


**Fig. 1.96** The JalView editor

13. The third '**Guide Tree**' tab displays by default displays the phylogram. The other buttons on the page include, "Download Guide Tree File", "Show as Cladogram Tree" and "Show Distances" (Fig. 1.97).

A Phylogram and Cladogram are branching diagram (tree) that are assumed to be an estimate of a phylogeny. In a Phylogram the length of the branch is proportional to the amount of inferred evolutionary change where as in a Cladogram do not show the amount of evolutionary "time" separating taxa as the branch length are equal and show a common ancestry. By right-clicking on the applet, the options can be controlled.

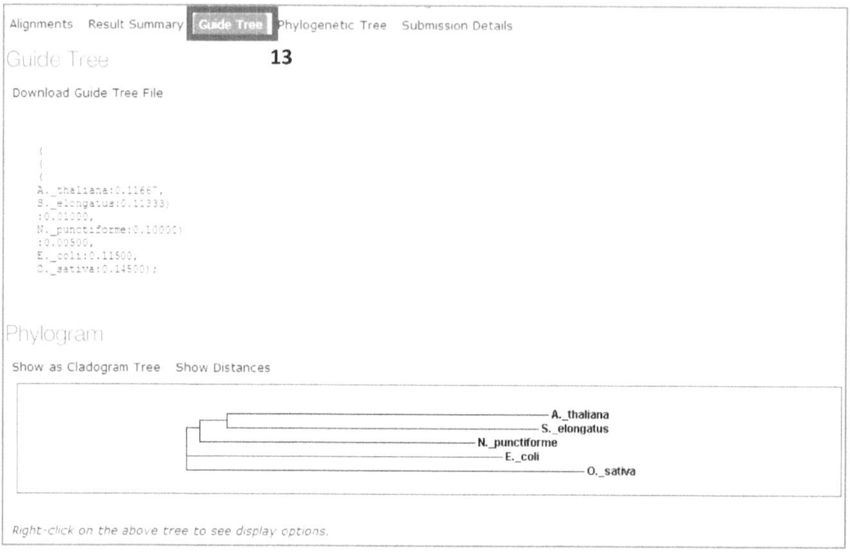


**Fig. 1.97** The guide tree generated during alignment process

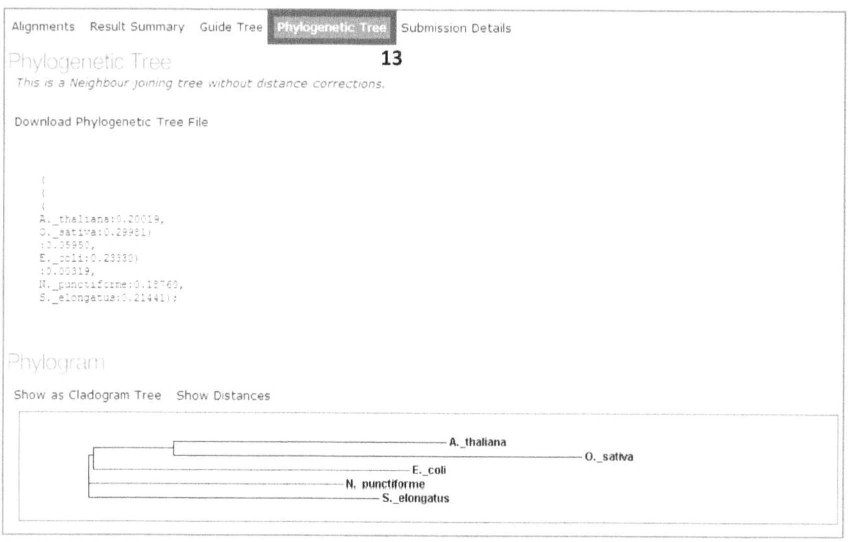


**Fig. 1.98** The neighbour-joining tree created for the sequences used

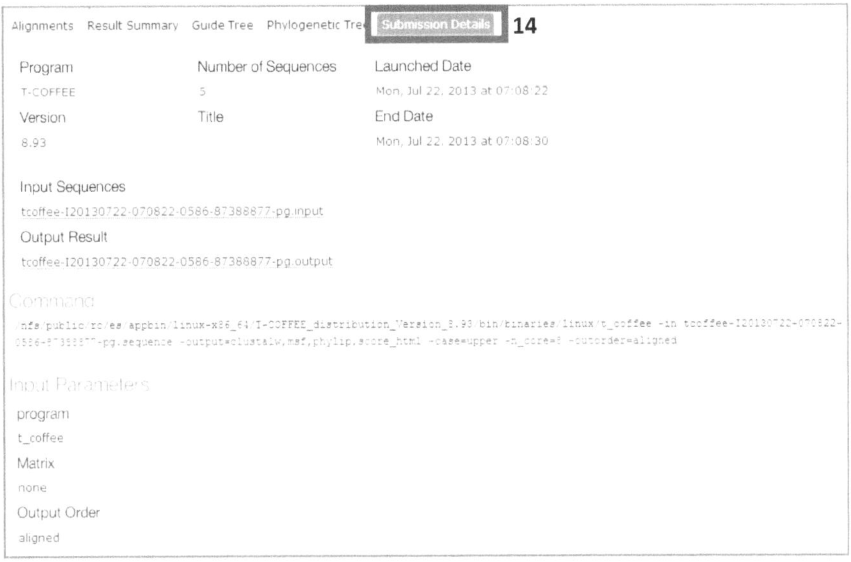


**Fig. 1.99** The submission details of the alignment

14. The fourth '**Phylogeny Tree**' tab displays the phylogenetic tree of the sequences used. It is actually a Neighbour-joining tree without correcting the distances (Fig. 1.98).
15. The fifth tab '**Submission Details**' displays the information regarding the program used, its version, input parameters etc. (Fig. 1.99).

The multiple sequence alignment basically indicates that all the sequences are moderately related. This can be inferred from the amount of '*', ':' and '.' observed throughout the sequence. Distantly related sequences will have sparse symbols like '*', ':' and '.', whereas a closely related one will have high occurrence of '*', ':' and '.'. This forms the basis for phylogenetic analysis, which requires the degree of similarity for choosing the method of tree construction.

It is also evident from the alignment that the active sites of all the FeSODs are highly conserved (Fig. 1.94).

## Exercise

1. Differentiate between conserved and semi-conserved substitutions.
2. What do you mean by a consensus sequence?
3. Differentiate between phylogram from cladogram.
4. What is guide tree and how is it different from a phylogenetic tree?
5. What do you mean by progressive algorithm?

## 1.18 Multiple Sequence Alignment—MUSCLE for Medium Alignments

**Introduction**

MUSCLE stands for **MU**ltiple **S**equence **C**omparison by **L**og-**E**xpectation. MUSCLE is accurate especially with proteins and faster than ClustalW2 or T-Coffee, depending on the options selected and is best suited for medium alignments.

MUSCLE algorithm includes fast distance estimation using *k*mer counting, progressive alignment using a new profile function called the log-expectation score, and refinement using tree-dependent restricted partitioning. Without any refinement, the speed and accuracy of MUSCLE are comparable with that of CLUSTALW, T-Coffee and MAFFT. MUSCLE is the fastest of the tested methods for aligning 5000 sequences having an average length 350 (Edgar 2004).

**Exploration Activity**

To do multiple sequence alignment of iron superoxide dismutase (FeSODs) of *Oryza sativa* subsp. *indica* (B8B2C9), *Arabidopsis thaliana* (P21276), *Escherichia coli* (P0AGD3), *Nostoc punctiforme* (B2IZB2) and *Synechococcus elongatus* strain PCC 7942 (P18655). The sequences are to be prepared in FASTA format as a single file (in Notepad).

1. Go to the MUSCLE multiple sequence alignment page by typing http://www.ebi.ac.uk/Tools/msa/muscle/ in your browser (Fig. 1.100).

Fig. 1.100 MUSCLE multiple sequence alignment interface at EBI

2. Input your sequences by pasting in the text boxes provided in **Step 1**. This tool reads three or more (protein or nucleotide). Input sequence can be in GCG, FASTA, EMBL, GenBank, PIR, NBRF or UniProtKB/Swiss-Prot format. But they do not accept partially formatted sequences. Generally, adding a return (pressing ENTER button on keyboard) to the end of the sequence may help, in most cases, understand the input by these programs. ***Note***: *Avoid using data directly from word processors as it may yield unpredictable results due to the presence of hidden/control characters. There is currently a limit of 500 sequences and 1 MB of data.*
3. Alternatively, file containing three or more valid sequences in any format (GCG, FASTA, EMBL, GenBank, PIR, NBRF or UniProtKB/Swiss-Prot) can be uploaded using the '**Choose File**' button (provided below the text box) as input. ***Note***: *Avoid using word processors files. Preferably generate a file with sequence in FASTA format and save the file with extensions .fas or .fasta. There is currently a limit of 500 sequences and 1 MB of data.*
4. Click on 'more options' button to set the alignment options in **Step 2**. Change the output format to '**clustalw**' to display the result in aln format. Default value is Pearson/FASTA [fasta]. Other output options are listed in Table 1.8.
   **Output Order**
   The order in which the sequences appear in the final alignment
   Default value is: aligned
5. Click '**Submit**' button for getting the results interactively in **Step 3**. If you wish to receive the results as link via. e-mail, check the box before 'Be notified by email'.

**Interpreting the Results**

6. The results are presented under different tabs (Fig. 1.101).
7. The '**Alignment**' tab (default) shows the alignment format specified during input.

**Interpreting the Results**

6. The results are presented under different tabs (Fig. 1.101).
7. The **Alignment** tab (default) shows the alignment format specified during input.

**Consensus Symbols**

By default an alignment will display the following symbols that denote the degree of conservation observed in each column:

**Table 1.8** Output options supported by MUSCLE program

| Output format | Description | Abbreviation |
|---|---|---|
| Pearson/FASTA | Pearson or FASTA sequence format | Fasta |
| ClustalW | ClustalW alignment format without base/residue numbering | Clw |
| ClustalW (strict) | Strict clustalW alignment format without base/residue numbering | Clwstrict |
| HTML | HTML format coloured alignment | Html |
| GCG MSF | GCG multiple sequence file (MSF) alignment format | Msf |
| Phylip interleaved | PHYLIP interleaved alignment format | Phyi |
| Phylip sequential | PHYLIP sequential alignment format | Phys |

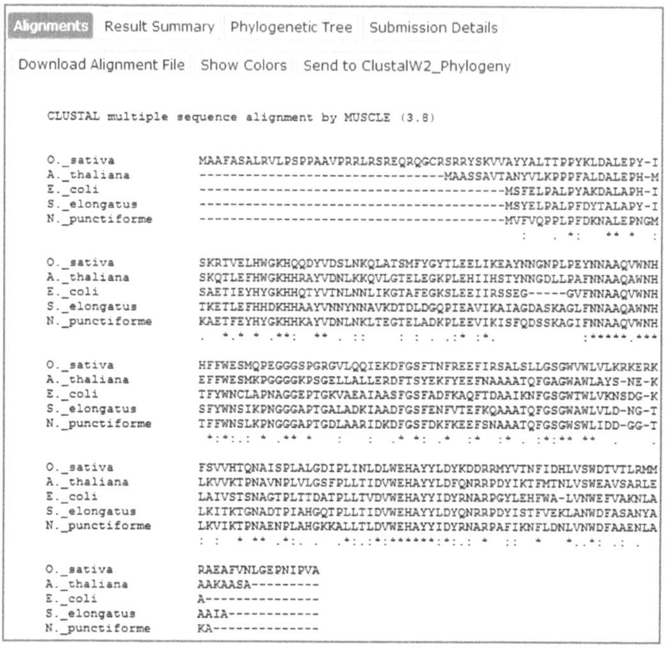


**Fig. 1.101** Multiple sequence alignment generated by MUSCLE

"*" refers to the residues or nucleotides in that column are identical in all sequences in the alignment.
":" indicates that conserved substitutions have been observed.
"." indicates that semi-conserved substitutions are observed.

8. Clicking the '**Download Alignment File**' will allow you to download the alignment in .aln format.
9. Also you can change the colour option in the output results. By clicking the button '**Show Colors**', the alignment will be shown in colour.
10. The second tab '**Result Summary**' displays the result files comprising the input sequences for the alignment (.input), tool output (.output), which is a log file created during the alignment, and alignment in CLUSTAL format (.clw)
11. The '**Start JalView**' button under JalView triggers JalView, a Java based editor in new window.
12. The third '**Phylogeny Tree**' tab displays the phylogenetic tree of the sequences used. It is actually a Neighbour-joining tree without correcting the distances
13. The fourth tab '**Submission Details**' displays the information regarding the program used, its version, input parameters, etc.

Using same sequence set, alignments from both T-Coffee and MUSCLE were generated for comparison purpose (Figs. 1.101 and 1.102). Both the programs use

**Fig. 1.102** Multiple sequence alignment generated by T-Coffee

different algorithms, which is clearly evident from the result generated by each program. It has to be noted that though alignments differ, conservation of amino acids at the active site are still retained. Again a phylogenetic tree construction is purely dependent on the alignment. Hence, one should take utmost care in performing MSA.

### Exercise

1. How MUSCLE is different from T-Coffee?
2. How MUSCLE program works?
3. Under what circumstances, one can choose MUSCLE than other programs?

## 1.19 Multiple Sequence Alignment—MAFFT for Large Alignments

### Introduction

MAFFT (**M**ultiple **A**lignment using **F**ast **F**ourier **T**ransform) is a high speed multiple sequence alignment program that uses Fast Fourier Transforms. This is suitable for making medium to large sequence alignments.

In MAFFT, the computational time is drastically reduced when compared to other existing methods in MSA. This is achieved by including two novel techniques in which first the homologous regions are identified quickly by the fast Fourier transform (FFT), where an amino acid sequence is converted to a sequence having volume and polarity information of every amino acid. Then a simplified scoring system is used for reducing computational time and increasing the accuracy of alignments. This is applicable even for sequences having large insertions or extensions as well as distantly related sequences of similar length. Apart from these, the program uses two different heuristic algorithms, the progressive method (FFT-NS-2) and the iterative refinement method (FFT-NS-i). It was observed that the computational time of FFT-NS-2 is drastically reduced as compared with CLUSTALW with comparable accuracy and FFT-NS-i is over 100 times faster than T-COFFEE even when the number of input sequences exceeds 60, without sacrificing the accuracy (Katoh et al. 2002).

**Exploration Activity**

To do multiple sequence alignment of iron superoxide dismutase (FeSODs) of *Oryza sativa* subsp. *Indica* (B8B2C9), *Arabidopsis thaliana* (P21276), *Escherichia coli* (P0AGD3), *Nostoc punctiforme* (B2IZB2) and *Synechococcus elongatus* strain PCC 7942 (P18655). The sequences are to be prepared in FASTA format as a single file (Notepad).

1. Go to the MAFFT multiple sequence alignment page by typing http://www.ebi.ac.uk/Tools/msa/mafft/ in your browser (Fig. 1.103).
2. Choose the type of sequence to be analyzed Protein or Nucleic acid from the dropdown menu.

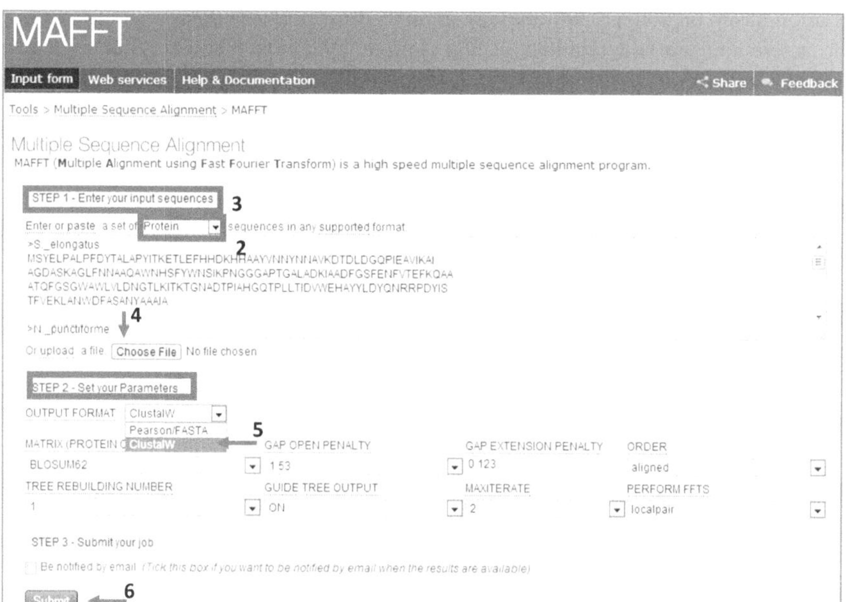


**Fig. 1.103** MAFFT multiple sequence alignment interface at EBI

3. Input your sequences by pasting in the text boxes provided in **Step 1**. This tool reads three or more (protein or nucleotide). Input sequence can be in GCG, FASTA, EMBL, GenBank, PIR, NBRF or UniProtKB/Swiss-Prot format. But they do not accept partially formatted sequences. Generally, adding a return to the end of the sequence may help, in most cases, understand the input by these programs. *Note*: *Avoid using data directly from word processors as it may yield unpredictable results due to the presence of hidden/control characters. There is currently a limit of 500 sequences and 1 MB of data.*
4. Alternatively, file containing three or more valid sequences in any format (GCG, FASTA, EMBL, GenBank, PIR, NBRF or UniProtKB/Swiss-Prot) can be uploaded using the '**Choose File**' button (provided below the text box) as input. *Note*: *Avoid using word processors files. Preferably generate a file with sequence in FASTA format and save the file with extensions .fas or .fasta. There is currently a limit of 500 sequences and 1 MB of data.*
5. Click on '**more options**' button to set the alignment options in **Step 2**. Change the output format '**clustalw**' (Default value is: Pearson/FASTA [fasta]). Change the Max Iterate value to '2' to change number of iterations for better alignment, if required. Leave other parameters to default settings:

   **Matrix**
   Protein comparison matrix to be used when adding sequences to the alignment.

   **Matrix (Protein Only)**
   Default value is: BLOSUM 62 [bl62]

   **Gap Open**
   Penalty for first base/residue in a gap.
   Default value is: 1.53

   **Gap Extension**
   Penalty for each additional base/residue in a gap.
   Default value is: 0.123

   **Order**
   The order in which the sequences appear in the final alignment
   Default value is: aligned

   **Tree Rebuilding Number**
   Tree Rebuilding Number
   Default value is: 1

   **Guide Tree Output**
   Generate guide tree file
   Default value is: ON [true]

   **Max Iterate**
   Maximum number of iterations to perform when refining the alignment
   Default value is: 0

| Order | Description | Abbreviation |
|---|---|---|
| None | | None |
| Localpair | | Localpair |
| Genafpair | | Genafpair |
| Globalpair | | Globalpair |

**Perform FFTS** (**F**ast Fourier Transform)
Default value is: localpair

6. Click '**Submit**' button for getting the results interactively in **Step 3**. If you wish to receive the results as link via. e-mail, check the box before '**Be notified by email**'.

### Interpreting MAFFT Result

7. The results are presented under different tabs (Fig. 1.104).

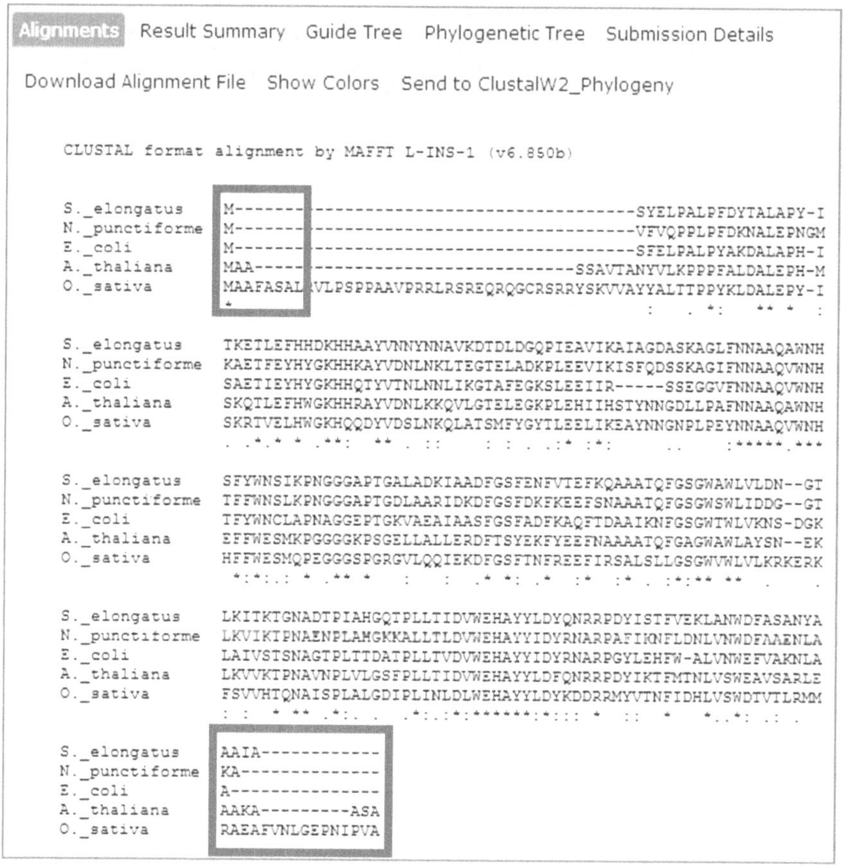


**Fig. 1.104** Multiple sequence alignment generated by MAFFT

8. The '**Alignment**' tab (default) shows the alignment format specified during input.
   **Consensus Symbols**
   By default an alignment will display the following symbols that denote the degree of conservation observed in each column:

   "*" refers to the residues or nucleotides in that column are identical in all sequences in the alignment.
   ":" indicates that conserved substitutions have been observed.
   "." indicates that semi-conserved substitutions are observed.

9. Clicking the '**Download Alignment File**' will allow you to download the alignment in .aln format.
10. Also you can change the colour option in the output results. By clicking the button '**Show Colors**', the alignment will be shown in colour.
11. The second tab '**Result Summary**' displays the result files comprising the input sequences for the alignment (.input), tool output (.output), which is a log file created during the alignment, and alignment in CLUSTAL format (.-clustalw) or alignment in FASTA format (.fasta)
12. The '**Start JalView**' button under JalView triggers JalView, a Java based editor in new window.
13. The third '**Guide Tree**' tab displays by default displays the phylogram. The other buttons on the page include, "Download Guide Tree File", "Show as Cladogram Tree" and "Show Distances".
    A Phylogram and Cladogram are branching diagram (tree) that are assumed to be an estimate of a phylogeny. In a Phylogram the length of the branch is proportional to the amount of inferred evolutionary change where as in a Cladogram do not show the amount of evolutionary "time" separating taxa as the branch length are equal and show a common ancestry. By right-clicking on the applet, the options can be controlled.
14. The fourth '**Phylogeny Tree**' tab displays the phylogenetic tree of the sequences used. It is actually a Neighbour-joining tree without correcting the distances
15. The fifth tab '**Submission Details**' displays the information regarding the program used, its version, input parameters, etc.

The N terminal alignment in the result generated by the MAFFT is similar to the one generated by the T-Coffee. But the alignment is in the middle and C terminal is entirely different from the T-Coffee or MUSCLE. Still it has to be noted that though alignments differ, conservation of amino acids at the active site are retained.

**Other Tools of MSA @ EBI**

**Clustal Omega**
New MSA tool that uses seeded guide trees and HMM profile-profile techniques to generate alignments (protein only). Suitable for medium-large alignments.

**DbClustal**
Create a Multiple Sequence Alignment from a protein BLAST result using the DbClustal program.
**MView**
Transform a Sequence Similarity Search result into a Multiple Sequence Alignment or reformat a Multiple Sequence Alignment using the MView program.
**WebPRANK**
The EBI has a new phylogeny-aware multiple sequence alignment program which makes use of evolutionary information to help place insertions and deletions.

**Exercise**

1. How MAFFT works and how it is different from T-Coffee?
2. What do you mean by Fast Fourier Transform?
3. What is meant by iteration?
4. What other programs are available for MSA?
5. What is the importance of MSA?

## 1.20 Multiple Sequence Alignment and Phylogenetic Analysis Using MEGA

**Introduction**
Traditionally, morphological (visible features) characters are used to classify organisms which can be collected from the organisms that are living and also from fossil records. But for microorganisms, fossils are essentially nonexistent, which makes it impossible to study phylogeny with this approach. In such cases, sequence data can be used (Molecular Phylogeny). Constructing a Phylogeny is tricky and many methods are proposed.

**Evolutionary Trees**
Evolutionary trees are two-dimensional graph showing evolutionary relationship among a set of items. The items can be organisms, genes, or sequences. Each unit is defined by a distinct branch.

**Phylogenetic Tree Construction**
It can be divided into five steps:

(1) Choosing molecular markers (DNA or protein sequence);
(2) Performing multiple sequence alignment;
(3) Choosing a model of evolution (Jukes–Cantor Model, Kimura Model, etc.);
(4) Determining a tree building method (Phenetic or Cladistic); and
(5) Assessing tree reliability (Bootstrapping, Jackkniffing, etc.).

**Methods of Tree Construction**

Tree constructing methods can be broadly classified into two main categories: Phenetic method and Cladistic method.

**Phenetic methods** are based on distance methods where trees are calculated by similarities of sequences and it is summarized using a hierarchical clustering algorithm such as UPGMA or Neighbour-Joining. The distance-based methods assume that all sequences involved are homologous and that tree branches are additive, meaning that the distance between two taxa equals the sum of all branch lengths connecting them. The tree thus generated is called a **Phylogram** and it might not reflect evolutionary relationships always. In the distance-based methods individual differences between pairs of sequences is compressed into or just represented a single number.

On the other hand, **Cladistic methods** use every position in the alignment as evolutionary information for building a tree and the resulting tree is known as a **Cladogram**. Here trees are generated by considering different possible pathways of evolution. Cladistic methods are also called as Character-based methods (or Discrete methods) that are based directly on the sequence characters rather than on pairwise distances. They count mutational events accumulated on the sequences and may therefore avoid the loss of information when characters are converted to distances. This preservation of character information means that evolutionary dynamics of each character can be studied and hence ancestral sequences can also be inferred. The two most popular character-based approaches are the Maximum parsimony (MP) and Maximum likelihood (ML) methods.

**Which Method to Choose?**

It depends upon the sequences that are being compared.

If sequence show strong similarity: Maximum parsimony
If sequence show clearly recognizable similarity: Distance methods
For all others: Maximum likelihood

It is best to choose at least two approaches and compare the results—if they are similar, you can have more confidence.

**Using MEGA**

**MEGA**—(**M**olecular **E**volutionary **G**enetics **A**nalysis) is an integrated tool for conducting automatic and manual sequence alignment, mining web-based databases, inferring phylogenetic trees, estimating rates of molecular evolution, and testing evolutionary hypotheses. It is available free of charge from http://www.megasoftware.net.

**Exploration Activity**

**Preparation of Sequence for Analysis**

Compile the sequences to be analyzed in FASTA format and save the file with an extension .fas or .fasta (Fig. 1.105).

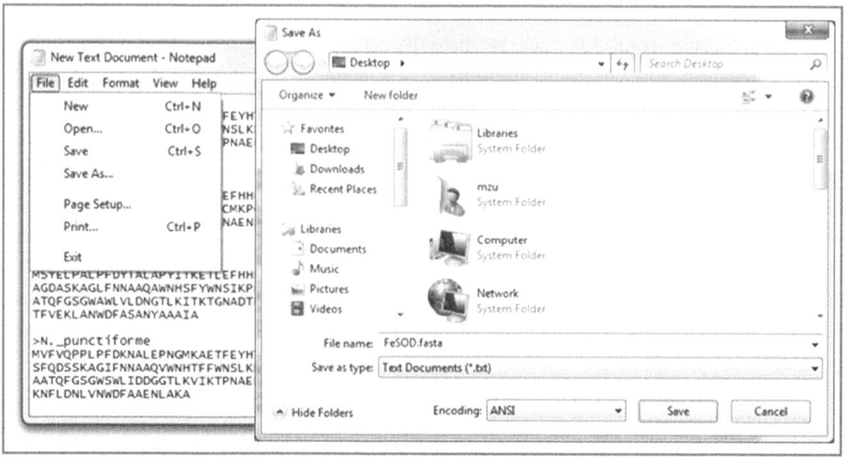


**Fig. 1.105** Preparing a sequence file for analysis

## Using MEGA for Multiple Sequence Alignment

1. With the '**Alignment Explorer**' tool it is possible to build and edit multiple sequence alignments in MEGA using either the ClustalW or Muscle algorithms.
2. Click and open MEGA program.
3. Click **File** and then **Open A File/Session**…
4. **Browse** the file you saved in FASTA format, earlier.
5. A pop up window will open asking for what purpose the file is to be opened.
6. Click '**Align**'.
7. A new window (**M5**: **Alignment Explorer**) will now open showing the sequences (DNA sequence and Translated Protein Sequence, if the DNA is a coding sequence, as tabs).
8. From main menu bar, under '**Alignment**' choose either **Align by ClustalW** or **Align by MUSCLE**.
9. If the program asks to select all the sequences, click '**OK**' (Fig. 1.106).

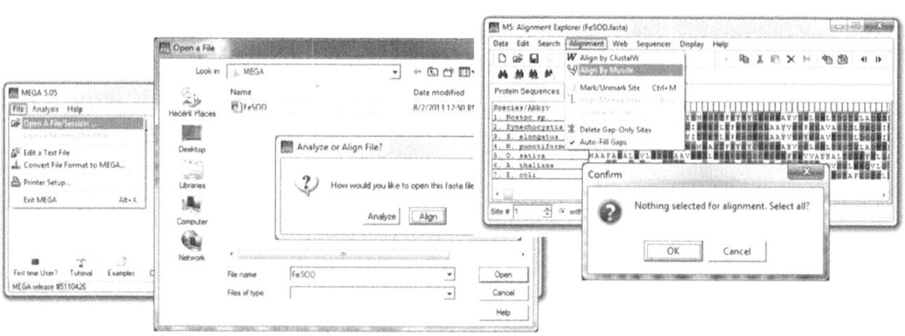


**Fig. 1.106** Making a multiple sequence alignment

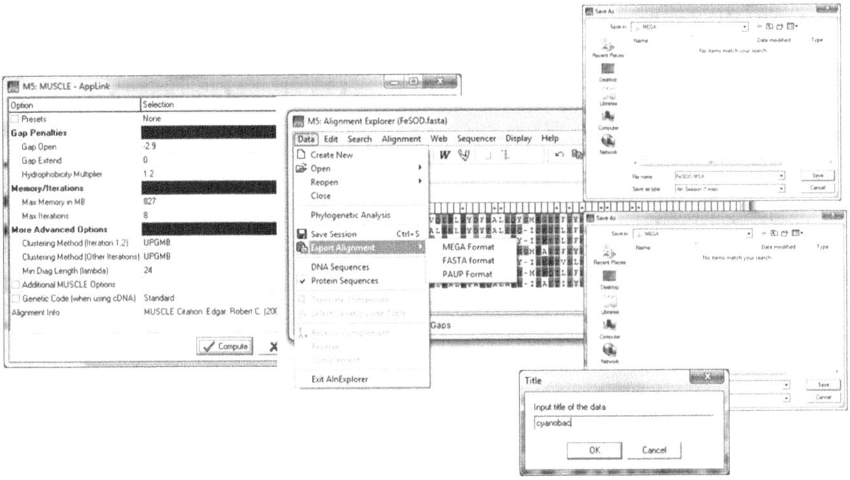


**Fig. 1.107** Saving alignment and exporting it to MEGA format

10. A new window displaying the parameters for multiple sequence alignment will now be displayed.
11. Click '**Compute**' button to accept default parameters (Fig. 1.107).
12. Multiple sequence alignment will now be displayed.

**Saving and Exporting the Session**

13. Save the session by clicking the **floppy disk button**.
14. The session will be saved as a file with extension **.mas**
15. For phylogenetic analysis, export the multiple sequence alignment session to MEGA format by clicking '**Data**', choose '**Export Alignment**' and select '**MEGA format**'.
16. Click '**Save**' and the file will be saved with extension '**.meg**'. This helps MEGA to read and interpret the data correctly.

**Computation of Evolutionary Parameters (Model Evaluation, Diversity, Distance, Mutation Rate, Test of Neutrality and Molecular Clock)**

**Computing Synonymous and Non-synonymous Mutations**

17. Open the .meg file you saved earlier (Fig. 1.108).
18. Click on the TA button (Fig. 1.109a).
19. The sequence will be open in a '**Sequence Data Explorer**' (Fig. 1.109b).
20. The information like Conserved sites [**C**], Variable sites [**V**], Parsimony-Informative sites [**Pi**] and Singleton sites [**S**] can be visualized by clicking appropriate buttons on the menu bar. A variable site generally contains at least two types of nucleotides or amino acids namely, singleton or parsimony-informative. Conserved site or constant site on the other hand is a

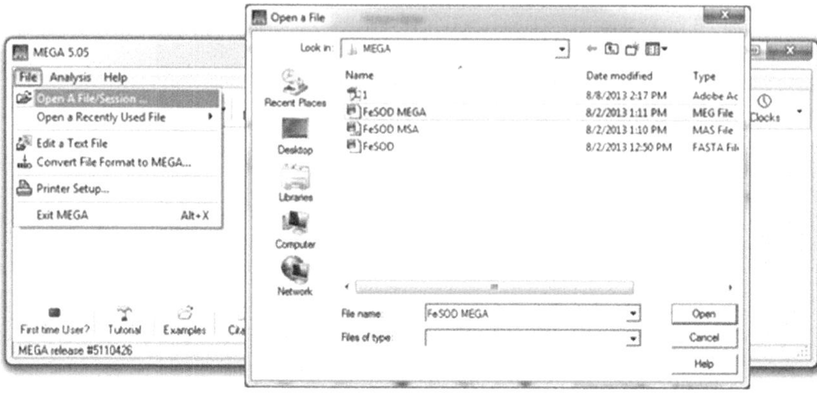


Fig. 1.108 Opening MEGA format file

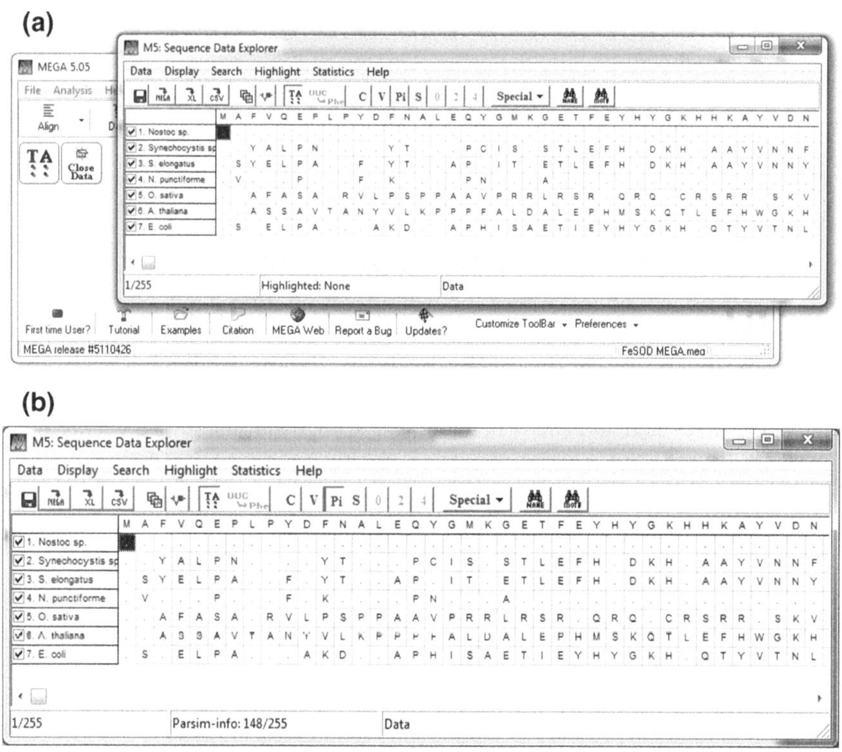


Fig. 1.109 Visualization of parsimony-informative sites

site that contains the same nucleotide or amino acid in all sequences. MEGA treats a site as constant only if at least two sequences contain unambiguous nucleotides or amino acids. A singleton site is a site that usually contains at least two types of nucleotides (or amino acids) one occurring multiple times. For MEGA a site is a singleton site if at least three sequences contain unambiguous nucleotides or amino acids. In case of Parsimony-informative site, at least two types of nucleotides (or amino acids) must occur with a minimum frequency of two.

21. The information will be displayed in the status bar at the bottom of the window.
22. For the provided data, Conserved sites: 2/255, Variable sites: 210/255, Parsimony-Informative sites: 148/255 and Singleton sites: 52/255 were observed.
23. Under '**Statistics**' choose '**Codon Usage**'. This provides details of Relative Synonymous Codon Usage (RCU) details. A RSCU value is mostly positive and a value >1 indicate a codon more frequently used and a value <1 indicates it is less frequently used (Sharp et al. 1986).

### Viewing Amino Acid Composition

24. Select '**Models**' from the main menu and select '**Compute Amino Acid Composition**' (Fig. 1.110).
25. It opens a window where you can specify the output options like the program in which the results has to be displayed, file name and location to be saved. Click '**OK**' button (Fig. 1.111).

The results will be displayed in Microsoft Excel (Fig. 1.112).

26. In Fig. 1.112, the amino acid Alanine was found to be higher.

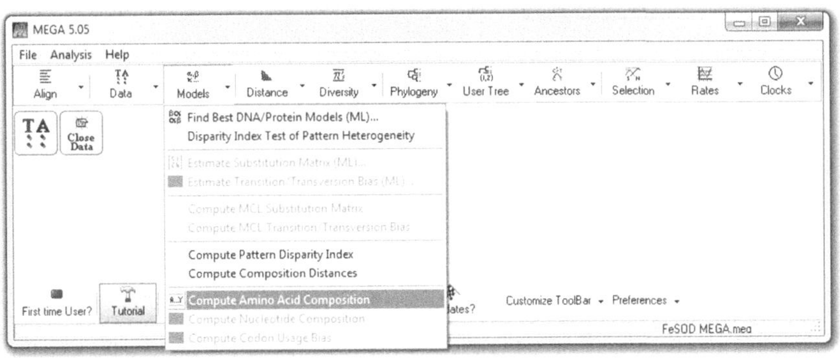


**Fig. 1.110** Computing amino acid composition

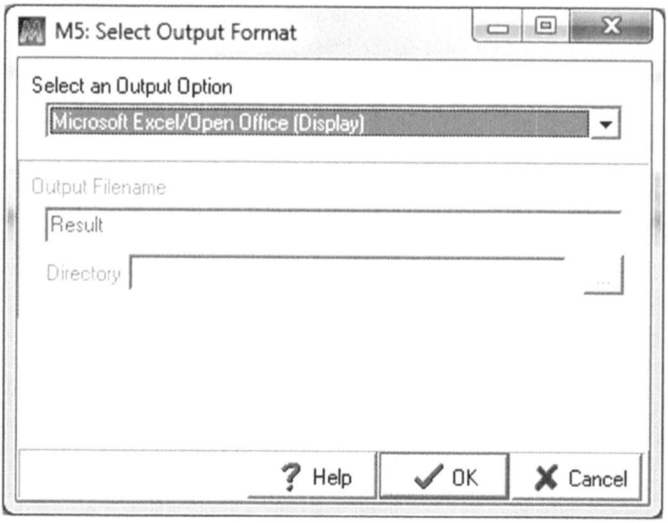


**Fig. 1.111** Selecting output option

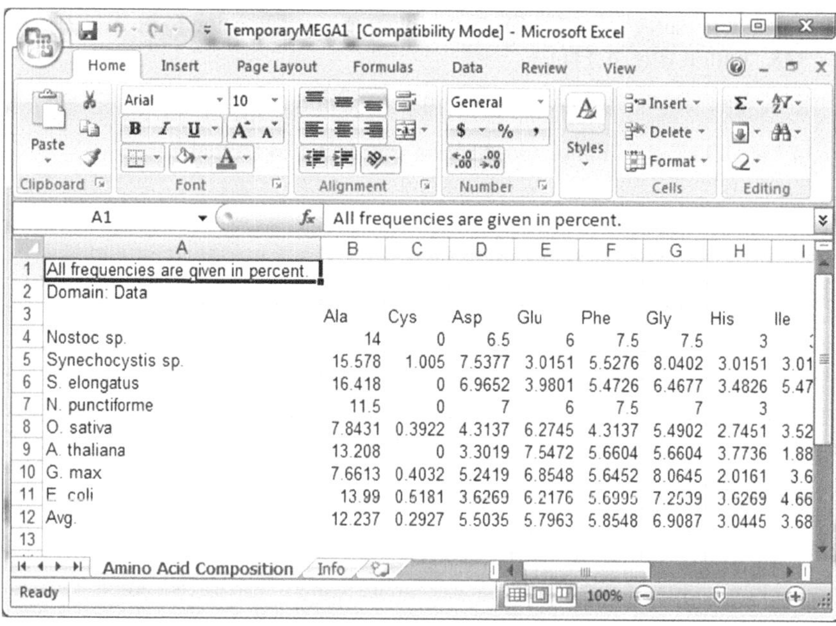


**Fig. 1.112** Results displayed in Microsoft Excel

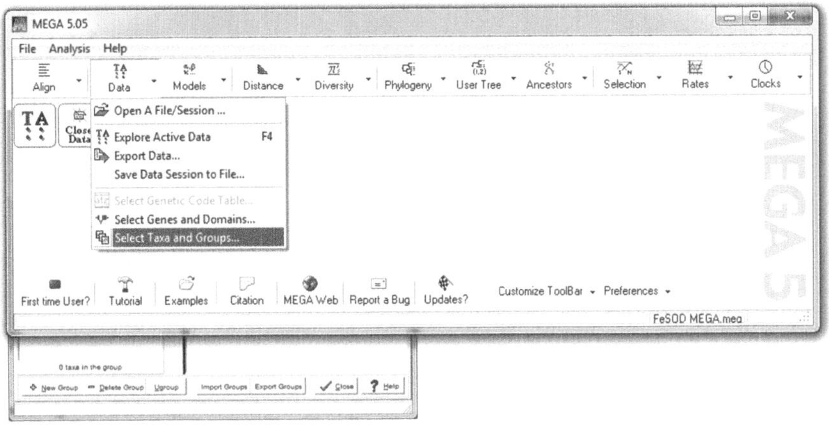


**Fig. 1.113** Initiating distance computing procedure

## Computing Distances

27. From the main window of MEGA, click on the '**Data**' button from the main menu and select '**Select Taxa and Groups** ...' (Fig. 1.113).
28. This opens a window to group the data.
29. Click on the '**+ New Group**' button at the bottom and give a name (Fig. 1.114).
30. Click on the arrow button to send the selected taxa (in the '**Ungrouped Taxa**') that belong to the first group. Make sure you have at least three taxa in one group.
31. Here two groups namely Bacteria and Plant are created.
32. Once done, click on the '**Close**' button at the bottom.
33. Now click on the '**Distance**' button on the main menu and select '**Compute Within Group Mean Distance**'. This displays the mean distance within group (Fig. 1.115).
34. Similarly you can compute the mean distance between group and net between group by selecting '**Compute Between Group Mean Distance**' and '**Compute Net Between Mean Distance**' from 'Distance' option on the main menu (Fig. 1.116).

## Computing Distance-Based Phylogenies

Distance-based phylogenetic reconstruction involves three steps. The first involves computing pairwise genetic distances between individuals which is then represented as tree (Figure) and finally evaluating the relevance of the representation. If there are n number of sequences, there are **n (n − 1)/2** distances. Thus if n = 5, then the genetic relationships between the samples are described by 5 × 4/2 = 10 pairwise distances.

The advantage of this method is that it is the fastest and flexible. It can also use a large number of substitution models for correcting distances. But the limitation of

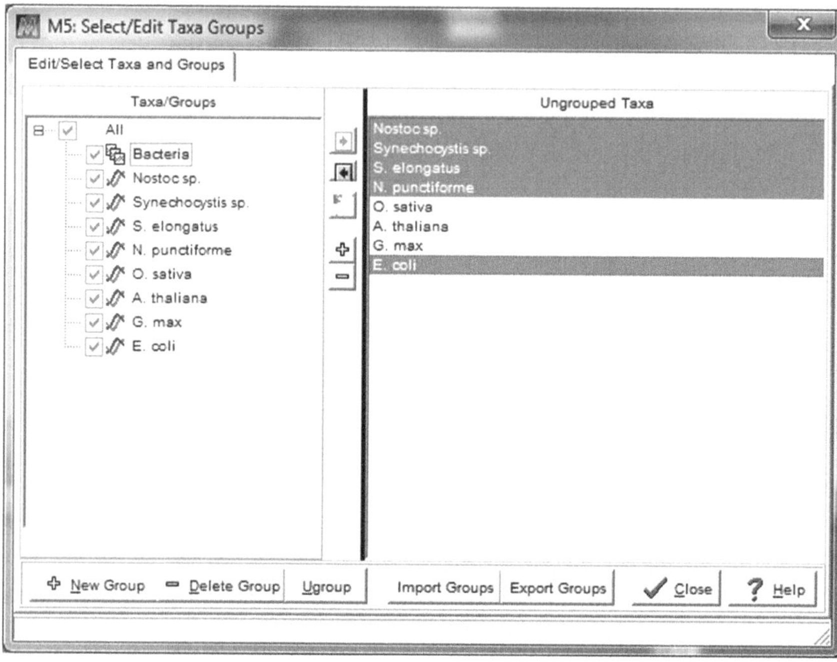


Fig. 1.114 Creating new group

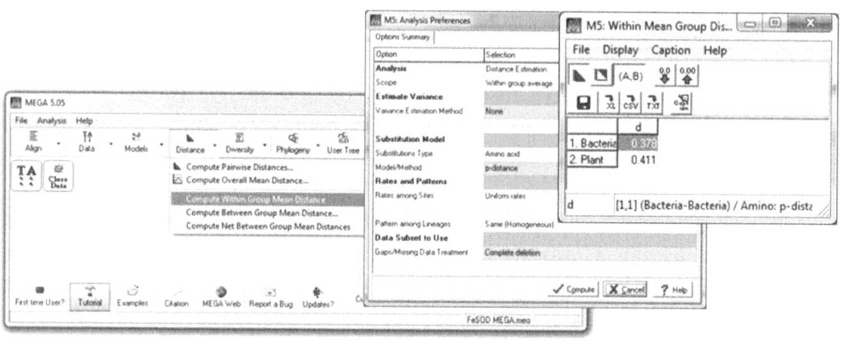


Fig. 1.115 Computing mean distance

this method is that when all the sequence variation is reduced to a single value, its actual sequence information is lost like ancestral sequences at internal nodes cannot be inferred. This method highly depends on the distance and clustering algorithm selected. Moreover tree construction may be inaccurate as there is no model comparison for testing the 'best' tree, or the 'best' model of evolution.

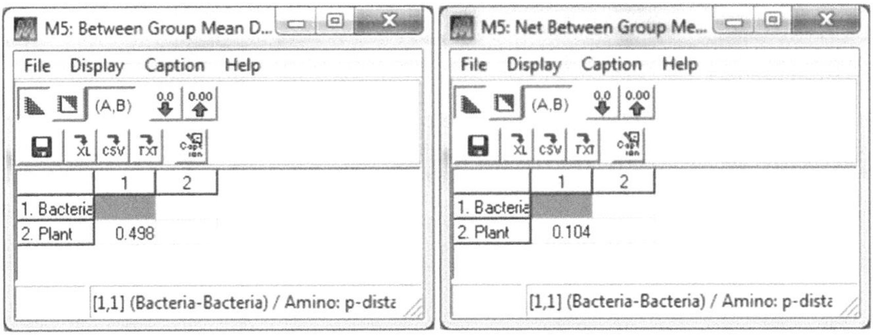


**Fig. 1.116** Computing 'between group mean distance' and 'compute net between mean distance'

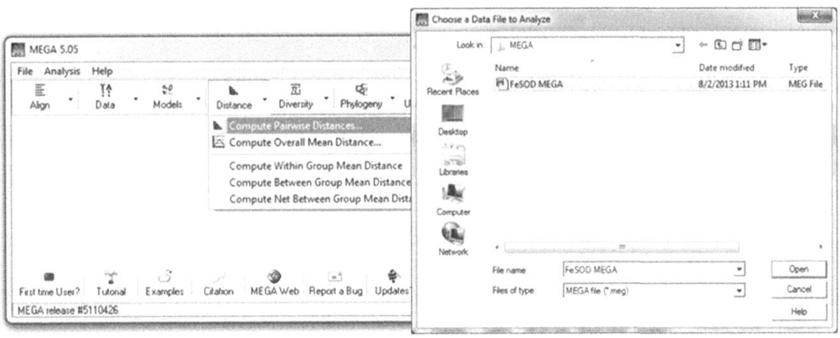


**Fig. 1.117** Computing pairwise distance

## Compute the Proportion of Amino Acid Differences

35. It is possible to estimate the evolutionary distance between sequences by computing the proportion of amino acid differences between each pair of the sequences.
36. Using options from menu bar **open** the file you exported earlier in MEGA format.
37. From the main MEGA window, click '**Distance**' and select '**Compute Pairwise Distances**' (Fig. 1.117).
38. This will display the Analysis Preferences window. Click the Model/Method pull-down and then select '**p-distance**'.
39. Click the '**Compute**' button to accept the default values for the rest of the options and begin the computation.
40. A results viewer window will be displayed, showing the distances in a grid format (Fig. 1.118).

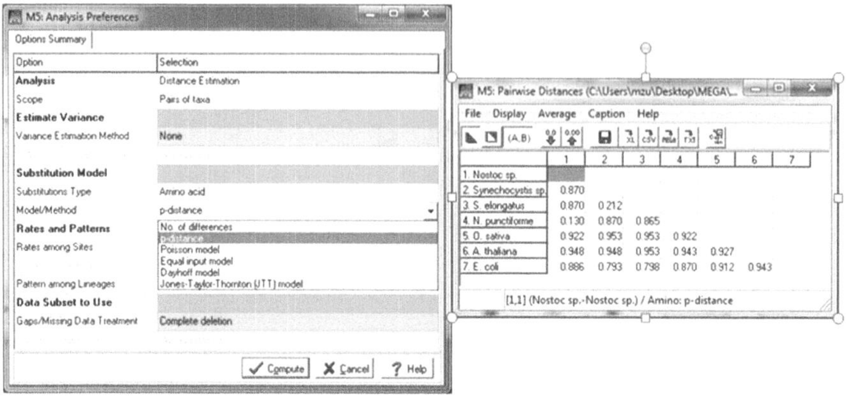


**Fig. 1.118** Results displayed in a grid format

41. By computing the proportion of difference of nucleotide between each pair of sequences, evolutionary distances between sequences can be estimated.
42. MEGA supports a wide collection of models for estimating evolutionary distances. You can calculate evolutionary distances by using different models and eventually compare them all.

**Constructing a Neighbour-Joining (NJ) Tree**

Though the trees represents the genetic distances between individuals in a better way, still it has to remembered that they are necessarily efficient representations of the original distances and there are chances that the vital information may be lost during the process.

The steps illustrate the basics of NJ phylogenetic tree re-construction using MEGA.

43. Using options from menu bar **open** the file you exported earlier in MEGA format.
44. From the main MEGA menu bar, click '**Phylogeny**' and select '**Construct/Test Neighbour-Joining Tree**' option.
45. In the Analysis Preferences window select '**Bootstrap method**' for the Test of Phylogeny option and set '**1000**' under '**No. of Bootstrap Replications**'. Then select the '**p-distance**' option from the Model/Method drop-down.
46. Click the '**Compute**' button to begin the computation with the default values for the rest of the options (Fig. 1.119).
47. The tree will be displayed in the Tree Explorer window.

**Interpreting a NJ Tree**

Assuming the left-side of the phylogeny as 'ancestral' and the right-side as 'recent' is the common mistake people make while interpreting a phylogenetic tree and might not be correct unless it is actually rooted. i.e. adding some external taxa, like

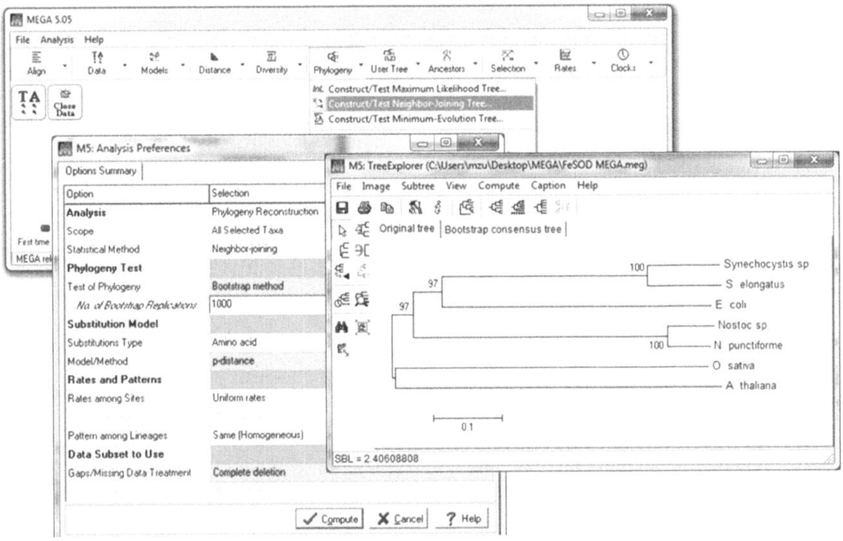


**Fig. 1.119** Constructing and visualizing a NJ tree

sister taxon, to define what is the most 'ancient' split in the tree. The following tree is not rooted, and should be designated as an unrooted tree (Fig. 1.120).

When you have an idea on the time of occurrence, any of the most ancient isolates can treated as a root. The rate of evolution of a given gene can be assessed with the help of rooted trees. Molecular clock represents the accumulation of mutations over time (Zuckerkandl and Pauling 1965). Performing a quantitative analysis is relatively easy as it requires just going back or backtracking the number of mutations from the root to the time it got diverged from the root.

Assigning the *E. coli* as the root, the tree will look like as shown in Fig. 1.121.

### Finding Best Model for Tree Construction

48. Open a previously saved multiple sequence alignment (Fig. 1.122).
49. On the program's main window, click '**Models**' and choose, '**Find Best DNA/Protein Models**' (Fig. 1.123).
50. Click '**Compute**' on the popup.
51. A table with all the best models will be ranked and displayed. Make a note of it (Fig. 1.123).

### Constructing a Maximum Likelihood (ML) Tree

Likelihood-based methods may be maximum likelihood and Bayesian methods. The former looks for the best tree and parameter values by maximizing the likelihood while the other derives samples of tree topologies and model parameters that are most consistent with the likelihood and also with the possible prior knowledge about the tree/parameters. Maximum likelihood phylogenetic reconstruction tries to

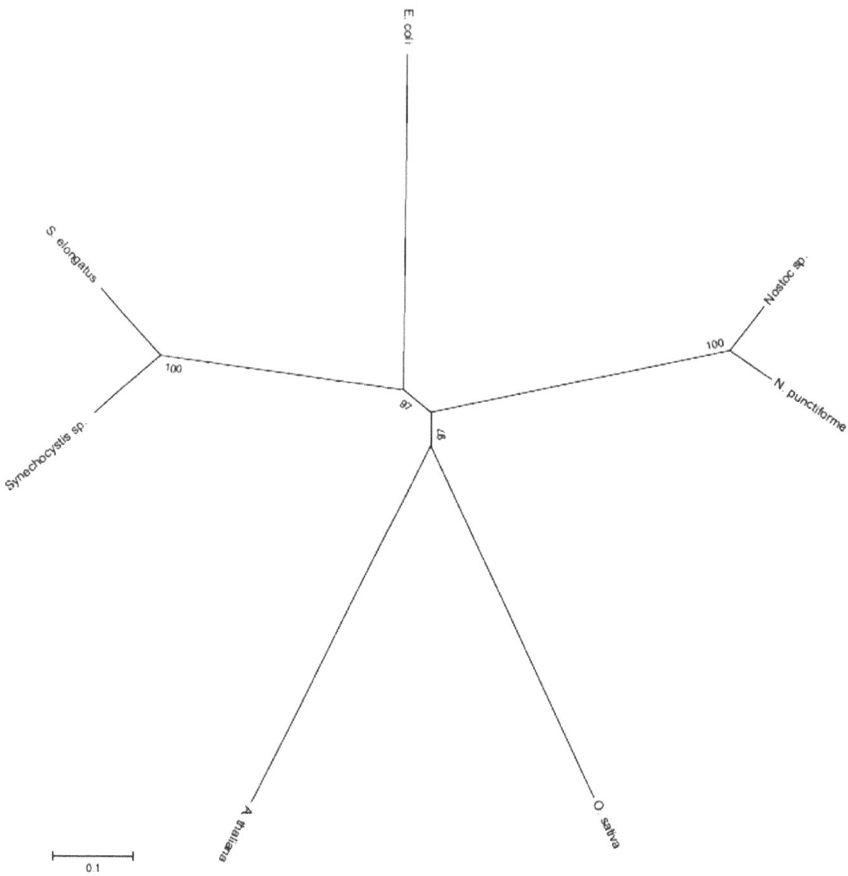


**Fig. 1.120** Unrooted NJ tree

evaluate the possible tree topologies for each position in an alignment and look for the 'best' tree. This method offers more flexibility where any model of sequence evolution can be considered. Thus for any model of evolution, the likelihood of a given tree is calculated and optimization procedures can be used to infer both the most likely tree topology and model parameters. Like distance-based methods, model-based phylogenetic reconstruction also requires prior knowledge on the parameters that should be considered in a model. Though this method is the slowest, it gives the best result and the most information about the tree constructed.
**Advantages**: It is flexible as it allows selection of any model of evolution, usually accurate, measure of uncertainty (in case of Bayesian approaches).
**Limitations**: It is the most rigorous among all approaches since ML uses the full sequence information, not just the informative sites.

Its accuracy depends on the substitution model used. Choosing an unrealistic substitution model may lead to an incorrect tree. Because of the exhaustive nature

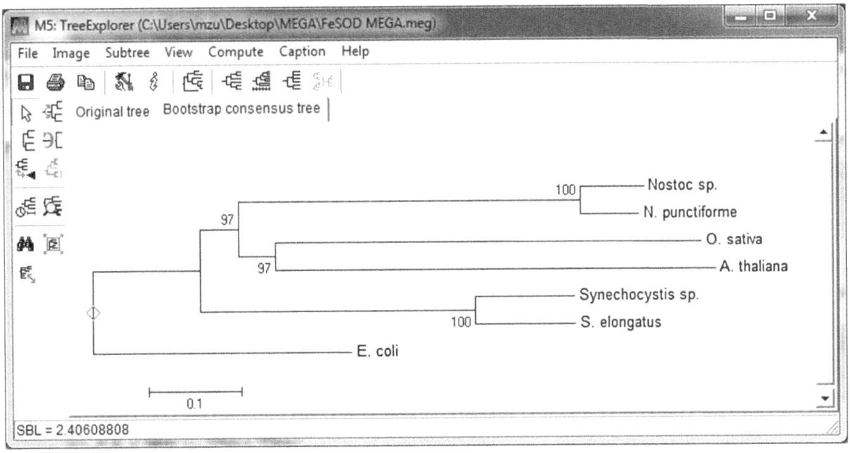


Fig. 1.121 A rooted tree

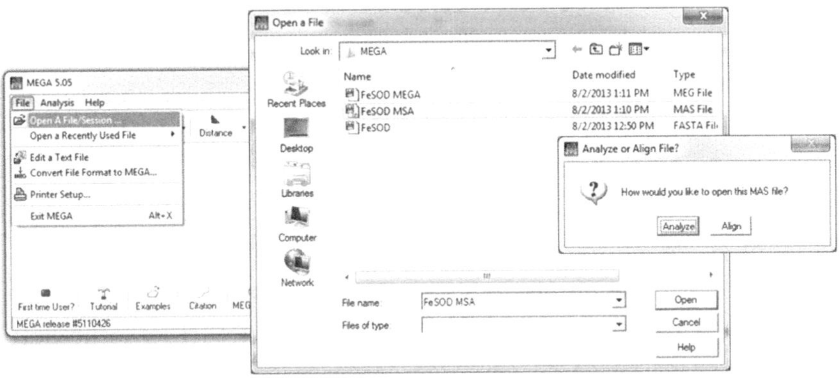


Fig. 1.122 Opening an alignment file for computing model

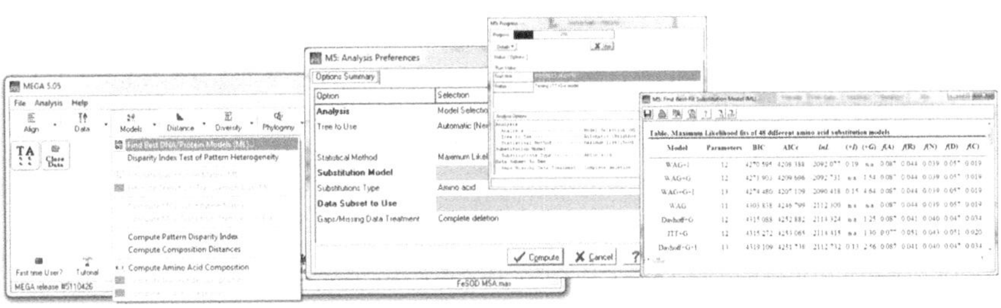


Fig. 1.123 Calculating best model from alignment

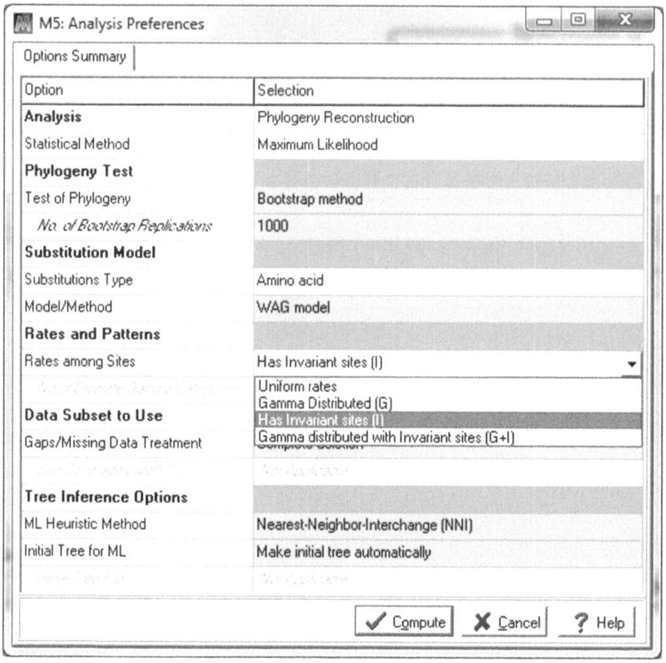


**Fig. 1.124** Parameter selection for Maximum Likelihood Tree construction

of the ML method, when the number of taxa increases to a modest size, it becomes impossible to use.

**Maximum Likelihood Phylogenetic Tree Re-construction Using MEGA**

52. Using options from menu bar, **open** the file you exported earlier in MEGA format.
53. From the main MEGA menu bar, click '**Phylogeny**' and select '**Construct/Test Maximum Likelihood Tree**' option.
54. In the Analysis Preferences window select '**Bootstrap method**' for the Test of Phylogeny option and set '**1000**' under '**No. of Bootstrap Replications**'.
55. Then select the '**WAG**' option from the Model/Method drop-down and under '**Rates Among Sites**' select '**Has Invariant Sites (I)**'. We got this information from our previous analysis of 'finding best model for constructing tree'.
56. Click the '**Compute**' button to accept the default values for the rest of the options and begin the computation (Fig. 1.124). The tree will be displayed in the Tree Explorer window (Fig. 1.125).

**Interpreting the ML Tree**

In Fig. 1.125, *O. sativa* has been chosen as the root, which is supposed to be wrong. Here the odd one out is the *E. coli*, which should be designated as the root. It can be

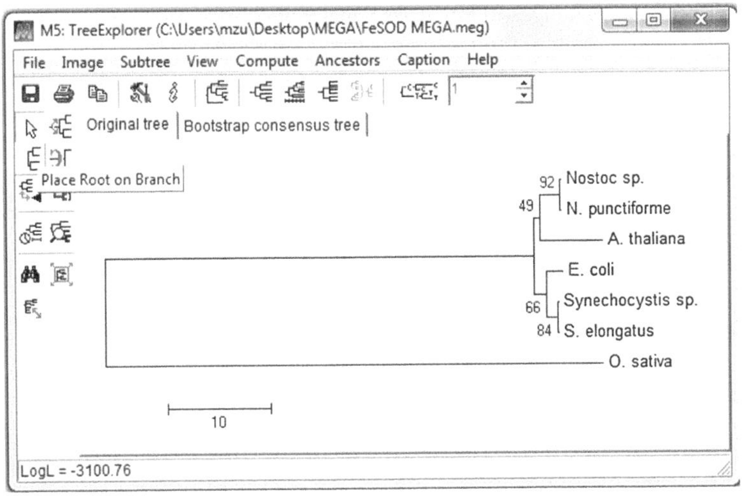


Fig. 1.125 Maximum Likelihood Tree with wrong root

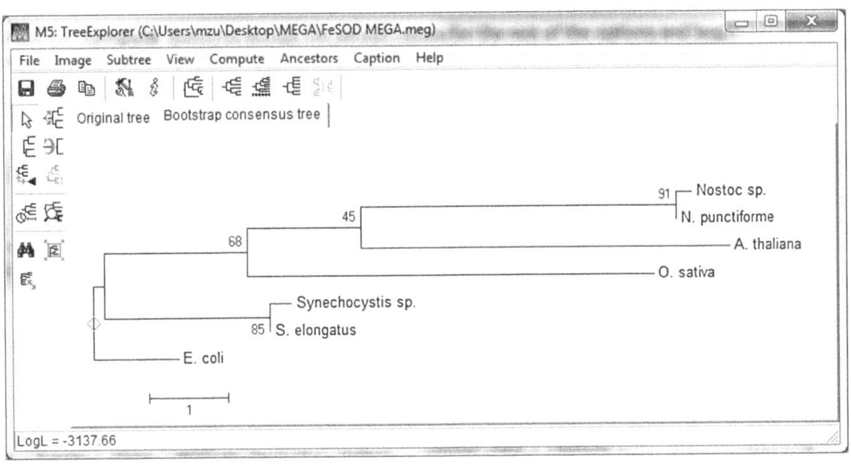


Fig. 1.126 Maximum Likelihood Tree with corrected root

done by clicking the *E. coli* branch and selecting the '**Place Root on the Branch**' option on the menu.

Thus the tree with *E. coli* as root will look like the one shown in the Fig. 1.126.

**Constructing a Maximum Parsimony Tree**

Phylogenetic reconstruction based on parsimony tries to find trees with minimum number of changes (substitutions) in total from ancestors to descendents. That is for every position in the alignment, all possible trees are evaluated and a score is given based on the number of evolutionary changes that are required to produce the observed sequence changes. Thus a tree the fewest evolutionary changes for all sequences derived from a common ancestor is supposed to be the most parsimonious tree. This view is justified by the fact that evolutionary changes are relatively rare within a reasonably short time frame. This implies that a tree with minimal changes is likely to be a good estimate of the true tree.

Though a lot of criticisms can be made to this approach, yet this is a simple way to infer phylogenies for sequences displaying low divergence like taxa differing from each other by only a few nucleotides and where the overall substitution rate is low. This method is more time-consuming method than the distance methods and works only well if the sequences have a strong sequence similarity.

**Advantages**: It tends to produce more accurate trees than the distance-based methods when sequence divergence is low.

**Limitations**: It is computational intensive. When sequence divergence is high, tree estimation by MP can be less effective. Estimation of branch lengths may also be erroneous because MP does not employ substitution models to correct for multiple substitutions. This drawback can become prominent when dealing with divergent sequences. In addition, MP only considers informative sites, and ignores other sites. Consequently, certain phylogenetic signals may be lost.

The following steps illustrate the basics of MP phylogenetic tree re-construction (by Close-Neighbour-Interchange (CNI) on Random Trees method) using MEGA.

57. Using options from menu bar, **open** the file you exported earlier in MEGA format.
58. From the main MEGA menu bar, click '**Phylogeny**' and select '**Construct/Test Maximum Parsimony Tree(s)**' option (Fig. 1.127).
59. In the Analysis Preferences window select '**Bootstrap method**' for the Test of Phylogeny option and set '**1000**' under '**No. of Bootstrap Replications**' (Fig. 1.127).

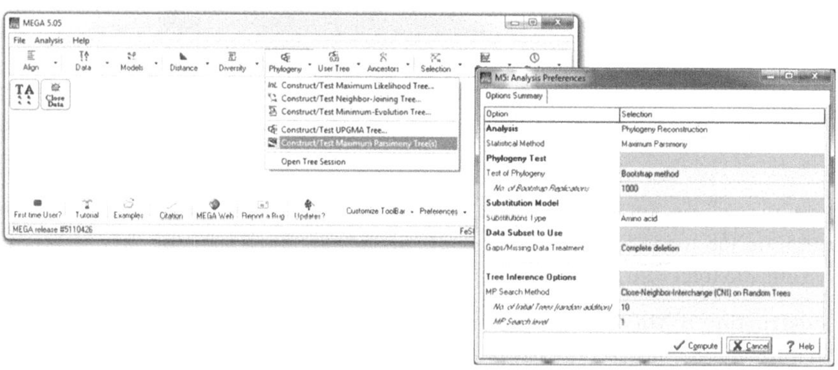


**Fig. 1.127** Choosing and specifying options for Maximum Parsimony Tree

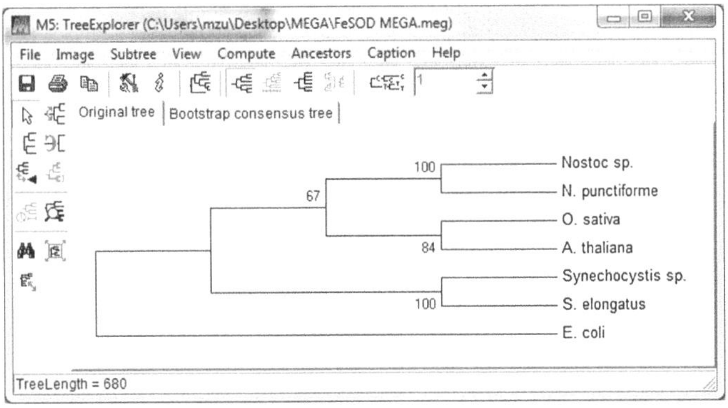


Fig. 1.128 Maximum parsimony tree

60. Under '**MP Search Method**' select '**Close-Neighbour-Interchange (CNI) on Random Trees**'. Click the '**Compute**' button to begin the computation with the default values for the rest of the options.
61. The tree will be displayed in the Tree Explorer window. Click on the '**Bootstrap consensus tree tab**' (Fig. 1.128).

**Interpreting Maximum Parsimony Tree**

The branching of the tree looks to be good. As expected *E. coli* is shown as the root and all other clustering are fine. What actually happened here is the sequence that appears at the end in the FASTA input file (and so in the MEG file) will be automatically considered as the outgroup. An outgroup might be a single or a collection of sister group or sister taxon to all other sequences in the dataset. If you change the order in the MEG file, you can see the effect. Thus one should be careful in this. If the outgroup is wrong, then it has to be selected and declared as outgroup explicitly.

**Exploring the *Tree Explorer* Window**

62. The Tree Explorer window displays the tree generated by any given method.
63. For selecting a branch, click on it with the left mouse button.
64. Clicking on a branch with the right mouse button, will shows options to Swap, Flip the branch or Place Root, etc. (Fig. 1.129).
65. If you wish to see how the cursor moves through the generated tree, click on a branch for selecting and then press the arrow keys **Up**, **Down**, **Left**, and **Right** from your keyboard.
66. You can change the branching style by choosing the **View > Tree/Branch Style** option from the main menu of Tree Explorer (Fig. 1.130). It has options to display it in Rectangular, Straight, Curved or as a Circle.

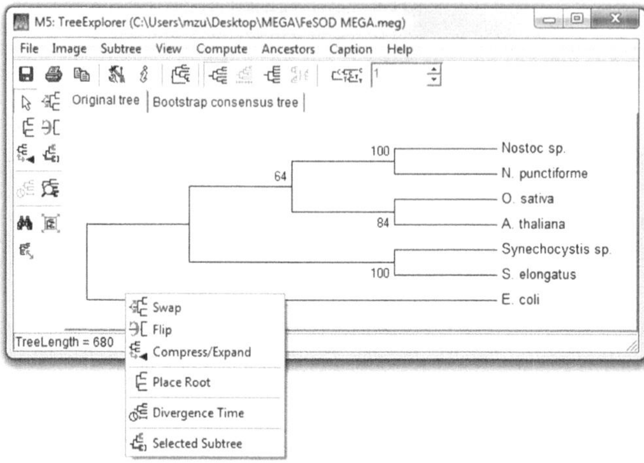


**Fig. 1.129** Swapping a branch

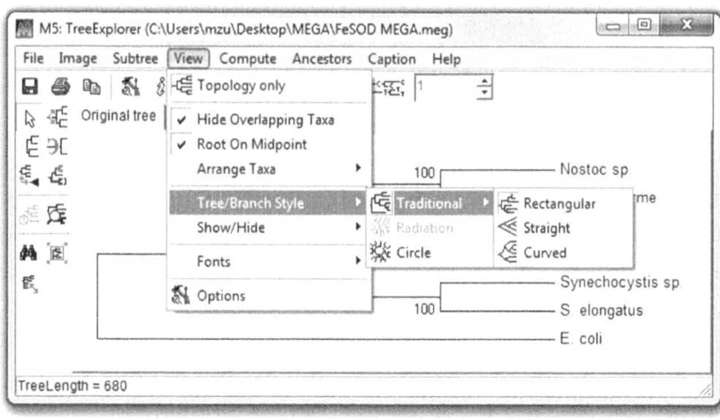


**Fig. 1.130** Changing the branch style

67. Click on the **View > Topology Only** option from the Tree Explorer main menu to toggle between **Cladogram** (showing only relationships) and **Phylogram** (branch length proportional to the amount of character change) representation.
68. The branch line thickness, font of the labels, displaying the branch lengths in the 'Topology Only' option can be done by selecting **View > Options** (Fig. 1.131).

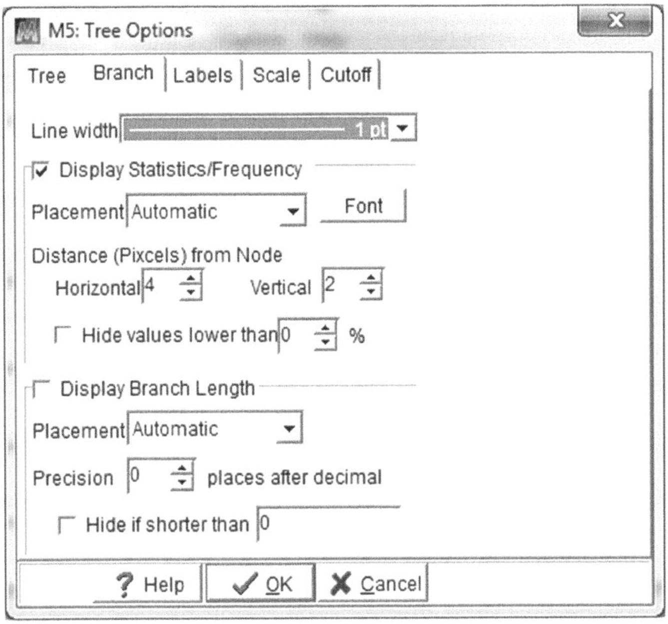


**Fig. 1.131** Managing other options in tree visualization

### Saving the Phylogenetic Tree Image

69. MEGA allows saving the tree generated in BMP, TIFF or in PDF format. To save in other formats, you can use the '**Copy to Clipboard**' option and paste it in an image editing software and can be saved in desired format.
70. This is done by clicking the '**Image**' option from the Tree Explorer main menu and selecting desired option (Fig. 1.132).

### Additional Analysis

### Computing Diversity

71. Click on the '**Diversity**' option from the main menu and select '**Compute Mean Diversity Within Subpopulation**' from the list. Accept the popup window with the default parameters by clicking '**Compute**' button. It displays the diversity within the subpopulation (Fig. 1.133).
72. Similarly you can compute the mean diversity in entire population and inter-population. You can also compute the coefficient of differentiation by selecting appropriate option from the list under '**Diversity**' in the main program window (Fig. 1.134).

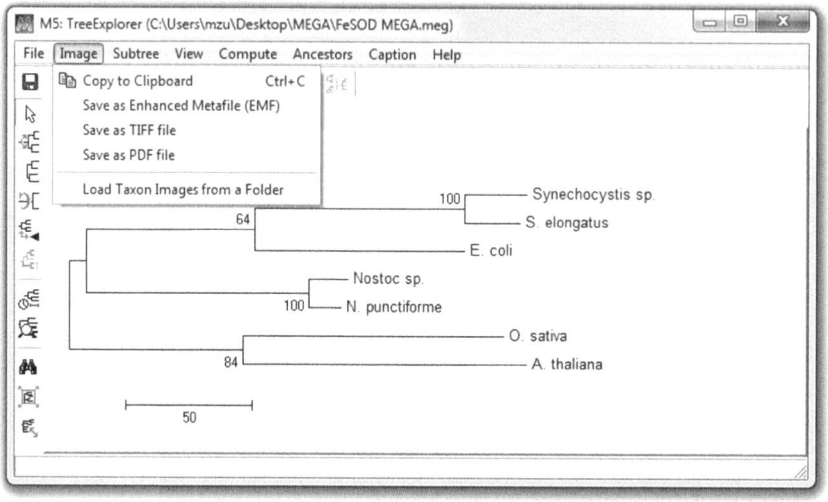


**Fig. 1.132** Exporting tree in for printing/publication

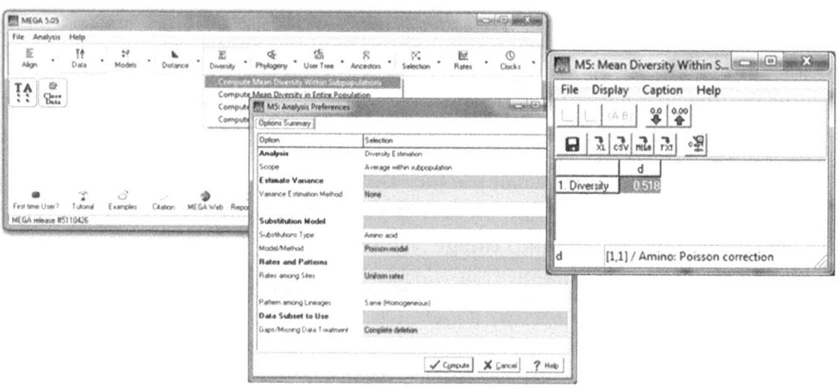


**Fig. 1.133** Computing mean diversity within subpopulation

## Computing Tajima's Test of Neutrality

73. Click on the '**Selection**' option from the main menu and select '**Tajima's Test of Neutrality**' from the list. Accept the popup window with the default parameters by clicking '**Compute**' button (Fig. 1.135).
74. The program displays the result Tajima's Test of Neutrality in new window (Fig. 1.136). Make a note of the values.

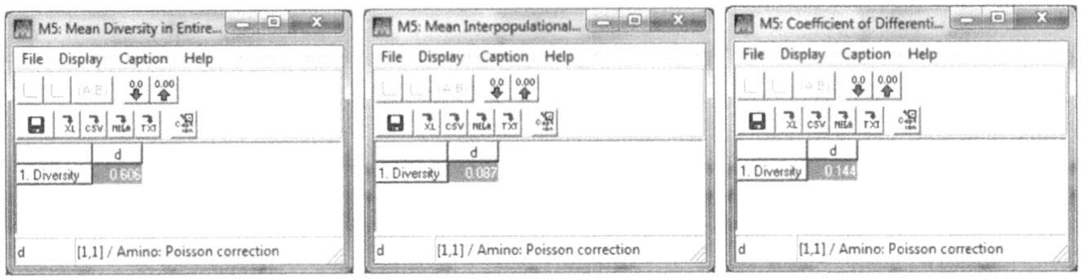


**Fig. 1.134** Computing mean diversity in entire population, interpopulation and coefficient of differentiation

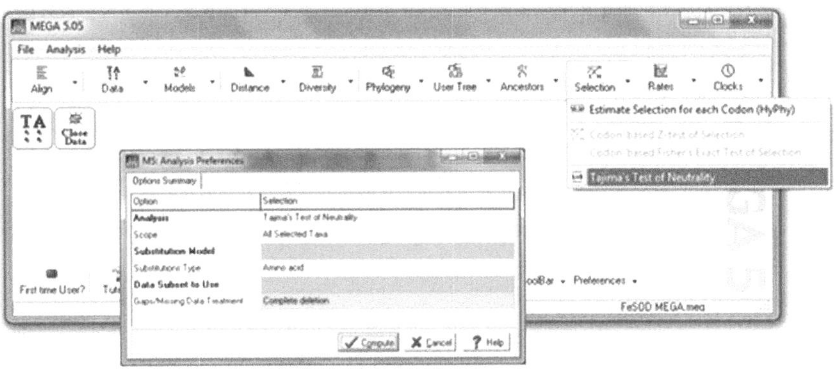


**Fig. 1.135** Computing Tajima's test of neutrality

## Maximum Likelihood Estimate of Gamma Parameter for Site Rates

75. Click on the 'Rates' option from the main menu and select '**Estimate Gamma Parameter for site rates**' from the list. Accept the popup window with the default parameters by clicking '**Compute**' button. After calculation the program displays the result (Fig. 1.137).
76. The results provide information on mean evolutionary rates which indicates how rapid the rate of evolution is. It also provides information about amino acid frequencies.

## Computing the Tajima's Relative Rate

77. Click on the '**Clocks**' option from the main menu and select '**Tajima's Relative Rate Test**' from the list. In the popup window specify any two taxon and one out group to be compared. Click '**Compute**' button (Fig. 1.138). After calculation the program displays the result. In this case, *E.coli* has been chosen as outgroup.
78. The result is displayed as a table indicating the unique differences, identical and divergent sites among the taxa compared (Fig. 1.139). Similarly you can calculate for other taxons and compile a table so as to compare the values.

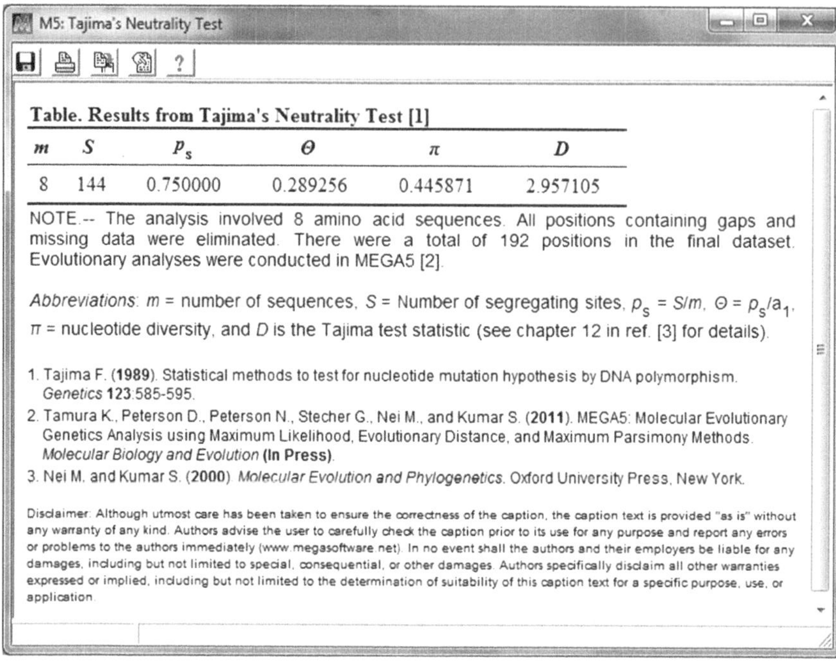


**Fig. 1.136** Results for Tajima's test of neutrality

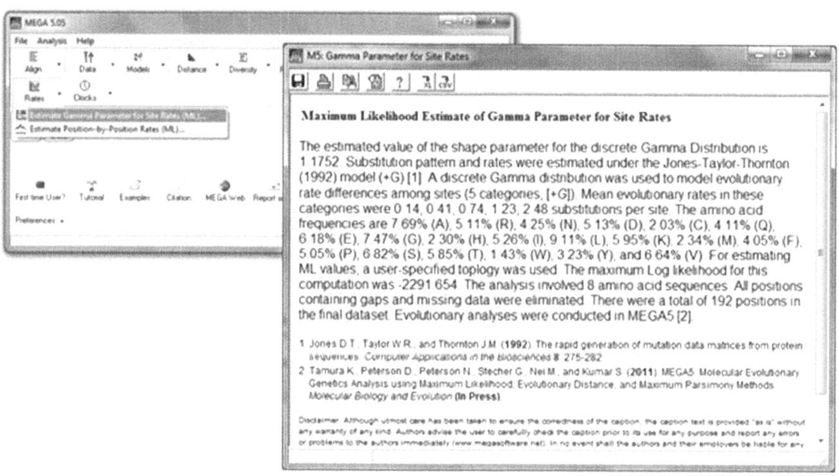


**Fig. 1.137** Estimating gamma parameter for site rates

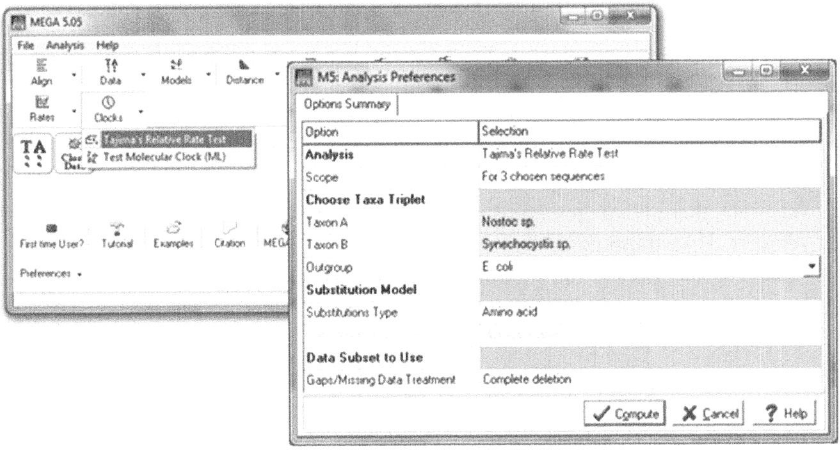


**Fig. 1.138** Computing the Tajima's relative rate

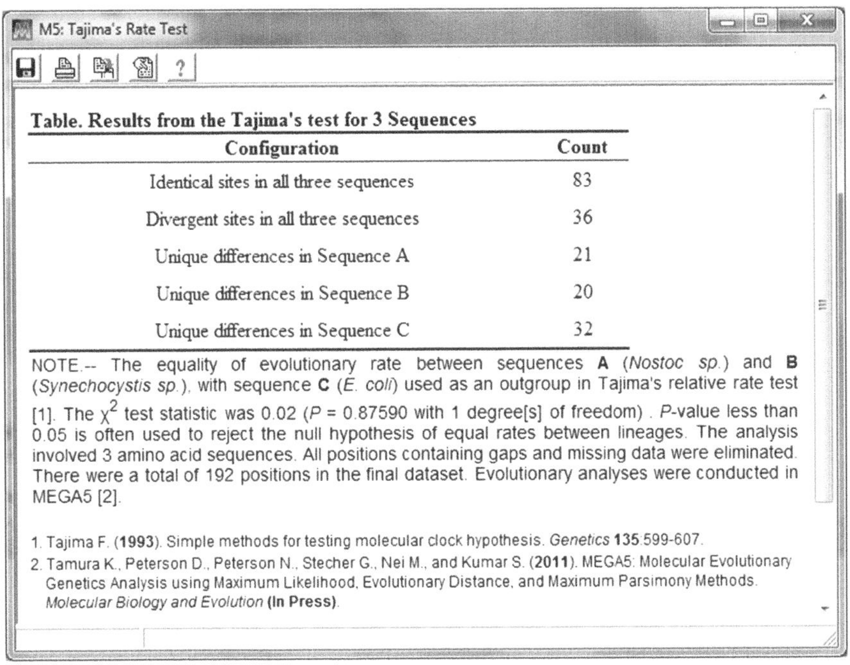
M5: Tajima's Rate Test

**Table. Results from the Tajima's test for 3 Sequences**

| Configuration | Count |
|---|---|
| Identical sites in all three sequences | 83 |
| Divergent sites in all three sequences | 36 |
| Unique differences in Sequence A | 21 |
| Unique differences in Sequence B | 20 |
| Unique differences in Sequence C | 32 |

NOTE.-- The equality of evolutionary rate between sequences **A** (*Nostoc sp.*) and **B** (*Synechocystis sp.*), with sequence **C** (*E. coli*) used as an outgroup in Tajima's relative rate test [1]. The $\chi^2$ test statistic was 0.02 ($P$ = 0.87590 with 1 degree[s] of freedom) . $P$-value less than 0.05 is often used to reject the null hypothesis of equal rates between lineages. The analysis involved 3 amino acid sequences. All positions containing gaps and missing data were eliminated. There were a total of 192 positions in the final dataset. Evolutionary analyses were conducted in MEGA5 [2].

1. Tajima F. (**1993**). Simple methods for testing molecular clock hypothesis. *Genetics* **135**:599-607.
2. Tamura K., Peterson D., Peterson N., Stecher G., Nei M., and Kumar S. (**2011**). MEGA5: Molecular Evolutionary Genetics Analysis using Maximum Likelihood, Evolutionary Distance, and Maximum Parsimony Methods. *Molecular Biology and Evolution* **(In Press)**.

**Fig. 1.139** Tajima's relative rate results

**Exercise**

1. How molecular phylogeny is different from traditional classification of organism?
2. Name the important steps involved in phylogeny construction.
3. What are phenetic and cladistic methods of phylogeny?
4. What is meant by conserved and variable sites in an alignment?
5. What are the advantages and limitations of distance-based phylogenies?
6. What is 'Boostrapping'?
7. How will you decide the method of phylogeny to perform?

# Chapter 2
# DNA Marker Analysis

Genetic markers are polymorphic genetic sequences, like RFLPs or microsatellites, that differ within chromosomal alleles. Rather than analysing the sequence directly, this gene is inferred through analysis of a genetic marker. Marker analysis approach is quite helpful in population biology and ecology studies that can be trace patterns in populations like plants, animals, humans, etc.

## 2.1 Genetic Analysis Using NTSYSpc (Numerical Taxonomy System)

**Introduction**

Numerical taxonomy deals with grouping taxonomic units by numerical methods, a concept first developed by Sokal and Sneath in 1963, divided the field into **phenetics** and **cladistics** in which the former was based on the patterns of overall similarities and in the latter classifications were based on the branching patterns of the estimated evolutionary history of the taxa.

NTSYSpc generate phenogram based on characters by performing various types of agglomerative cluster analysis with similarity or dissimilarity matrix like standardizing data matrix and computing distance coefficients among the columns of the matrix, using UPGMA to cluster the distance matrix, computing cophenetic-value (ultrametric) matrix and also computing the cophenetic correlation as a measure of goodness of fit, and then finally plotting the results in the form of a phenogram and generating the distance matrix also as an output. Matrix which is the actual input can be generated based on the presence (1) or absence (0) of characters of the variables.



K. Syed Ibrahim et al., *Bioinformatics—A Student's Companion*,
DOI 10.1007/978-981-10-1857-2_2

**Exploration Activity**

1. Based on the gel picture (Fig. 2.1), identify the distinct bands and prepare a table (Table 2.1). Enter 1 for the presence of band.

**Using NTedit**

2. Click and activate the NTedit program in your computer.
3. You can create a table of desired rows and columns by entering appropriate values in '**No. Rows**' and '**No. Cols**' (Fig. 2.2).
4. You can edit the row label and column label by clicking '**Row Lab**' and '**Col Lab**' button and editing accordingly.
5. Once the matrix has been prepared, enter the values as you have tabulated earlier. Enter 1 for the presence of band and 0 for the absence of band.

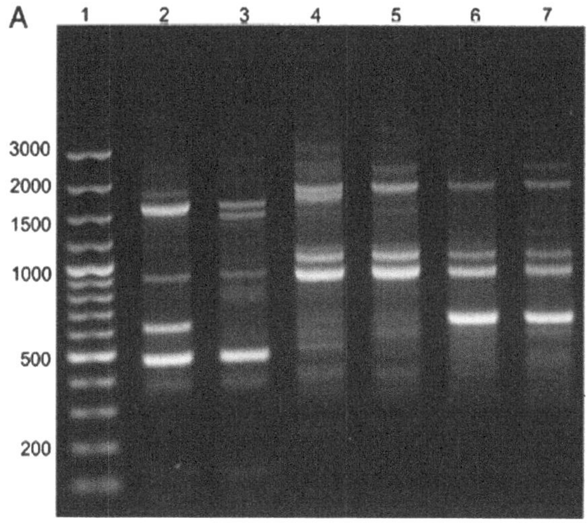


**Fig. 2.1** The RAPD gel picture of six different samples

**Table 2.1** Table prepared with according to the gel picture where one indicates presence of bands

| Marker (MW) | Sp1 | Sp2 | Sp3 | Sp4 | Sp5 | Sp6 |
|---|---|---|---|---|---|---|
| 3000 | | | 1 | 1 | | |
| 2500 | | | 1 | | | |
| 2400 | | | 1 | 1 | | |
| 2000 | | | 1 | 1 | 1 | 1 |
| 1600 | 1 | 1 | 1 | | | |
| 1500 | | 1 | 1 | 1 | | |
| 1000 | | | 1 | 1 | 1 | 1 |
| 700 | | | | | 1 | 1 |
| 600 | 1 | | | | | |
| 500 | 1 | 1 | | | | |
| 400 | 1 | 1 | | | | |
| 120 | | 1 | | | | |

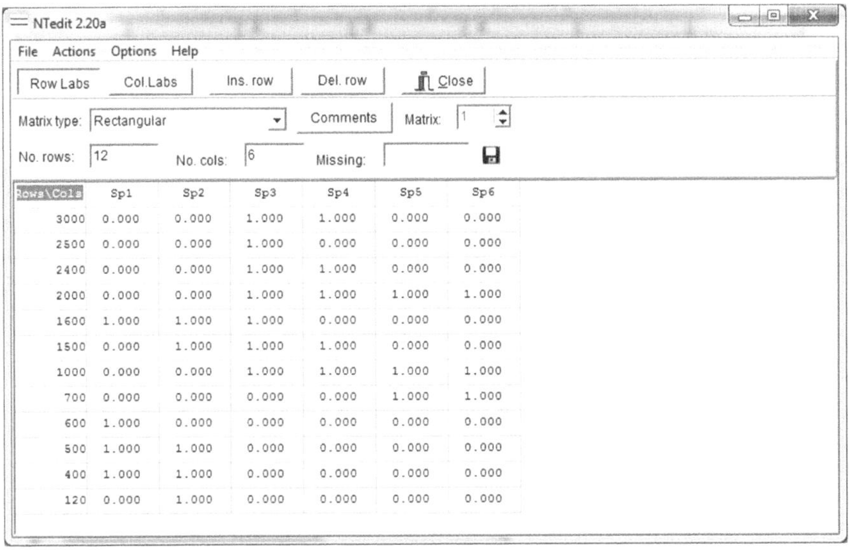


**Fig. 2.2** The NTedit program for preparing data

6. Save the file by choosing '**Save**' option from '**File**' in the menu bar or by clicking the floppy icon (data.NTS).

### Using NTSYSpc

7. Now click and activate the NTSYSpc program from your computer.
8. The options and the menus of the program have been listed on the left side of the program window (Fig. 2.3).
9. Click on the '**Transformation**' option (Fig. 2.4) and select '**Standardization**' (Stand).
10. Specify the file which you prepared and saved earlier using NTedit as input file (Data.NTS). And, specify a name for the output file (Stand.NTS).
11. Click '**Compute**' button.
12. The program generates a '**Report listing**' window (Fig. 2.5) whose contents can be copied and saved for future requirements, if necessary.
13. You can open the output file to visualize the distance matrix (Fig. 2.6).

### Constructing Dissimilarity Tree

14. Click on the '**Dis/similarity**' menu and select '**Genetic Distance**' (Simgend) option. Specify the output file you generated earlier in Standardization step as input file (Stand.NTS). Then, specify a new output file (simgend_out.NTS) (Fig. 2.7).
15. Click '**Compute**' button.

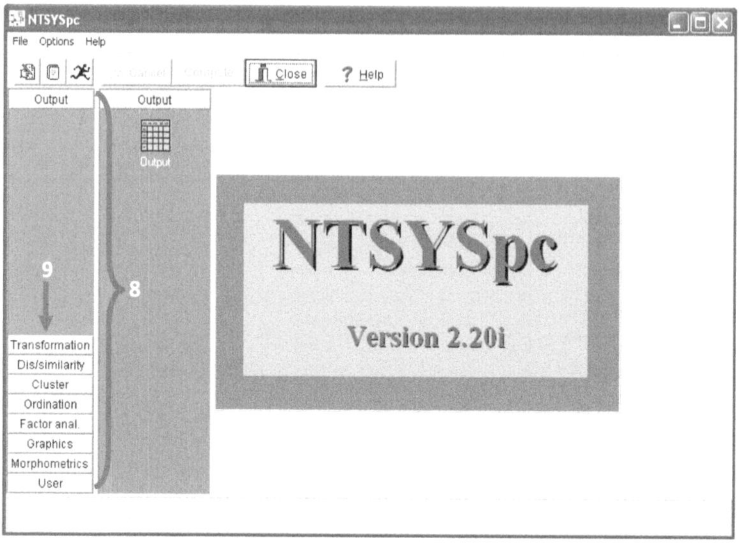


Fig. 2.3 The NTSYSpc program interface

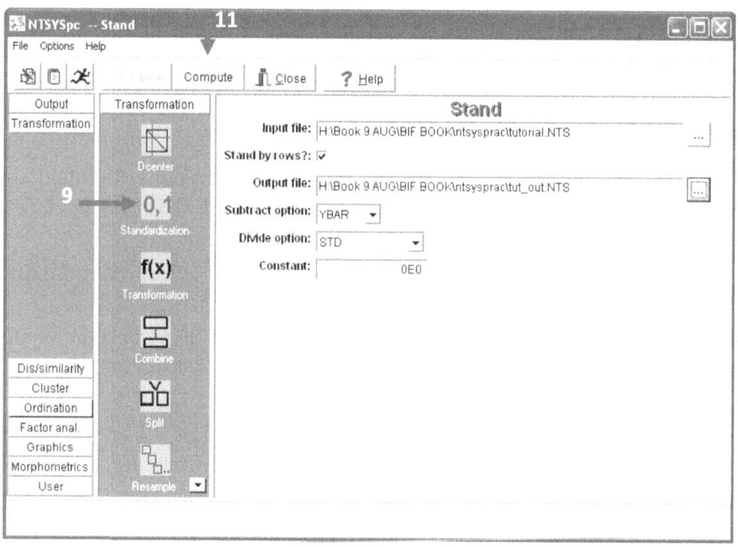


Fig. 2.4 Performing standardization under transformation

16. The program generates a '**Report listing**' tab whose contents can be copied and saved for future requirements, if necessary.
17. You can open the output file to visualize the distance matrix (Fig. 2.8).

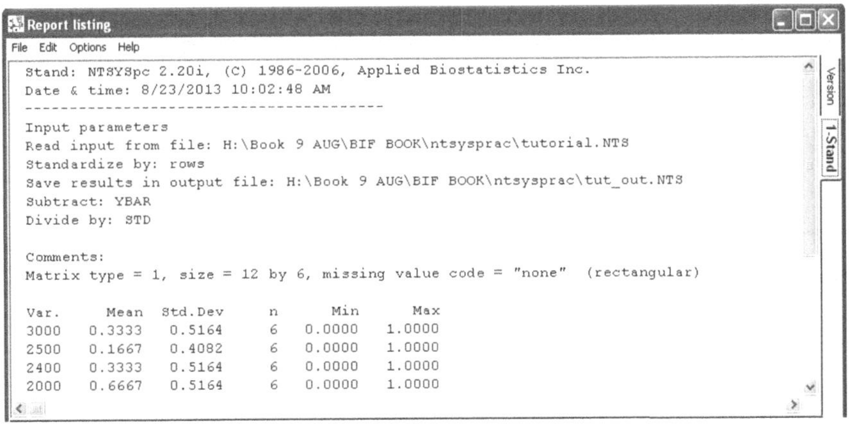


**Fig. 2.5** The report listing window

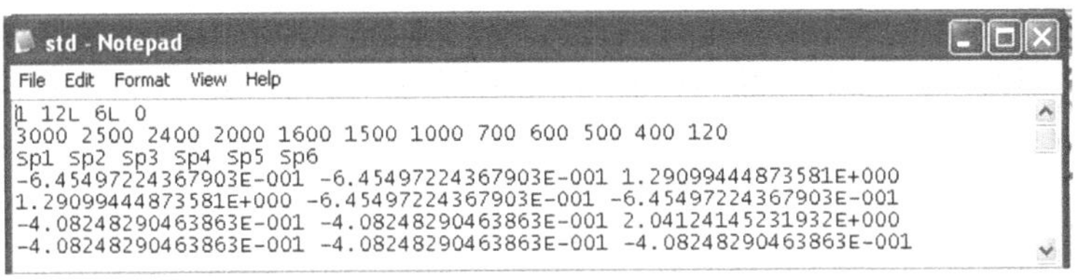


**Fig. 2.6** The content of output file displaying the distance matrix

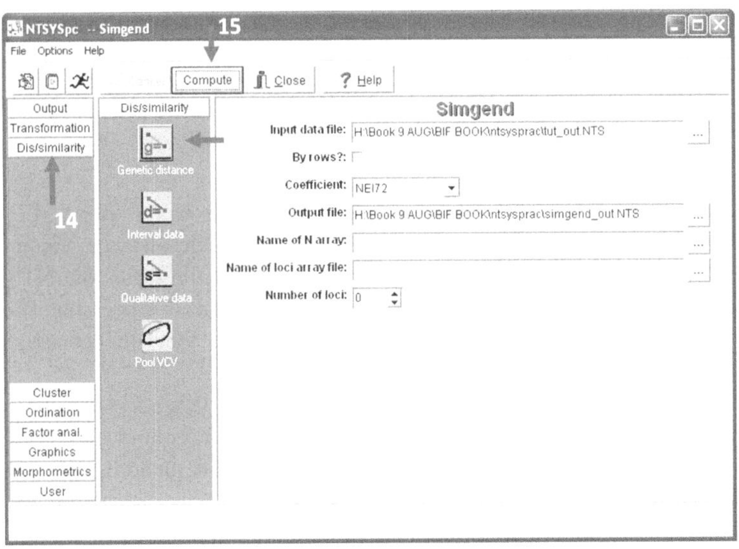


**Fig. 2.7** Calculating genetic distance from standardized data

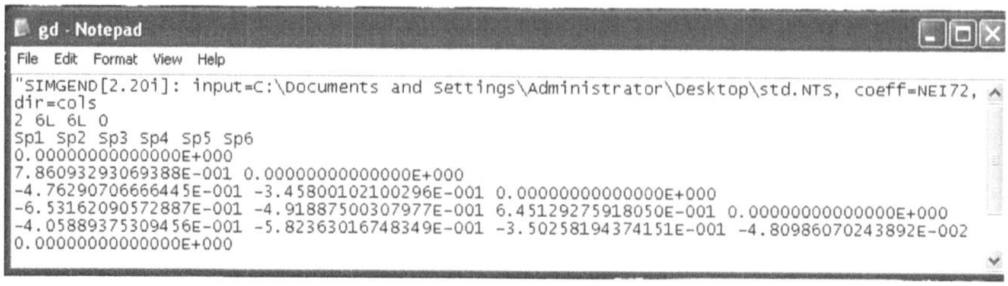


**Fig. 2.8** Simgend report

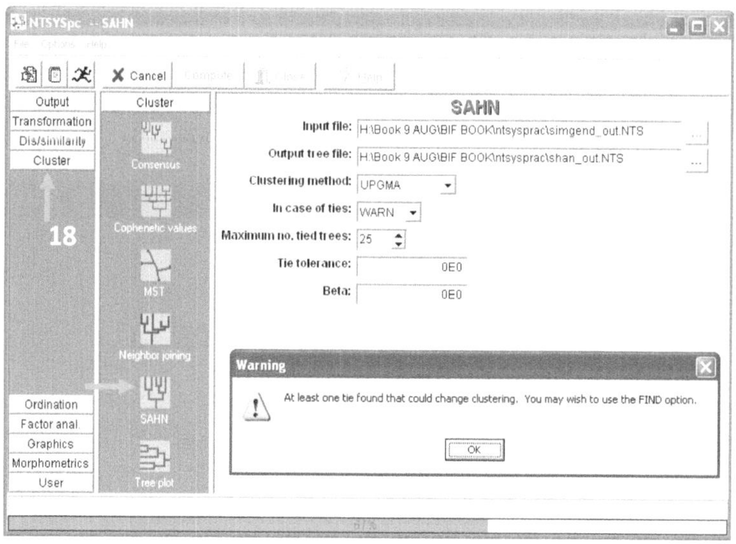


**Fig. 2.9** Preparing SAHN tree

18. Click '**Cluster**' from the menu and select '**SAHN**' from the option (Fig. 2.9). Specify the output file you generated earlier in Genetic distance step as input file (simgend_out.NTS). Then, specify a new output file (sahn_out.NTS).
19. Click '**OK**' on the warning popup generated to proceed processing. You can see an icon at left corner which can be clicked to view the generated tree (Fig. 2.10).
20. The tree is displayed on a new window '**Tree plot**' (Fig. 2.11).
21. The display options like flipping the tree, changing the font or thickness of the line can be changed by clicking the '**Options**' from the menu bar and selecting '**Plot options…**' (Fig. 2.12).
22. You can view the UPGMA file content (sahn_out.NTS) by opening the output file (Fig. 2.13).

**Fig. 2.10** The button to display the generated tree

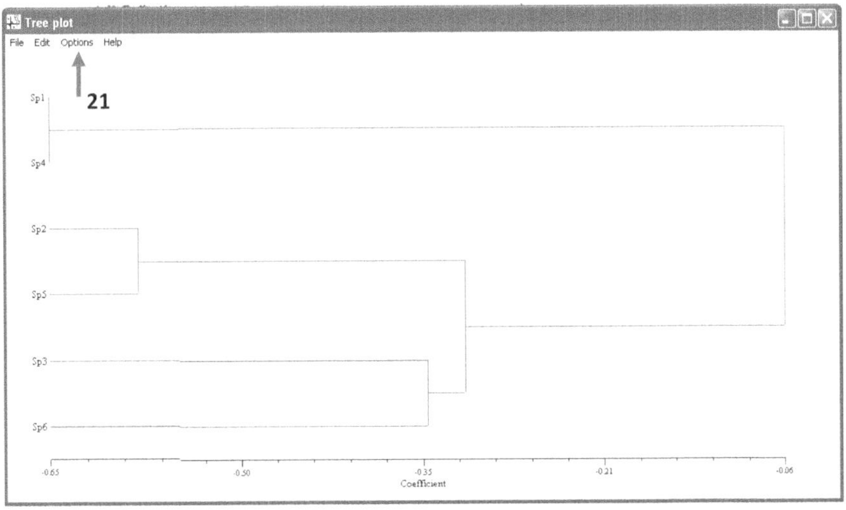


**Fig. 2.11** The dissimilarity tree

**Fig. 2.12** The tree plot options

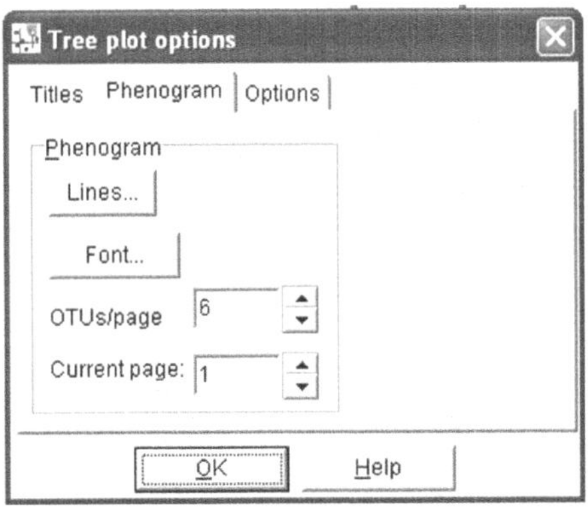

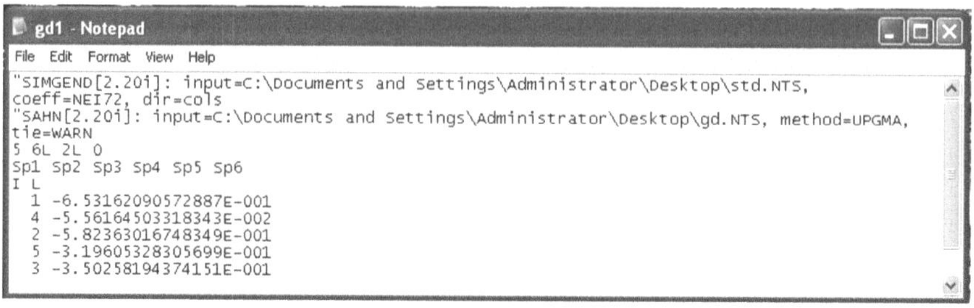


**Fig. 2.13** UPGMA tree generated by SAHN

### Constructing Similarity Tree

23. Click on the '**Dis/similarity**' menu and select '**Interval data**' (Simint) or '**Qualitative data**' (Simqual) option. Specify the output file you generated earlier in Standardization step as input file (stand.NTS). Then, specify a new output file (simint_out.NTS or simqual_out.NTS).
24. Click '**Compute**' button.
25. The program generates a '**Report listing**' tab and an output file whose contents can be copied and saved for future requirements, if necessary.
26. Click '**Cluster**' from the menu and select '**SHAN**' from the option to generate tree.

### Exercise

1. What are the applications of NTSYSpc?
2. Define the following terms:
    1. Ordination
    2. Factorial analysis
    3. Cluster analysis
    4. SAHN
    5. UPGMA
    6. Neighbour Joining

## 2.2 Principal Coordinate Analysis (PCOORDA) Using NTSYSpc

**Principal component analysis (PCA)** is a mathematical procedure invented in 1901 by Karl Pearson and later developed by Harold Hotelling in the 1930s. This method used orthogonal transformation for converting a set of possibly correlated variables into a set of linearly uncorrelated variables named principal components.

PCA can be done by Eigen value decomposition of a data matrix, usually after mean centring (and normalizing or using Z-scores) the data matrix for each attribute. The results of a PCA are usually discussed in terms of Eigen value. Largest Eigen values usually correspond to the principal components that are associated with most of the co-variability among a number of observed data.

PCOORDA is an alternative to PCA which will give the same results as PCA. When there are fewer points than variables, the computation time will be much less than for the PCA. The data matrix is standardized by variables (rows), a matrix of distances between the objects is computed, the double-centred distance matrix is then factored and a plot is drawn showing the objects in a three-dimensional space.

1. Now click and activate the NTSYSpc program from your computer.
2. The options and the menus of the program have been listed on the left side of the program window (Fig. 2.14).

**Preparation of Similarity Matrix Using Qualitative Data**

3. Click on the '**Dis/similarity**' option and select '**Qualitative data**' (Simqual).
4. Specify the file which you prepared and saved earlier using NTedit as input file (**Data**). And for the **output file** specify a name (**DJ**) (Fig. 2.15).
5. Choose '**J**' under '**Coefficient**'.
6. Leave other options as default (Positive code 1 and Negative code 0).
7. Click '**Compute**' button.

**Preparation of Double-Centred Matrix**

8. Click '**Transformation**' and select '**Dcenter**' (Fig. 2.16).

**Fig. 2.14** The NTSYSpc program interface

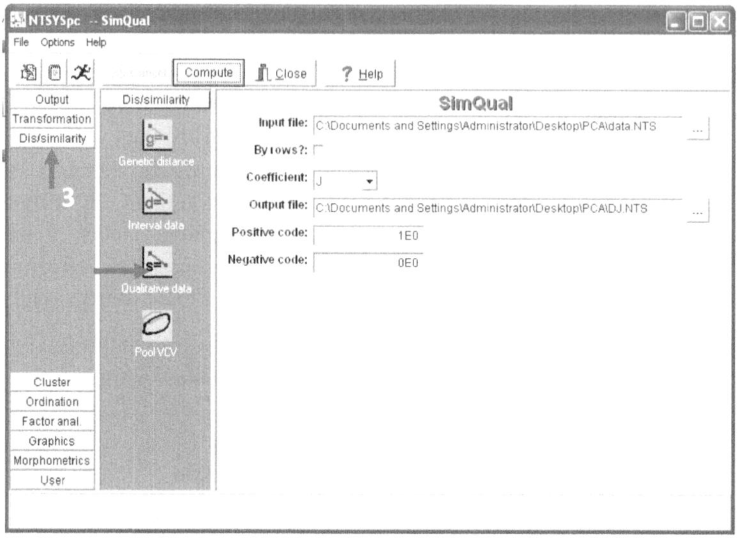


**Fig. 2.15** Calculating similarity matrix using qualitative data

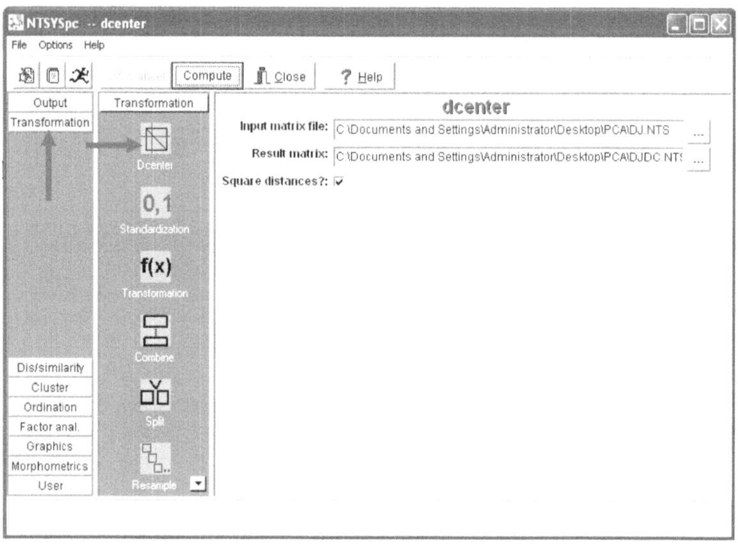


**Fig. 2.16** Calculating Double-centred matrix

9. Specify the file which you prepared and saved earlier (**DJ**) as Input matrix file. And for the **Result matrix** (output file) specify a name (**DJDC**).
10. Click '**Compute**' button.

**Fig. 2.17** Calculating Eigen values

### Calculation of Eigen Value

11. Click '**Ordination**' and select '**Eigen**' (Fig. 2.17).
12. Specify the file which you prepared and saved earlier (**DJDC**) as input matrix file.
13. Leave **no. of dimensions** as default (4).
14. For the **output Eigenvector file** specify a name (**DJDCVec**) and for **output Eigenvalue file** a name (**DJDCVal**).
15. Leave other options as default (sample size and degrees of freedom = 0).
16. Click '**Compute**' button.
17. The Eigen values are displayed on the '**Report listing**' window (Fig. 2.18).

### Plotting Coordinate Graph

18. Click '**Graphics**' and select '**Mod 3D Plot**' (Fig. 2.19).
19. Specify the file which you prepared and saved earlier (DJDCVec) as input file. Deselect '**Plot by rows**'.

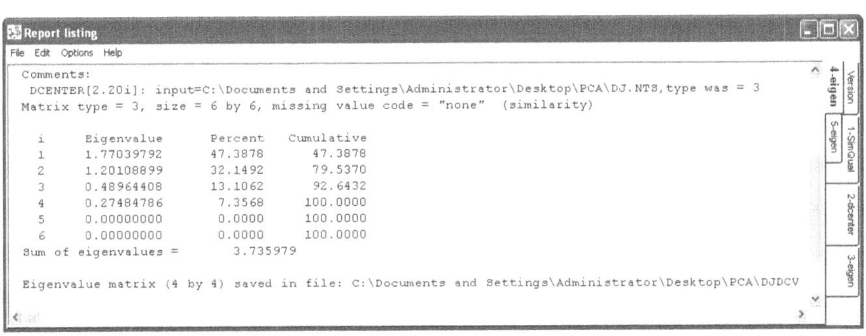


**Fig. 2.18** Eigen values displayed on the report listing window

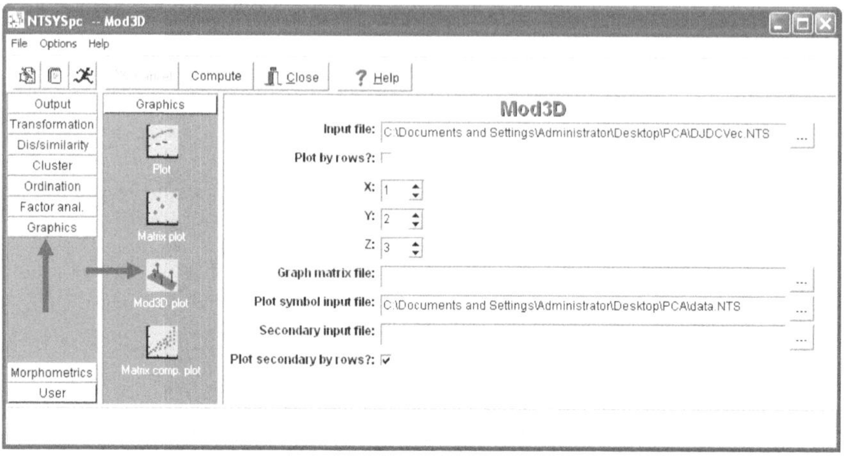


**Fig. 2.19** Preparing plot

20. For **Plot symbol input file** (dominant symbol), specify the matrix file you created with NTedit (Data).
21. Leave '**Graph matrix file**' as blank.
22. Click '**Compute**' button.
23. The graph is displayed by which can be rotated by mouse. The display options can be changed by changing the options in '**Plot options**' (Fig. 2.20).

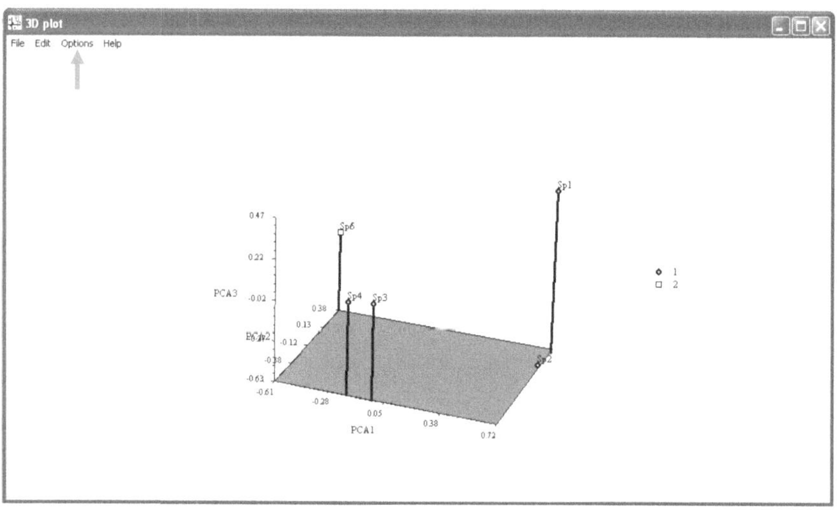


**Fig. 2.20** The plot in 3D. *Diamond symbol* indicates one species (1, 2, 3 and 4) and the *square symbol* indicates 2 species (5 and 6)

**Exercise**

1. What do you mean by multivariate analysis and PCA?
2. How PCA is different from PCOORDA?
3. Define the following

   A. Eigen value
   B. PCA 1 and PCA 2
   C. SVD.

## 2.3 Population Genetic Analysis Using PowerMarker

**Introduction**

PowerMarker is a comprehensive collection of both new and traditional statistical methods that can be employed for population genetic analysis which is basically designed for SSR/SNP data analysis (Liu and Muse 2005).

**Preparation of Matrix Data Based on Banding Pattern**

1. Open a new Excel (xls) file and create a data file containing **1 for the presence of band**, **0 for the absence of band** and **9 for missing data** (Fig. 2.21).
2. Find and replace all 0 to **Z**, 1 to **Y** and 9 to **Q** (Fig. 2.22).
3. Copy the whole table and paste in a Notepad (Fig. 2.23). Now replace all the Z to **0/0**, Y to **1/1** and Q to **9/9** and save the file (Data.txt).

| | A | B | C | D | E | F | G | H |
|---|---|---|---|---|---|---|---|---|
| 1 | popn | accession | p1000 | p800 | p600 | p400 | p200 | p100 |
| 2 | p1 | A1 | 1 | 0 | 1 | 0 | 9 | 1 |
| 3 | p1 | A2 | 1 | 1 | 0 | 0 | 0 | 0 |
| 4 | p1 | A3 | 1 | 0 | 0 | 0 | 1 | 0 |
| 5 | p1 | A4 | 1 | 1 | 0 | 1 | 1 | 1 |
| 6 | p1 | A5 | 1 | 0 | 1 | 1 | 1 | 1 |
| 7 | p2 | A6 | 0 | 0 | 1 | 1 | 0 | 1 |
| 8 | p2 | A7 | 9 | 1 | 1 | 1 | 0 | 9 |
| 9 | p2 | A8 | 1 | 0 | 0 | 1 | 1 | 1 |
| 10 | p2 | A9 | 0 | 1 | 0 | 0 | 1 | 0 |
| 11 | p2 | A10 | 0 | 0 | 1 | 0 | 0 | 0 |
| 12 | p3 | A11 | 0 | 1 | 0 | 9 | 0 | 0 |
| 13 | p3 | A12 | 1 | 0 | 9 | 0 | 1 | 1 |
| 14 | p3 | A13 | 0 | 0 | 9 | 1 | 1 | 1 |
| 15 | p3 | A14 | 1 | 0 | 0 | 1 | 0 | 1 |
| 16 | p3 | A15 | 1 | 9 | 1 | 1 | 9 | 1 |
| 17 | | | | | | | | |

**Fig. 2.21** Preparing data on excel sheet

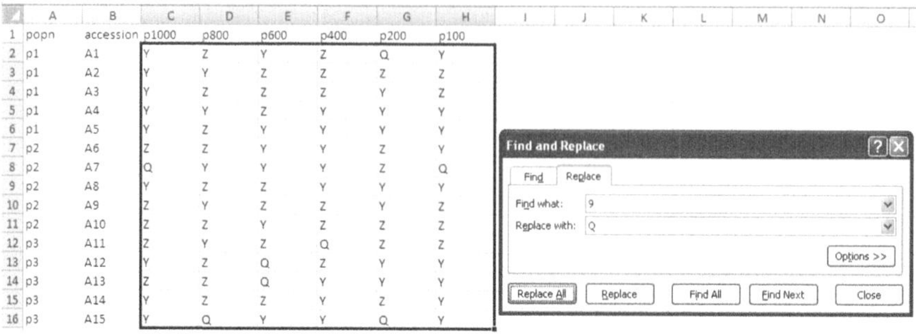


Fig. 2.22 Finding and replacing 0's and 1's

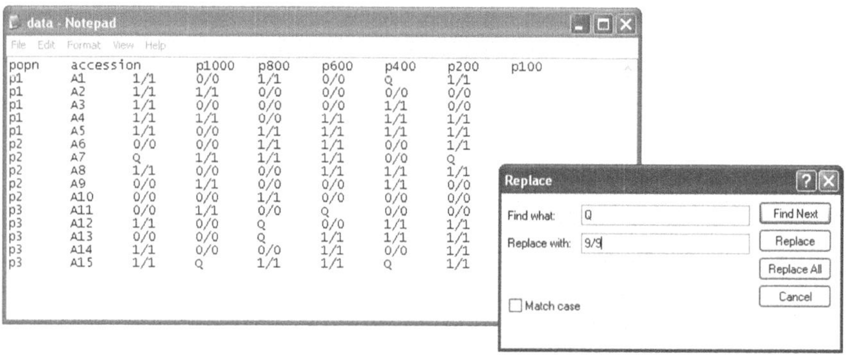


Fig. 2.23 Modifying data on a notepad

## Using Power Marker

1. Click and initiate PowerMarker program from your computer.
2. Choose '**File**' and select '**Close all Projects**' from the menu to close all the default projects.
3. Create a new project by choosing '**File**' and selecting '**New Project**' by giving a name (PMTut).
4. Click '**File**' and select Import and Dataset. A dataset import wizard window now opens. Click on the browse and select the 'Data.txt' you saved earlier. Click '**Next**' button (Fig. 2.24).
5. Change the 'popn' and 'accession' to categorical type by clicking the column item and clicking the '**Categorical**' link at the bottom of the table (Fig. 2.25).
6. Then under '**Hierarchy**', change the Level-1 to '**accession**' and level-2 to '**popn**' by clicking on the dropdown menu and selecting. Click '**Next**' button (Fig. 2.25).
7. Change the '**Missing allele**' and '**Missing category**' values to **9**. Click '**Next**' button (Fig. 2.26).

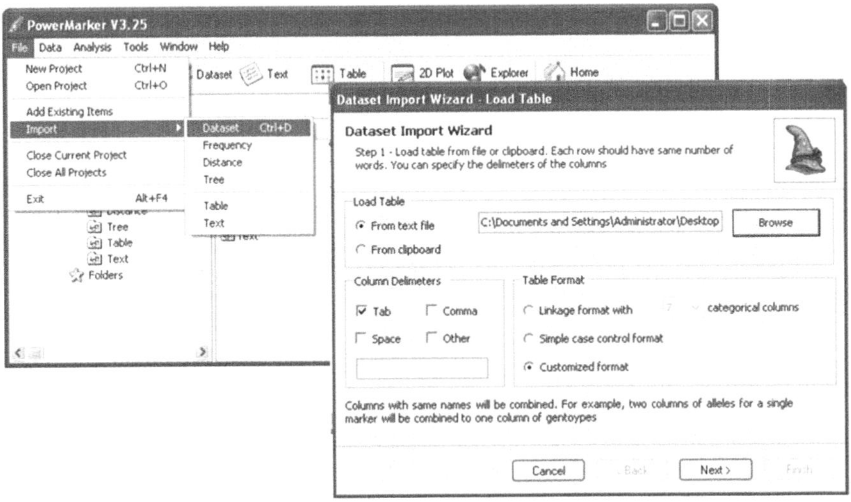


**Fig. 2.24** Importing dataset into PowerMarker

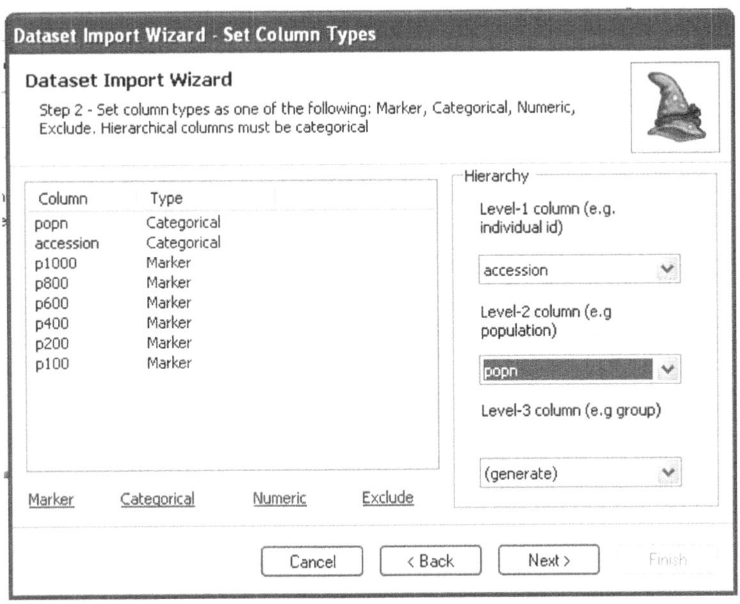


**Fig. 2.25** Choosing options in the dataset import wizard

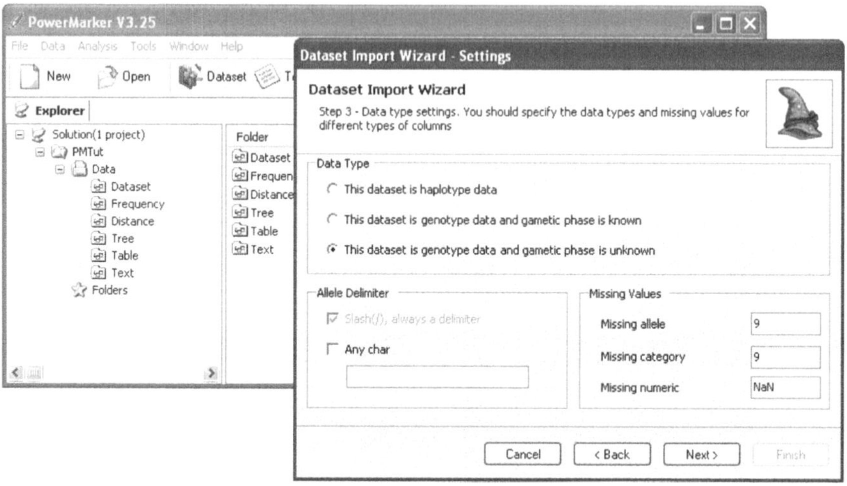


Fig. 2.26 Specifying the data types and missing values

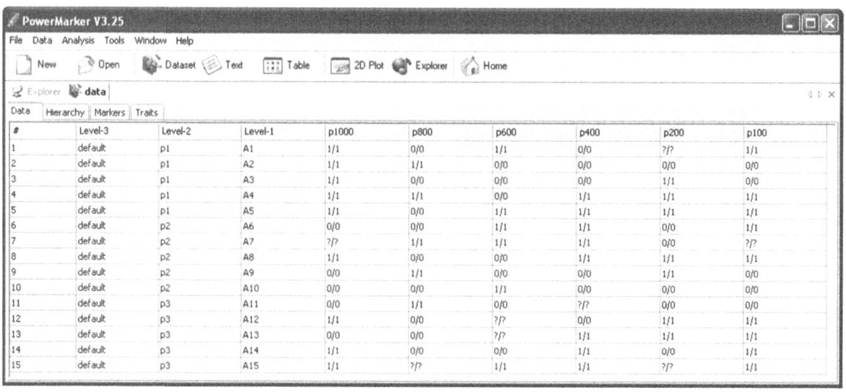


| # | Level-3 | Level-2 | Level-1 | p1000 | p800 | p600 | p400 | p200 | p100 |
|---|---|---|---|---|---|---|---|---|---|
| 1 | default | p1 | A1 | 1/1 | 0/0 | 1/1 | 0/0 | ?/? | 1/1 |
| 2 | default | p1 | A2 | 1/1 | 1/1 | 0/0 | 0/0 | 0/0 | 0/0 |
| 3 | default | p1 | A3 | 1/1 | 0/0 | 0/0 | 0/0 | 1/1 | 0/0 |
| 4 | default | p1 | A4 | 1/1 | 1/1 | 0/0 | 1/1 | 1/1 | 1/1 |
| 5 | default | p1 | A5 | 1/1 | 0/0 | 1/1 | 1/1 | 1/1 | 1/1 |
| 6 | default | p2 | A6 | 0/0 | 0/0 | 1/1 | 1/1 | 0/0 | 1/1 |
| 7 | default | p2 | A7 | ?/? | 1/1 | 1/1 | 1/1 | 0/0 | ?/? |
| 8 | default | p2 | A8 | 1/1 | 0/0 | 0/0 | 1/1 | 1/1 | 1/1 |
| 9 | default | p2 | A9 | 0/0 | 1/1 | 0/0 | 0/0 | 1/1 | 0/0 |
| 10 | default | p2 | A10 | 0/0 | 0/0 | 1/1 | 0/0 | 0/0 | 0/0 |
| 11 | default | p3 | A11 | 0/0 | 1/1 | 0/0 | ?/? | 0/0 | 0/0 |
| 12 | default | p3 | A12 | 1/1 | 0/0 | ?/? | 0/0 | 1/1 | 1/1 |
| 13 | default | p3 | A13 | 0/0 | 0/0 | ?/? | 1/1 | 1/1 | 1/1 |
| 14 | default | p3 | A14 | 1/1 | 0/0 | 0/0 | 1/1 | 0/0 | 1/1 |
| 15 | default | p3 | A15 | 1/1 | ?/? | 1/1 | 1/1 | ?/? | 1/1 |

Fig. 2.27 The final matrix

8. It will display a final matrix that is ready to be imported to the project. Click '**Finish**' button (Fig. 2.27).

## Statistical Analysis

9. Click on the '**Analysis**' from the main menu to see the list of analyses that can be performed for the provided data (Fig. 2.28). The type of analyses that can be performed is listed in Table 2.2.
10. Important features to look in '**Summary**' (Table 2.2).
11. Important features (Fig. 2.29) to look in '**Structure**' (Table 2.3).
12. The list of analyses that can be done with PowerMarker are provided in the developer's website.

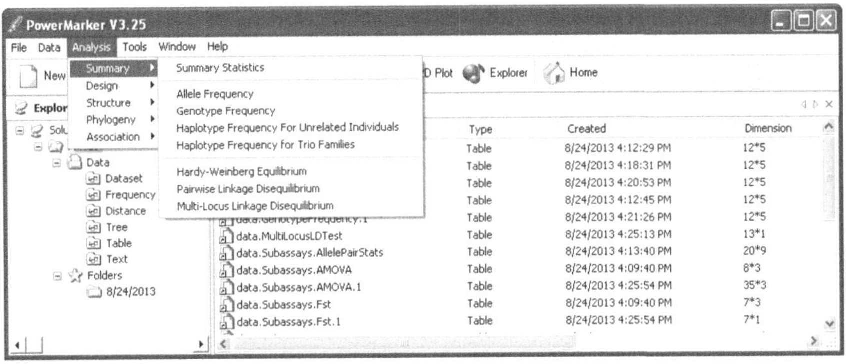


**Fig. 2.28** The Analysis options in PowerMarker

**Table 2.2** Options under 'summary statistics'

| | |
|---|---|
| Summary statistics | Major allele, Gene diversity, Heterozygosity, PIC |
| Allele frequency | Marker, variance |
| Genotype frequency | Covariance, marker |
| Hardy-Weinberg equilibrium | Chi square, p value |
| Pairwise linkage disequilibrium | Multi allelic D and N allelic D |

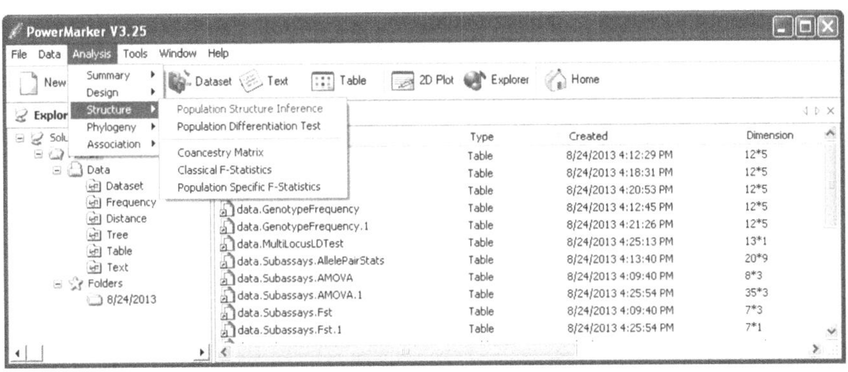


**Fig. 2.29** The structure options under analysis

**Table 2.3** Classical F statistics

| | |
|---|---|
| Classical F-statistics | Select 'ANOVA table' |
| Population specific F-statistics | Mean SD values of beta, alpha and covariance |

**Exercise**

1. Define the following
    1. Covariance
    2. Heterozygocity
    3. PIC
    4. Hardy Weinberg equilibrium
    5. Exact Test
    6. Mantel Test
2. Explain the basic steps involved in AMOVA and state their applications.

## 2.4 Dissimilarity Analysis—DARwin5 (Dissimilarity Analysis and Representation for Windows)

**DARwin** offers both traditional and more original approaches and focuses mainly on diversity structure description which is based on the distance methods.

**Exploration Activity**

1. DARwin program can be downloaded from http://darwin.cirad.fr/Download.php for free after registration.
2. Prepare the data in excel sheet and save it in a notepad (data.txt) (Fig. 2.30). Type Unit in column A. Row may be accession, species or variety. Each column represents the marker/bands.
3. Activate the Darwin5 program from your computer (Fig. 2.31).

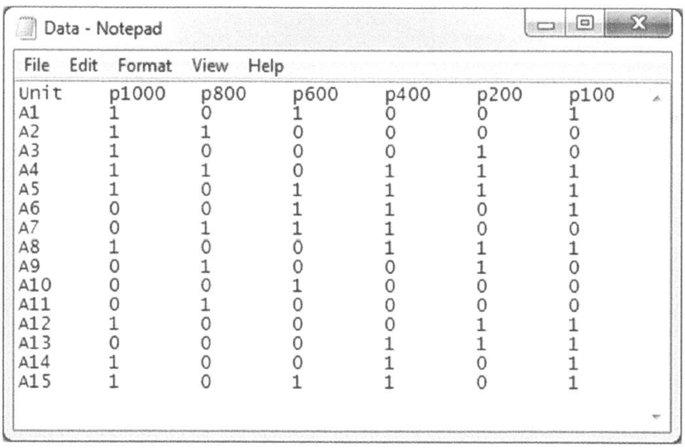


| Unit | p1000 | p800 | p600 | p400 | p200 | p100 |
|---|---|---|---|---|---|---|
| A1 | 1 | 0 | 1 | 0 | 0 | 1 |
| A2 | 1 | 1 | 0 | 0 | 0 | 0 |
| A3 | 1 | 0 | 0 | 0 | 1 | 0 |
| A4 | 1 | 1 | 0 | 1 | 1 | 1 |
| A5 | 1 | 0 | 1 | 1 | 1 | 1 |
| A6 | 0 | 0 | 1 | 1 | 0 | 1 |
| A7 | 0 | 1 | 1 | 1 | 0 | 0 |
| A8 | 1 | 0 | 0 | 1 | 1 | 1 |
| A9 | 0 | 1 | 0 | 0 | 1 | 0 |
| A10 | 0 | 0 | 1 | 0 | 0 | 0 |
| A11 | 0 | 1 | 0 | 0 | 0 | 0 |
| A12 | 1 | 0 | 0 | 0 | 1 | 1 |
| A13 | 0 | 0 | 0 | 1 | 1 | 1 |
| A14 | 1 | 0 | 0 | 1 | 0 | 1 |
| A15 | 1 | 0 | 1 | 1 | 0 | 1 |

**Fig. 2.30** Preparation of data

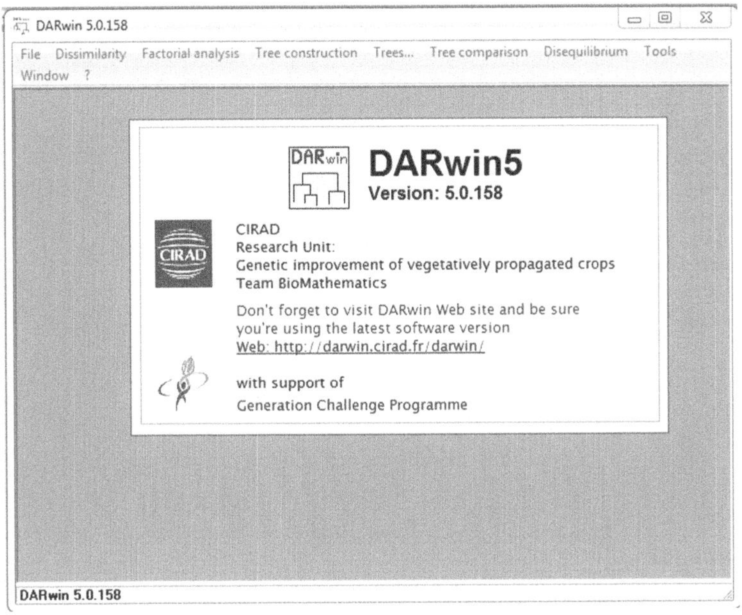


**Fig. 2.31** The DARwin5 interface

4. Click on '**File**' and select '**Import Data Matrix**'. Locate the data file you prepared earlier in notepad (data.txt). You can now see the name on the imported file under the button '**Txt File to Import**'.
5. Then click on the '**Save Data as…**' button and save the file (data.var). Once done you can see the file name under the button '**Save Data as…**'.
6. Click the check box next to 'View file when done' to see the file.
7. Click the '**OK**' button.
8. Click on the '**Dissimilarity**' option on the main menu and select '**Calculate from Single Data**'. Now locate the var file you saved earlier (Data.var) (Fig. 2.32).
9. In case there are missing data in your data file you can click the '**Options**' button in the '**Missing Data**' column, tick the '**Some data are missing**' and under '**Pairwise variable deletion**' choose 50%. Click '**OK**' (Fig. 2.33).
10. If you don't have any missing data, you can skip the earlier step.
11. Under '**Dissimilarity index**' choose '**Jaccard**'.
12. Click '**Save dissimilarity as…**' button on the top to save (Data.dis).
13. Click the '**OK**' at bottom of the program window to execute the program.

**Factorial Analysis**

Principal Coordinate analysis produces Euclidean plans which preserve at best the distances between units through graphical representations.

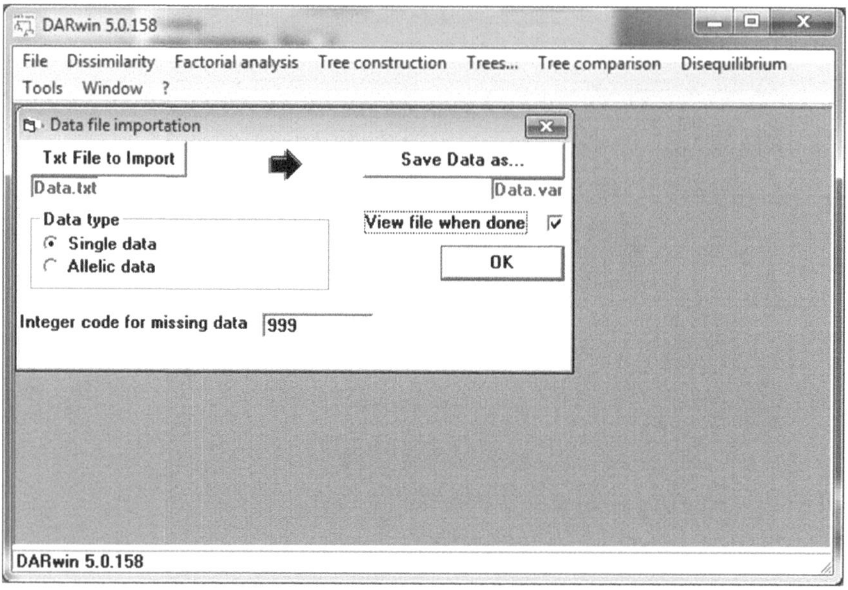


Fig. 2.32 Importing the data file and converting it to var file

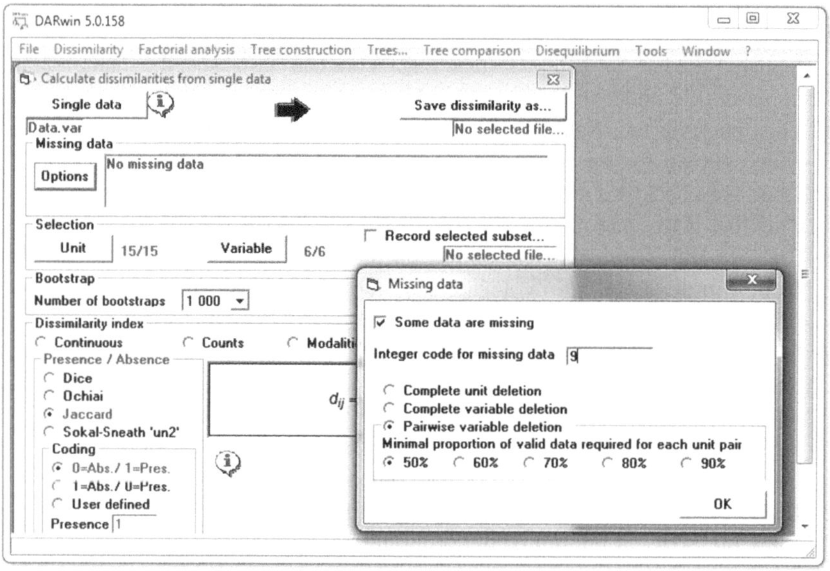


Fig. 2.33 Preparing dissimilarity file

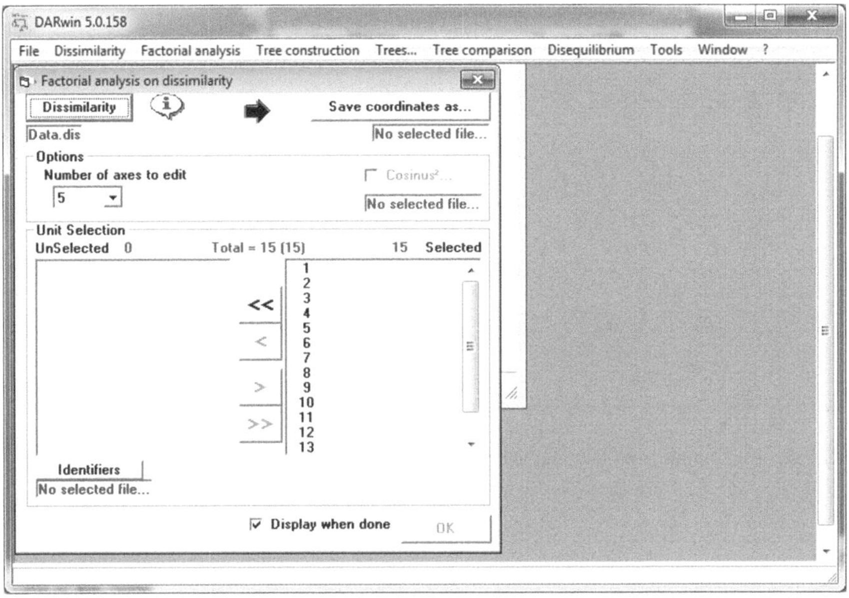


**Fig. 2.34** Performing factorial analysis

14. Click on the '**Factorial analysis**' option on the main menu and select '**Analysis**'. Now locate the dis file you saved earlier (Data.dis).
15. Once loaded, now click on the '**Save coordinates as…**' button on the top of the program interface and save the file Factorial coordinate file (Data.AFT). Click '**OK**' button at the bottom to execute the procedure (Fig. 2.34).
16. The program now displays the plot and a report on the Factorial analysis done (Fig. 2.35).
17. To save the report as text file (FA.txt) click on the floppy disk icon on the '**AFTD sur**' window.

**Tree Construction by Neighbour-Joining Method**

Tree construction can be done by methods like aggregation criteria, Neighbour Joining or by Scores. The first two methods can be either weighted or unweighted. Methods like NJTree and Scores produce more reliable trees by attempting to reduce sensibility to data error. More over bootstrapping, a statistical measure, in NJTree can be employed to estimate how well the tree is supported by the data.

18. Click on the '**Tree construction**' option on the main menu and select '**Neighbour joining**'. Now locate the dis file you saved earlier (Data.dis). Check the '**Bootstrap analysis**' and '**Display when done**' options at the bottom (Fig. 2.36).

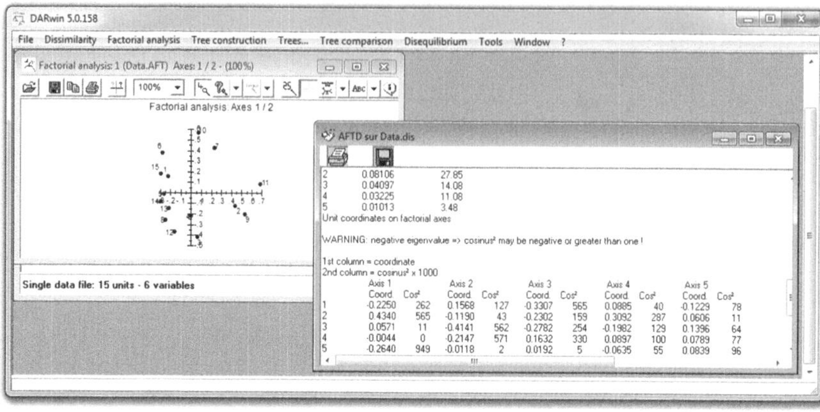


Fig. 2.35 The output from factorial analysis

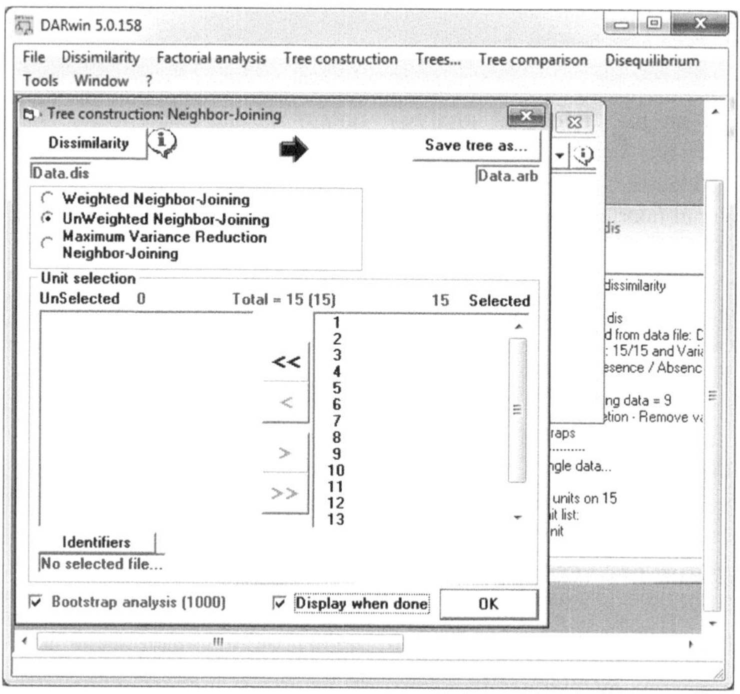


Fig. 2.36 Tree construction using neighbour-joining method

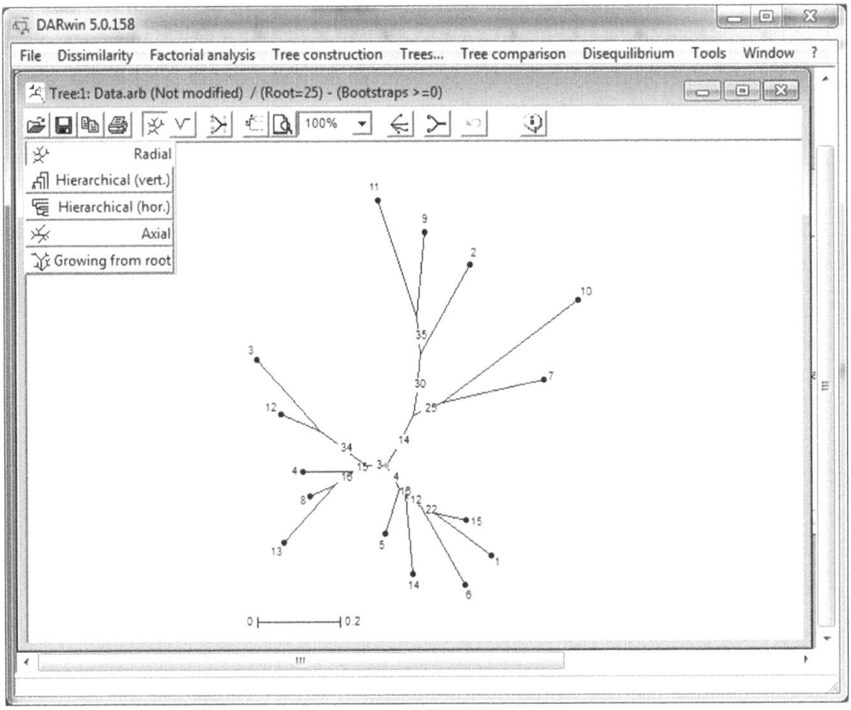


**Fig. 2.37** The neighbour-joining tree

19. Select '**Unweighted Neighbour joining**' from the options and click '**Save tree as…**' button on the top of the program interface to save the tree file (Data.arb).
20. By default radial tree is displayed. The tree can be displayed in other forms by clicking '**Tree representation**' button from the menu bar (Fig. 2.37).

### Tree Representation

21. The menu bar provides options to change the display of the tree. For example to specify a root, click on the '**Root selection tool**' and specify or to edit the tree, click on the '**Edition**' button.
22. To save the final tree, click on the floppy disk icon on the menu.

### Exercise

1. What do you mean by dissimilarity analysis?
2. What do you mean by weighted and unweighted Neighbour-Joining method?
3. Define Jaccard Index of dissimilarity.

# Chapter 3
# RNA Analysis

RNA besides playing a major role in gene regulation they are essential for health and understanding disease. It can form complex structures by base-pairing with itself and other nucleic acids. Hence it is the complex folding of RNA molecules that holds the information content rather than its linear nucleotide sequences (Spitale et al. 2014). Identification of enormous variety of functional RNAs acting besides or along with proteins in complex biological functions inside cells have increased bioinformatics analysis of RNA sequences and structures in recent years.

## 3.1 Predicting RNA Secondary Structure

**Introduction**

The Vienna RNA package provides a wide range of tools that offers algorithms for comparing RNAs, RNA folding, and predicting interactions between RNAs. Besides that there is also a set of comprehensive programs to provide better insights into RNA secondary structure analysis.

RNAfold, a component of Vienna RNA package, uses dynamic programming algorithm that calculates minimum free energy (mfe) structure for the user provided RNA sequences and produces the mfe structure in bracket notation along with its calculated free energy. The program can also compute partition function (pf), base pairing probability matrix, free energy of the thermodynamic ensemble, frequency of the mfe structure in the ensemble and the ensemble diversity (Lorenz et al. 2011). This program can be downloaded free or can be accessed from Vienna RNA Web Servers.

**Exploration Activity**

1. The RNAfold—Vienna RNA web server can be accessed from http://rna.tbi.univie.ac.at/cgi-bin/RNAfold.cgi.



K. Syed Ibrahim et al., *Bioinformatics—A Student's Companion*,
DOI 10.1007/978-981-10-1857-2_3

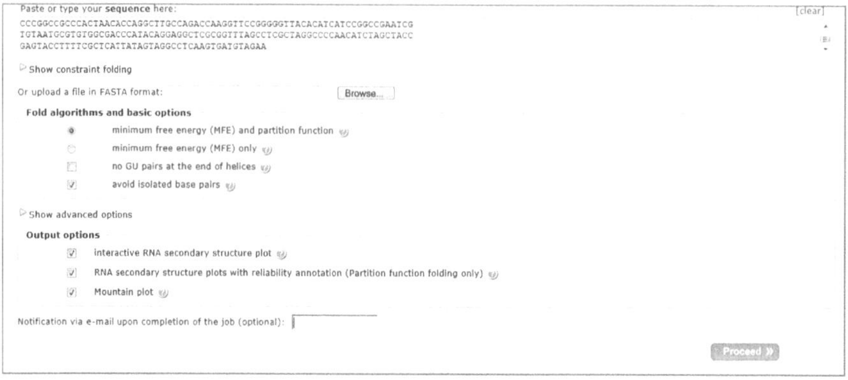


**Fig. 3.1** The RNA fold—Vienna RNA interface of the Vienna RNA web suite

2. Paste the RNA or DNA sequence in plain text or FASTA format. Here, *Anopheles jeyporiensis* ITS2 (Acc. No. JN643728) has been used as input (Fig. 3.1).
3. Select the options '**minimum free energy (MFE) and partition function**' and '**avoid isolated base pairs**' under '**Fold Algorithms and basic options**'.
4. Select all options under '**Output options**'.
5. You can also specify your email (optionally) if you wish to be notified by email about the completion of the job.
6. Click on the '**Proceed** >>' button at the bottom of the page.
7. The output gives a detailed description on the thermodynamic based on the loop-based energy model, displaying results under three headings. Since RNA structure prediction is always error-prone, the program usually provides predicted structures with the reliability information.
8. The first '**Results for minimum free energy prediction**' displays the predicted optimal secondary structure in the dot-bracket (Newick's format) along with the calculated minimum free energy value (Fig. 3.2).

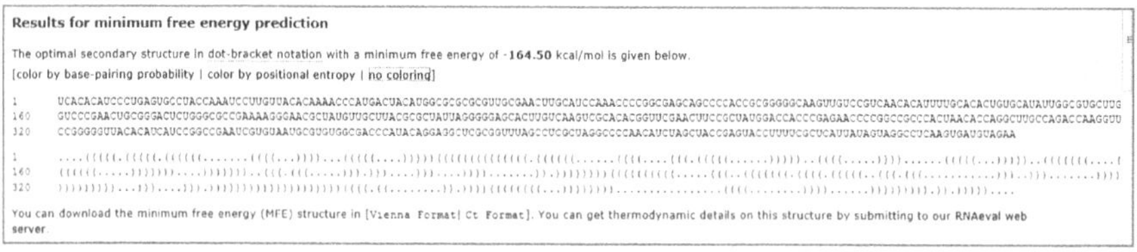


**Fig. 3.2** Results for minimum free energy prediction

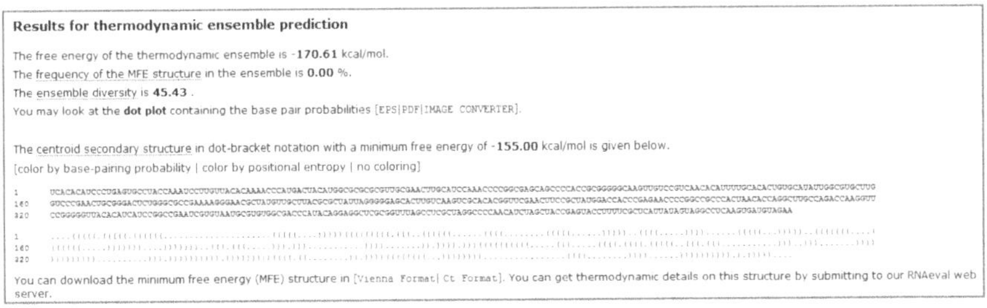


**Fig. 3.3** Results for thermodynamic ensemble prediction

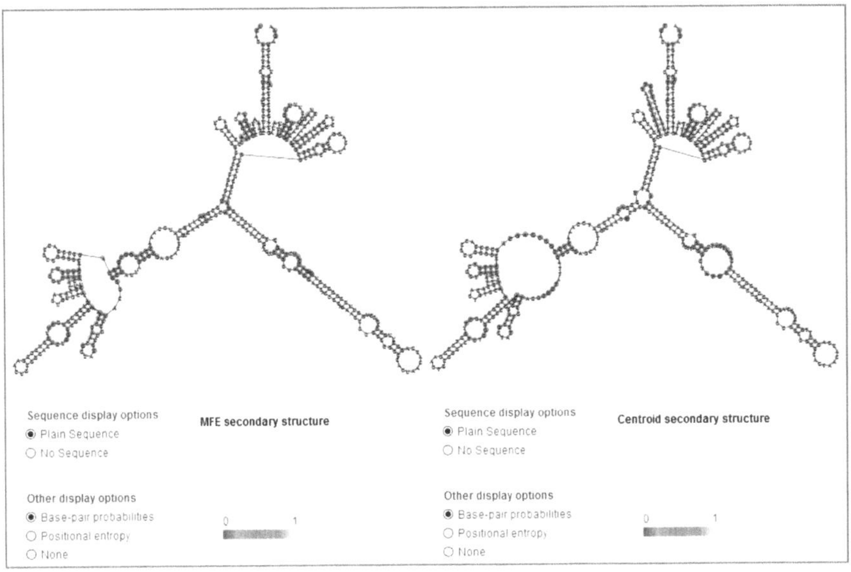


**Fig. 3.4** The graphical output

9. The second part '**Results for thermodynamic ensemble prediction**' displays the calculated thermodynamic ensemble value, the frequency of MFE structure, ensemble diversity and the centroid secondary structure in Newick's format (Fig. 3.3).
10. The third part '**Graphical output**' displays the minimum free energy (MFE) and Centroid secondary structures in formats like EPS and PDF for publication purposes (Fig. 3.4).
11. The centroid structure is normally annotated with colour using local (per base) reliability measures and is the structure that has statistically significant minimal

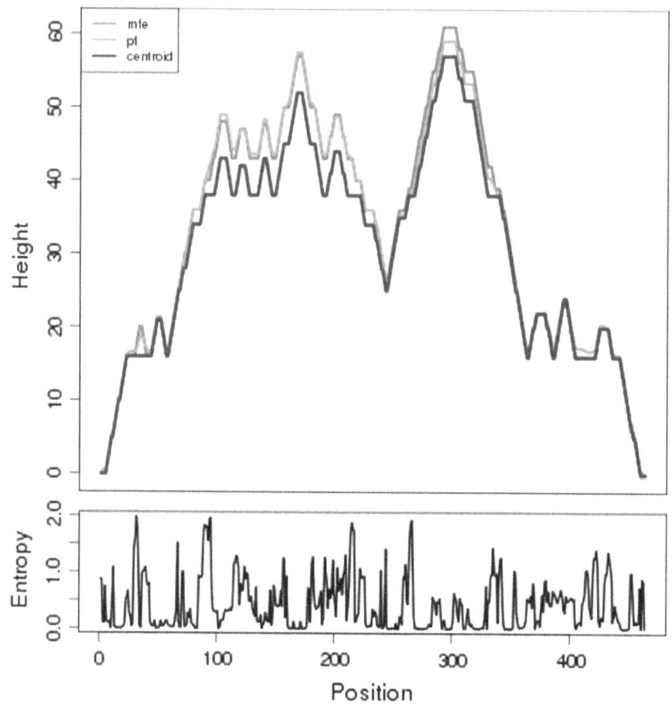


**Fig. 3.5** The mountain plot

base-pair distance compared to all other structures in the thermodynamic ensemble.

12. It is then followed by the '**Mountain plot**' which shows the plotting of sequence position with number of base pairs enclosing that position (Fig. 3.5).

## Interpreting the Results

13. The colour of the bases in the structure denotes the probability, i.e. in case of nucleotides that are base paired, it denotes the probability of being base paired and in case of unpaired nucleotides, it denotes the probability of being unpaired.
14. The colour annotation key in the result details provides information about the confidence in the prediction of a specific pair.
15. The red as highest probabilities ($\geq$ 99%), orange (99% > probability $\geq$ 95%), yellow (95% > probability $\geq$ 90%), dark green (90% > probability $\geq$ 80%),

light green (80% > probability $\geq$ 70%), light blue (70% > probability $\geq$ 60%), dark blue (60% > probability $\geq$ 50%), and purple ($\leq$ 50%).

16. It has been shown that more probable pairs are more likely to be correctly predicted (Mathews 2004).

**Exercise**

1. What is the significance of RNA structure?
2. What are the secondary structure elements of RNA?
3. What is 'minimum free energy'?
4. What is Newick's format?
5. What is mountain plot?

## 3.2 Finding Repeats

**Introduction**

A repeat is a substring of DNA sequence that occurs 'very often' in a genome and its prediction is becoming important as they cause genomic instability also participate in a number of regulatory functions. Eukaryotic repeats are two types, tandem or dispersed within the genome. Computational prediction involves applying exact or approximate string matching algorithm from a dictionary of known repeat sequences. When these repeats get mutated excessively, their prediction becomes challenging as their observable periodicity is diminished.

Spectral Repeat Finder (SRF) predicts repeats by analysing of the power spectrum of a given DNA sequence (Sharma et al. 2004). A repeat is the repeated occurrence of a segment of particular set of nucleotides within a DNA sequence as tandem or dispersed. SRF uses ab initio technique as no prior assumptions are made or holds any information regarding either the repeat length, its fidelity, or whether the repeats are in tandem or dispersed.

**Exploration Activity**

1. Retrieve any RNA sequence from NCBI e.g: *Anopheles jeyporiensis* ITS2 (AccNo. JN643728).
2. Spectral Repeat Finder can be accessed from http://www.imtech.res.in/raghava/srf/.
3. Click on the link '**Submit sequence**' from the list of menus in the left side of the page (Fig. 3.6).
4. You can paste the sequence as plain text, EMBL, FASTA, GENBANK, etc. Or alternatively you can also upload the sequence.

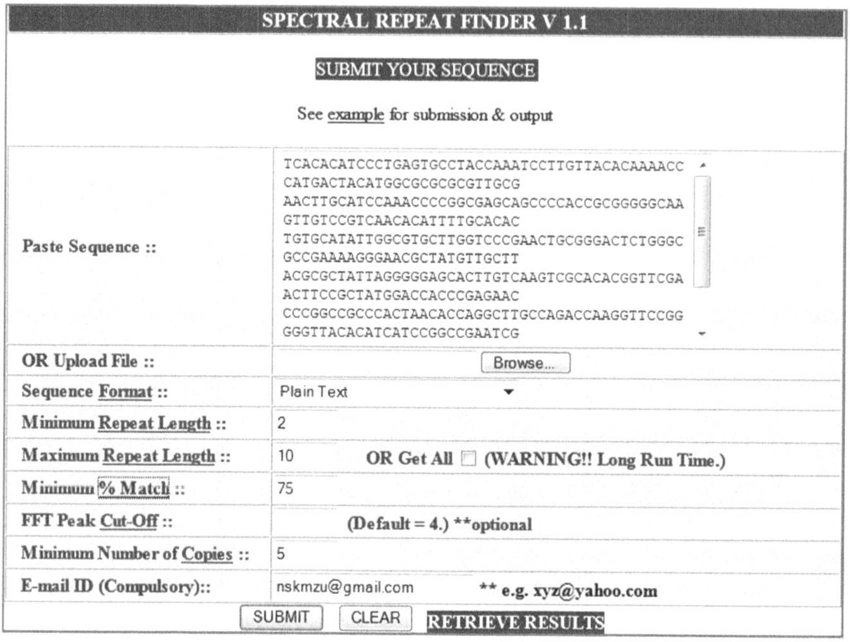


**Fig. 3.6** The spectral repeat Finder at Institute for Microbial Technology (IMTECH), India

5. Accordingly, specify the format you are using under '**Sequence Format**'.
6. Here, *Anopheles jeyporiensis* ITS2 (Acc. No. JN643728) has been used as input.
7. Leave other options as default. By default minimum repeat length: 2 and Maximum Repeat Length: 10. This program can search up to 300 bps (maximum repeat length). Minimum of 75% Match is needed to consider as repeats.
8. Results can be retrieved only via email and hence provide a valid e-mail address.
9. Click on the '**Submit**' button at the bottom of the page.
10. Upon submission you are provided with a job id which can be used later for retrieving the results after computation (Fig. 3.7). It also provides information like the length of the query, minimum and maximum repeat size. The program expects a minimum % of identity a DNA segment so as to qualify as a repeat.
11. If you have a job id already then results can be retrieved by clicking '**Retrieve results**' link.

**OUTPUT OF SPECTRAL REPEAT FINDER**

**YOUR QUERY IS IN QUEUE!!**

JobID Number is: **31466**

This page will refresh in 30 seconds.

You can look at results by
feeding the Job Identification number
given above to the retrieve results page

| **Your Query Details** | |
|---|---|
| **Length of Sequence** | 464 bp |
| **Minimum Length of Repeat Unit** | 2 bp |
| **Maximum Length of Repeat Unit** | 10 bp |
| **Minimum Percentage of Identity** | 75 % |
| **Minimum Number of Copies** | 5 |

**Fig. 3.7** The intermediate result page displaying the job id and input sequence characters

12. The program's analysis report is tabulated with information like pattern searched, the region, consensus pattern, its copy number and the score in identity percentage (Fig. 3.8).
13. The value 1 in the 'Score' denotes a 100% identity.
14. Links to view the detailed analysis of any particular repeat unit and also the Fourier spectrum is also available in the results.
15. The Fourier spectrum plots *Power* verses *Frequency* of different mers and any region where a repetitive sequence structure is present, a peak above '*threshold*' for the mer that is repeated is observed.

**OUTPUT OF SPECTRAL REPEAT FINDER**

| Pattern Searched | Region | Consensus | Copy No | Score |
|---|---|---|---|---|
| View Window Spectrum for 2 mer | | | | |
| CA | 1::352 | CA | 49 | 1.000 |
| View Window Spectrum for 4 mer | | | | |
| CTGG | 1::166 | CTGG | 14 | 0.786 |
| CTAG | 1::166 | CTAG | 14 | 0.786 |
| AGCT | 1::166 | AGCT | 14 | 0.786 |
| View Window Spectrum for 6 mer | | | | |
| GGCTGT | 1::191 | GGCTGT | 12 | 0.722 |
| View Window Spectrum for 7 mer | | | | |
| AGGCTGT | 1::180 | AGGCTGT | 7 | 0.776 |
| View Window Spectrum for 9 mer | | | | |
| ACACACATC | 158::393 | ACACACAt/cC | 11 | 0.747 |

**Fig. 3.8** The SRF result for *Anopheles jeyporiensis* ITS2

**Exercise**

1. What are repeats and how it functions in a genome?
2. What are the two major groups of repeats?
3. What is 'copy number' with respect to repeat prediction?
4. What are the applications of SRF in diversity research?

# Chapter 4
# Protein Sequence Analysis

## 4.1 Protein Sequence Retrieval from UniProtKB

**Introduction**

ExPASy (Expert Protein Analysis System) is a Bioinformatics Resource Portal from Swiss Institute of Bioinformatics that offers bioinformatics support like accessing scientific databases and software tools for the research in life sciences.

UniProtKB is a resource for protein sequence and functional information from ExPASy. This basically consists of sections Swiss-Prot that offers manually annotated and reviewed protein sequences, whereas TrEMBL has automatically annotated and the protein sequences **not** reviewed.

**Exploration Activity**

1. The ExPASy homepage can be accessed from http://www.expasy.org/.
2. Select UniProtKB option in the dropdown box. This tells the program which database you wish to search (Fig. 4.1). Alternatively, you can also click on the UniProtKB link on the right side of the page.
3. In search text box type in the name of the protein (query) as the search field.
4. Click on the "**Search**" button to execute search.
5. The program will provide a list of the search results (Fig. 4.2). If the query was not specific, the list will be quite big.
6. To narrow down the search hit, follow the links next to "**Restrict term**". The program will list all results of the new search.
7. To search with multiple keywords, for example, serine protease, type in the respective keywords in the following supported query syntax (Table 4.1) in the search field. For example in (Table 4.1).
8. To view the information click on the link (Fig. 4.2).
9. The page displays the protein information in tabulated form (Fig. 4.3).
10. To view the sequence click on the '**Sequences**' link or scroll down.
11. To view the sequence in FASTA format click on the **FASTA** link (Fig. 4.4).



K. Syed Ibrahim et al., *Bioinformatics—A Student's Companion*,
DOI 10.1007/978-981-10-1857-2_4

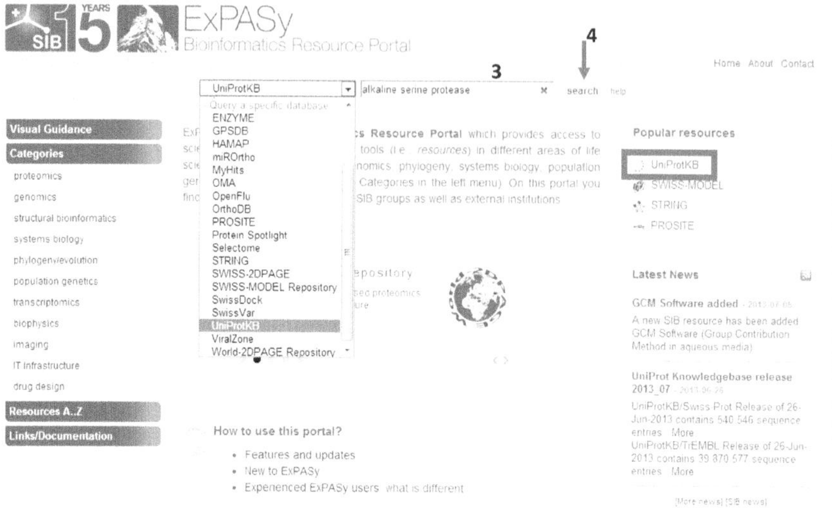


**Fig. 4.1** ExPASY home page

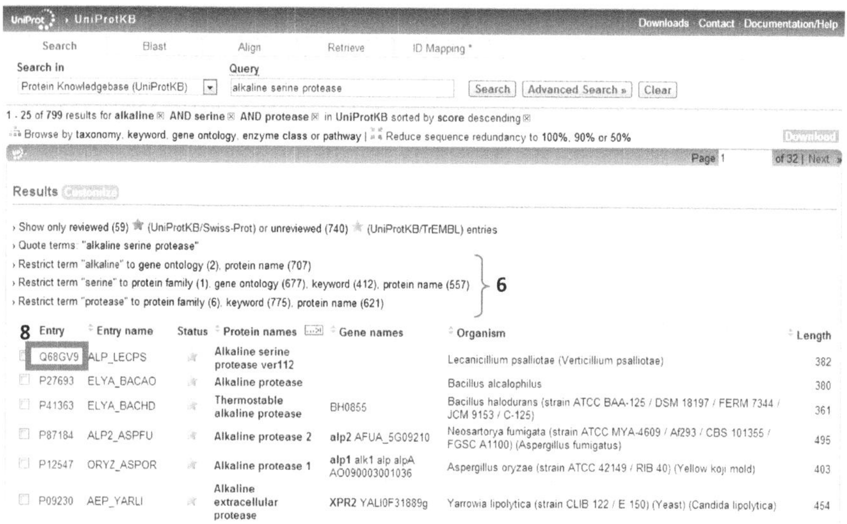


**Fig. 4.2** The UniProtKB search result page

12. The sequence can be parsed to sequence analysis programs like BLAST, ProtParam, Compute pI/MW, ProtScale, PeptideMass or PeptideCutter by clicking the dropdown box and choosing appropriate program and clicking **'go'** button (Fig. 4.4).

**Table 4.1** The query syntax of UniProtKB

| | |
|---|---|
| serine protease | All entries containing both terms |
| serine AND protease | |
| serine && protease | |
| "serine protease" | All entries containing both terms in the exact order |
| serine–protease | All entries containing the term serine but not protease |
| serine NOT protease | |
| serine ! protease | |
| serine OR protease | All entries containing either term |
| serine \|\| protease | |
| protease AND (human OR mouse) | Using parentheses to override boolean precedence rules |
| ser* | All entries containing terms starting with ser. Asterisks can be used anywhere. Note: *If the terms starts with an asterisk or a single letter followed by an asterisk slows down the searching process considerably* |
| author: Chow* | To search publications having author name starting with Chow |
| length: [75 TO *] | All entries with a sequence of at least 75 amino acids |
| citation: (author: John author: Steve) | All entries with a publication that was co-authored by two specific authors |

*Note* If you wish to search with characters having special meaning in the query syntax, use a backslash, e.g. use K\(2\)2XY for searching the gene name K(2)2XY

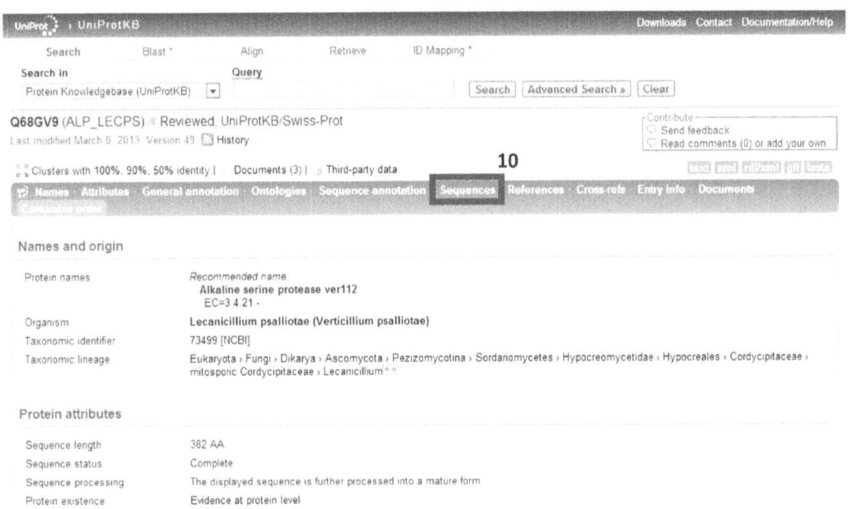


**Fig. 4.3** The UniProt/swiss-prot entry

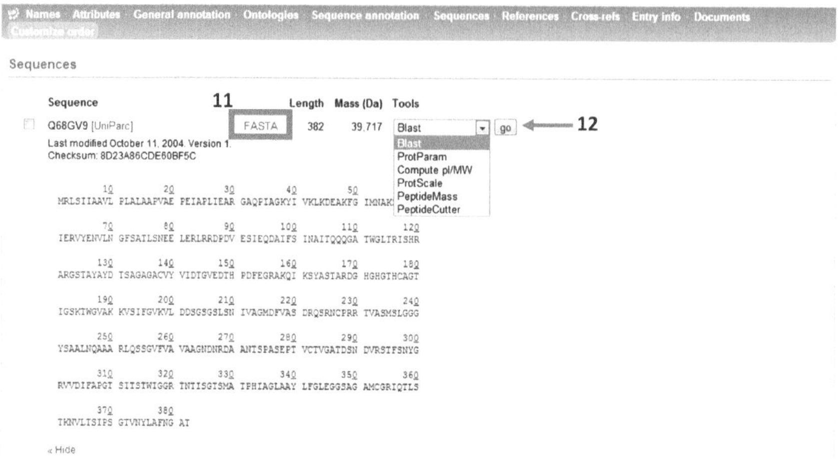


Fig. 4.4 The sequence section in the entry

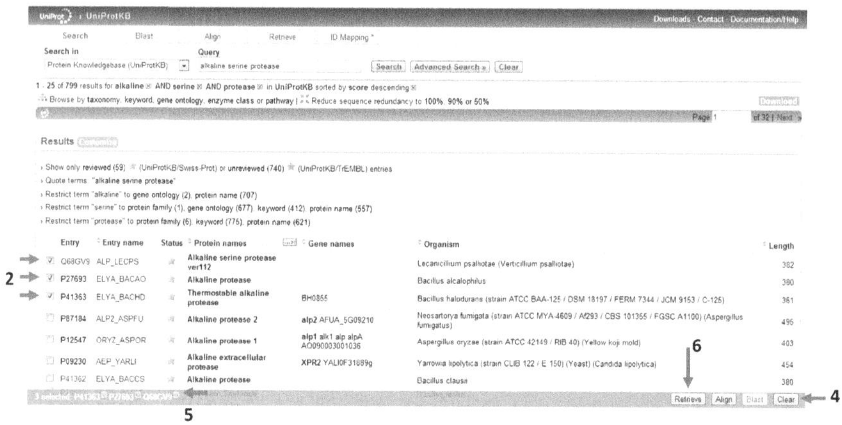


Fig. 4.5 The search result page

## To Download Multiple Sequence

1. Perform the steps 1–7 as mentioned above.
2. To retrieve multiple sequences, select on the check boxes of each sequence (Fig. 4.5).
3. Once one or more sequences have been marked, the *Retrieve* button in the green bar becomes available.
4. To clear selection completely, click 'Clear' button.

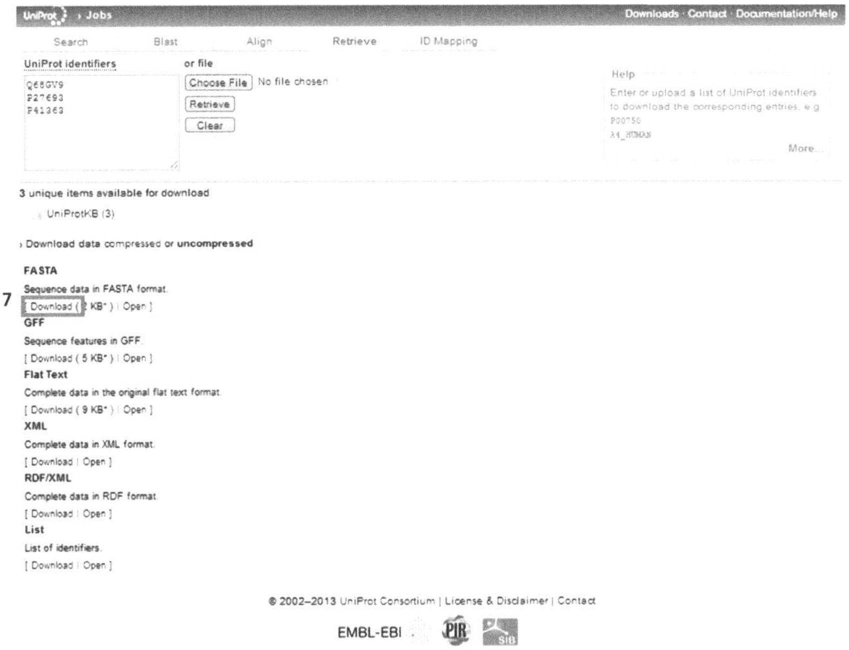

Fig. 4.6 The download options in various file formats

5. To clear only a selected sequence click on the x symbol over the ID.
6. To retrieve the sequences click on the retrieve button. The sequences can be different formats like FASTA, GFF, Flat Text, XML, RDF/XML or only the LIST of identifiers (Fig. 4.6).
7. Click on the corresponding '**Download**' link to download the sequences in FASTA file format.

## Exercise

1. What is the difference between primary and secondary database?
2. What is the difference in retrieving a protein sequence from NCBI's protein database and UniProtKB database?
3. How will you retrieve multiple sequences from the UniProtKB database?
4. Name a few other secondary databases?
5. What is meant by annotation?
6. Use key words 'amylase', 'alpha amylase' and 'beta amylase' separately as a search query and try to answer the following:

a. How many hits are obtained?
b. Retrieve the protein with id **Q9LIR6**. How many amino acids it has, what is its function and where is it located?
c. Retrieve the protein with id **P19961**. Find out the source, and mention the amino acids that participate in the metal binding.

## 4.2 Visualization of Features in a Multiple Sequence Alignment

**Introduction**

Multiple sequence alignment becomes more effective only when the features are highlighted in the alignment. This can be done on the fly when doing alignment in the ExPASy. In this exercise, leptin protein is chosen as an example.

**Exploration Activity**

1. The ExPASy homepage can be accessed from http://www.expasy.org/.
2. Select UniProtKB option in the dropdown box. This tells the program which database you wish to search (Fig. 4.6). Alternatively, you can also click on the UniProtKB link on the right side of the page.
3. In search text box, type in the name of the protein leptin as the search field.
4. Click on the "Search" button to execute search.
5. The program will provide a list of the search results (Fig. 4.7).
6. Select the leptin from Human, Bovine, Pig, Rat, Horse and Chicken from the list.

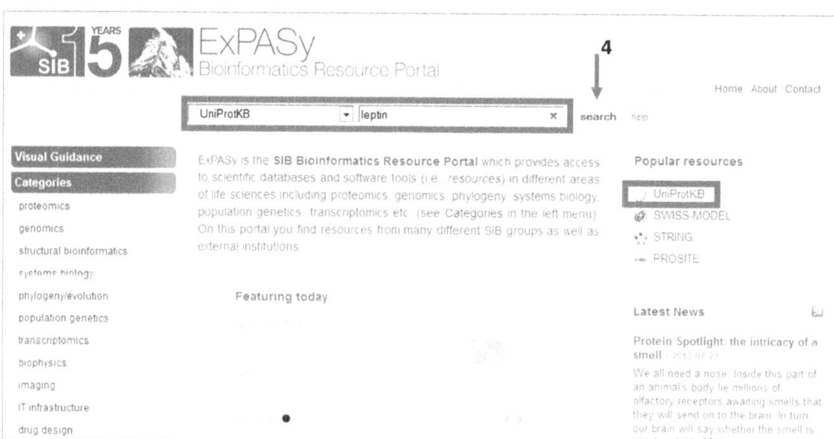


**Fig. 4.7** Searching leptin protein in UniProtKB database

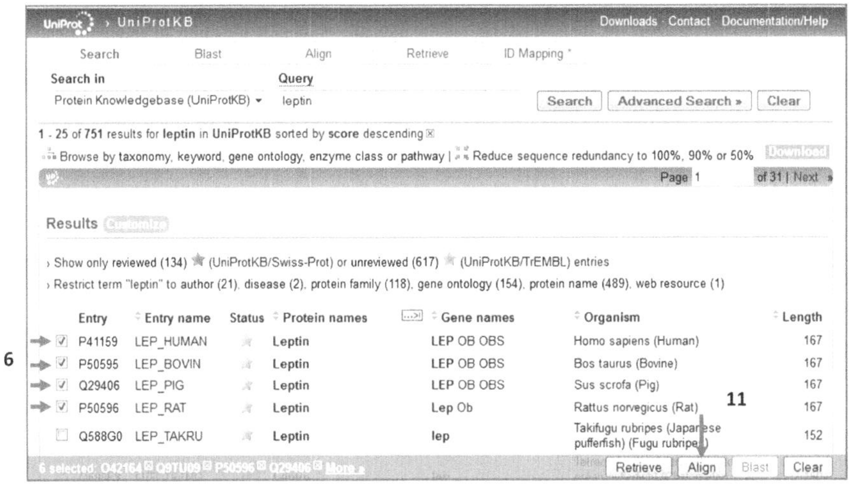


**Fig. 4.8** Selecting leptin protein from the list

7. To retrieve multiple sequences, select on the check boxes of each sequence (Fig. 4.8).
8. Once one or more sequences have been marked, a green bar becomes available at the bottom.
9. To clear selection completely, click '**Clear**' button.
10. To clear only a selected sequence click on the X symbol over the ID.
11. Click '**Align**' button to do an alignment for the selected sequence.
12. The MSA for the selected sequence appears.

## Highlighting the Features in the Alignment

13. To the right side of the alignment, features for annotation and amino acid properties are listed along with a check box.
14. In the annotation section, select on the Disulphide bond, Signal peptide and Helix to visualize the features (Fig. 4.9).
15. To visualize the amino acid properties of hydrophobicity and serine threonine amino acids, select the respective boxes (Fig. 4.10).

## Exercise

1. How are amino acids classified?
2. What is a signal peptide?
3. Name the amino acid that involves in disulphide bridge.
4. Name important secondary structural elements of proteins.

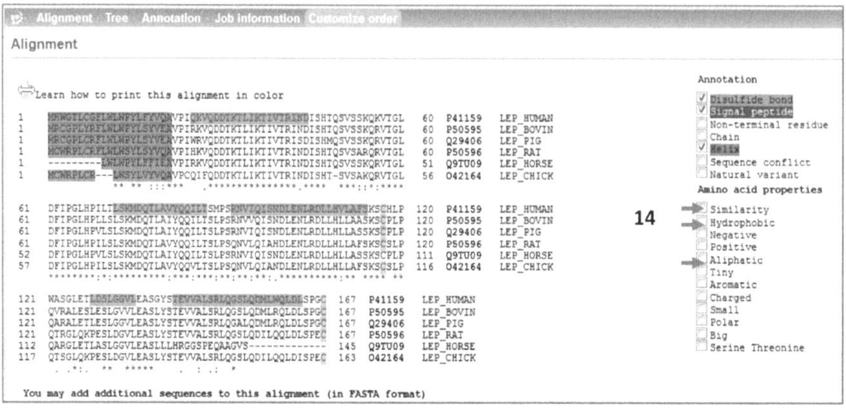


**Fig. 4.9** Annotating the disulphide bond, signal peptide and helix region in alignment

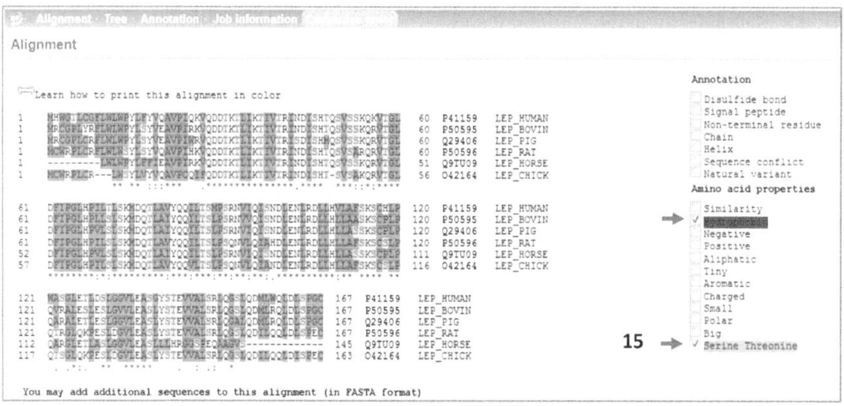


**Fig. 4.10** Highlighting the amino properties like hydrophobicity and serine threonine amino acids in the alignment

## 4.3 Predicting Signal Peptides in Proteins Using SignalP 4.1 Server

### Introduction

### Signal Peptides

The signal peptide is a chain of approximately 3–20 amino acids, essential to direct the protein where it should go. If the protein has to be secreted, signal peptide plays important role in directing the protein to endoplasmic reticulum where it will grow further and get matured for secretion. Once the protein reaches its final destination, the signal peptide is chopped off by the enzyme called signal peptidases.

The signal peptides are basically positively charged n-region, followed by a hydrophobic h-region and a neutral but polar c-region (Mølhøj and Degan 2004).

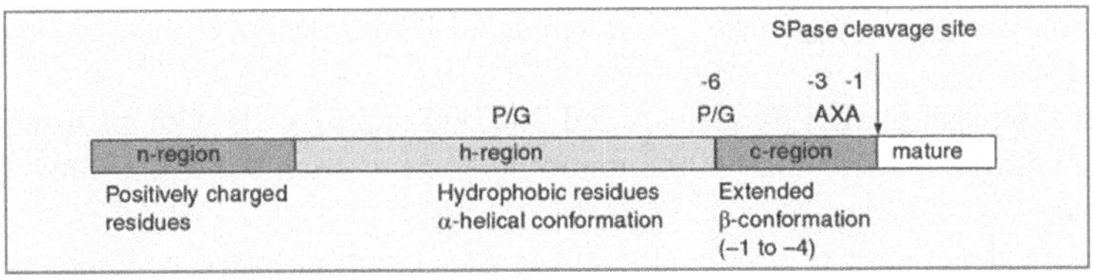


**Fig. 4.11** The regions of a signal peptide

According to the (−3, −1)-rule, it states that the residues at positions −3 and −1 (relative to the cleavage site) should be small and neutral in charge for cleavage to occur properly (Fig. 4.11).

**SignalP 4.1 Server**

SignalP 4.1 server helps in predicting the presence and location of signal peptide cleavage sites in the user-provided amino acid sequences from Gram-positive prokaryotes, Gram-negative prokaryotes and eukaryotes based on a combination of several artificial neural networks (Petersen et al. 2011).

SignalP has been constantly updated from earlier versions. The version 4.1 has been updated with the facility in choosing the D-score cut-off and a customizable minimum length for the signal peptide to be predicted.

**Exploration Activity**

1. SignalP 4.1 server can be accessed from the link http://www.cbs.dtu.dk/services/SignalP/.
2. Copy and paste the valid amino acid sequence of the protein to be analysed. In case a number of sequences have to be analysed then paste in FASTA format into the upper window of the main server page (Fig. 4.12).

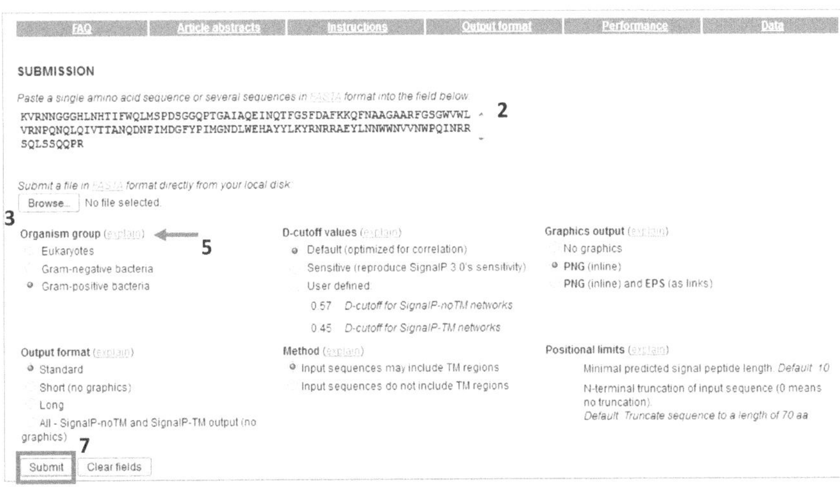


**Fig. 4.12** SignalP interface of Centre for Biological Sequence Analysis at Technical University of Denmark

3. Alternatively, the sequence can be uploaded from your local disk by browsing the disk.
4. In a submission, 2000 sequences and 200,000 amino acids with each sequence containing not more than 6000 amino acids have been set as the limit by the server.
5. Choose the organism group, like eukaryote, Gram-negative or Gram-positive (here Nostoc—Gram-positive).
6. Leave other parameters as default.
7. Click the '**Submit**' button.

### Interpreting Results

- Under standard output format, the result page displays a graphical image plotting the scores *C*, *S* and *Y*, for each position in the sequence obtained during the prediction (Fig. 4.13).
- It is then followed by the maximal values of the three scores as summary (Fig. 4.14).
  - *C-score* (raw cleavage site score)—which is trained to distinguish signal peptide cleavage sites from everything else.
  - *S-score* (signal peptide score)—which is trained to differentiate the positions within signal peptides from the proteins' mature part and from proteins devoid of signal peptides.

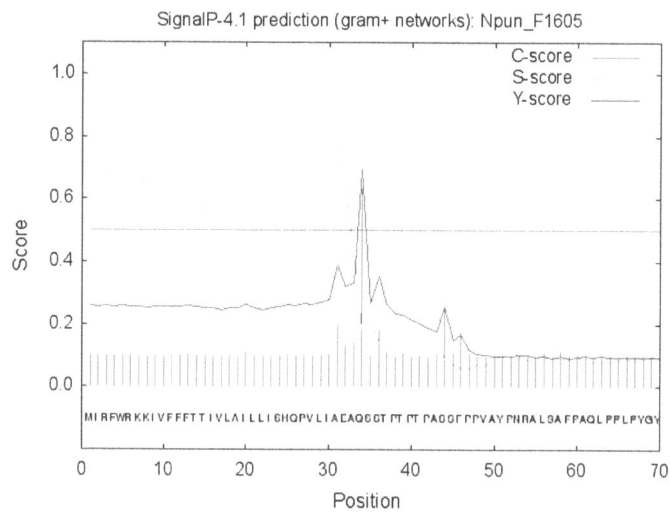


**Fig. 4.13** Graphical image obtained by plotting the *C*, *S* and *Y* scores

```
# Measure  Position  Value    Cutoff   signal peptide?
  max. C    34       0.628
  max. Y    34       0.695
  max. S    30       0.886
  mean S     1-33    0.700
       D     1-33    0.697   0.450    YES
Name=Sequence   SP='YES' Cleavage site between pos. 33 and 34: AEA-QS D=0.697 D-cutoff=0.450 Networks=SignalP-TM
# data
# gnuplot script

Signal peptides: 1
# processed fasta entries
# gff file of processed entries
```

**Fig. 4.14** The *C*, *S* and *Y* scores and the status of the predicted signal peptide

- *Y-score* (combined cleavage site score)—that combines the geometric average of the C-score and the S-score of the slope, to make a better prediction than with the raw C-score alone.
- If any cleavage site is found, then it is indicated by its position.
- For **non-secretory proteins**, SignalP should report ideally low scores, which is closer to the negative target value of 0.1.
- Links to the '**data**' and '**gnuplot script**' are also provided, where these two files can be used to customize the plot using free graphics program, gnuplot on your computer.
- At the bottom of the output, links for downloading the mature protein alone and signal peptide features in GFF format is provided via. 'processed fasta entries' and 'gff file of processed entries', respectively.

**Other Results**

The following examples are presented so as to give a clear picture how other predictions look like:

Transmembrane proteins without signal peptides (Fig. 4.15).
Non-transmembrane proteins with signal peptides (Fig. 4.16).
Non-transmembrane proteins without signal peptides (Fig. 4.17).
Transmembrane proteins with signal peptides (Fig. 4.18).

**Exercise**

1. For the following proteins, identify whether they have signal peptides and if so specify how many amino acids are present in the signal peptide. Prepare the results as a table.

   a. **P19961**
   b. **P22963**
   c. **P22963**
   d. **Q2KTB3**
   e. **P19515**

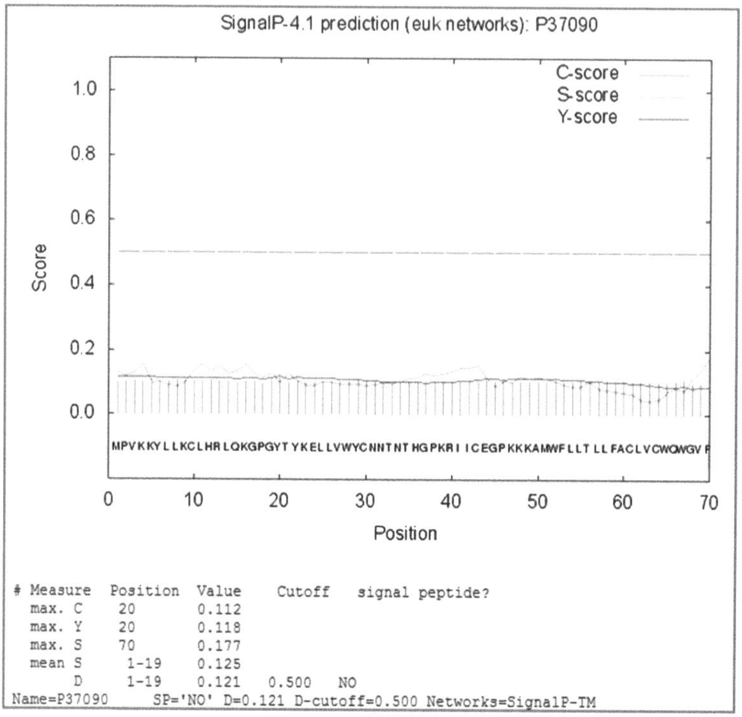


**Fig. 4.15** Amiloride-sensitive sodium channel subunit beta (P37090) from rat

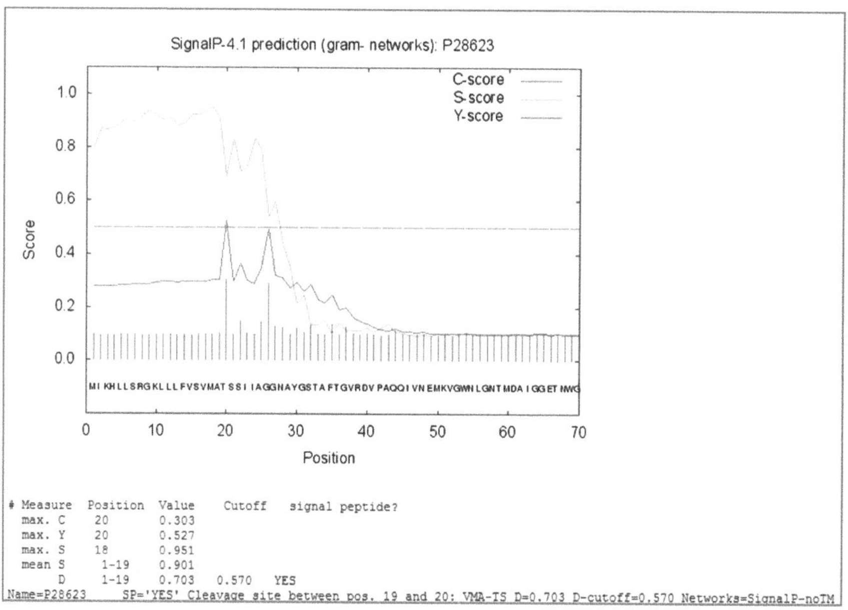


**Fig. 4.16** Endoglucanase D (P28623) from *Clostridium cellulovorans* (strain ATCC 35296)

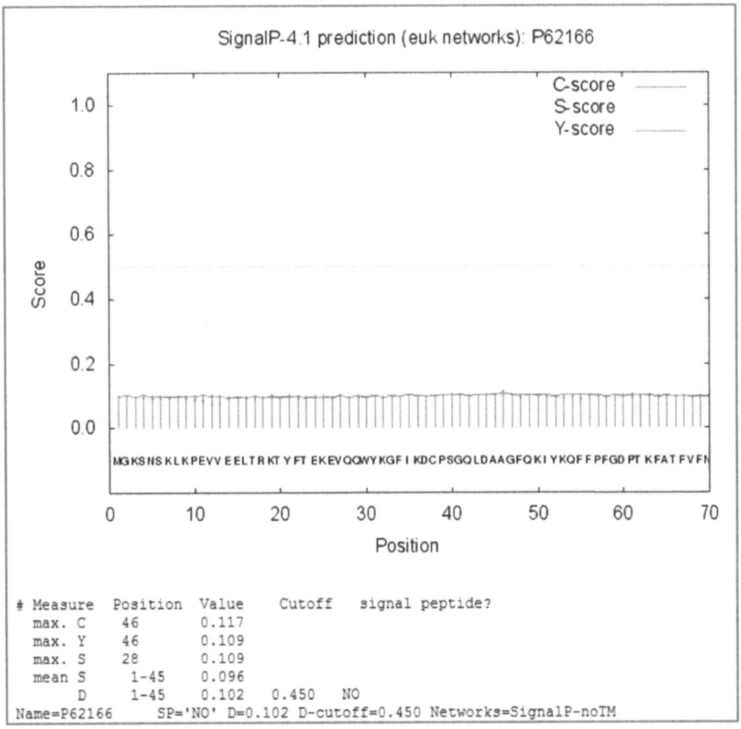


**Fig. 4.17** Neuronal calcium sensor 1 (P62166) from *Homo sapiens* (Human)

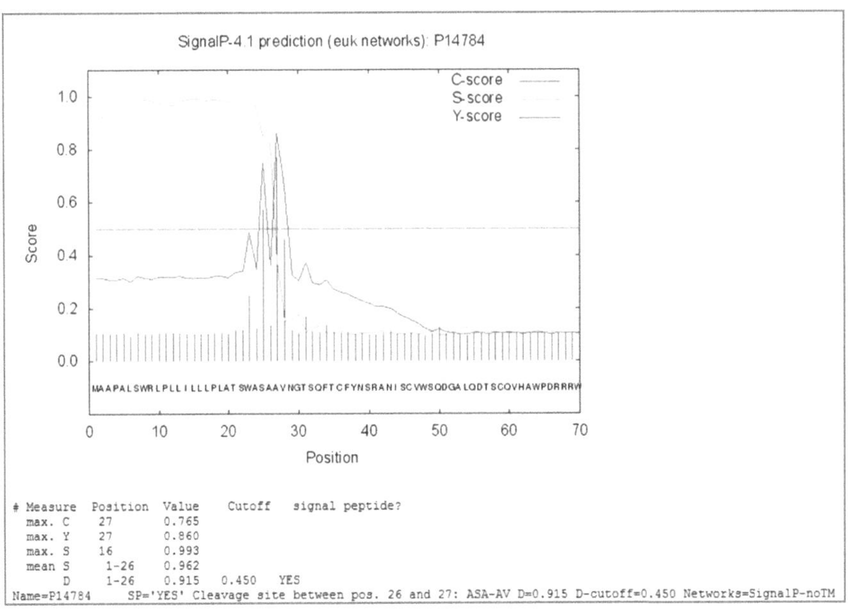


**Fig. 4.18** Interleukin-2 receptor subunit beta (P14784) from *Homo sapiens* (Human)

## 4.4 Predicting Transmembrane Segments and Signal Peptides in Proteins Using Phobius

### Introduction

The high similarity between the signal peptide and hydrophobic regions of a transmembrane helix poses inherent problem in predicting transmembrane protein topology and signal peptide. This leads to cross reaction between the two types of predictions. In order to improve predictions, this predictor aims to discriminate between the two classes. In addition, topology information can be used to successfully predict a signal peptide preceding transmembrane protein, as it shows that the N-terminus of the mature protein should be on the non-cytoplasmic side of the membrane.

Phobius is a program that can predict both transmembrane protein topology and signal peptide predictor. It is based on a Hidden Markov Model (HMM) which models the different portions of a signal peptide and the transmembrane protein in a series of interconnected states (Käll et al. 2007). The program is trained on newly assembled and curated datasets. The Phobius is reported to substantially reduce the errors arise from cross prediction between signal peptides and transmembrane segments when compared to TMHMM and SignalP.

### Exploration Activity

1. Phobius can be accessed from http://phobius.sbc.su.se/index.html.
2. Copy and paste the valid amino acid sequence of the protein (*N. punctiforme* Mn and Fe superoxide dismutase Npun_F1605) to be analysed in plain or in FASTA format into the sequence window of the main server page (Fig. 4.19).
3. Alternatively, the sequence can be uploaded from your local disk by clicking '**Browse**' button.
4. Under '**Select output format**' choose '**Long with Graphics**' (default).
5. Press '**Submit Query**' to start analysis.

### Interpreting Results

1. There are two output formats: Long and short depending on the choice you have made during submission.
2. By default the program generates a **long format** (Fig. 4.20).
3. The first part of Phobius provides information on the **location** of the predicted transmembrane helices, the intervening loop regions and the signal peptide.
4. If the whole sequence is labelled as cytoplasmic or non-cytoplasmic, then it indicates that there are no transmembrane helices in the sequence. Note: *In this case do not take it as a prediction of location.*
5. The prediction gives the most probable location and orientation of transmembrane helices in the sequence (if any). It is found by an algorithm called N-best (or 1-best in this case) that sums over all paths through the model with the same location and direction of the helices.

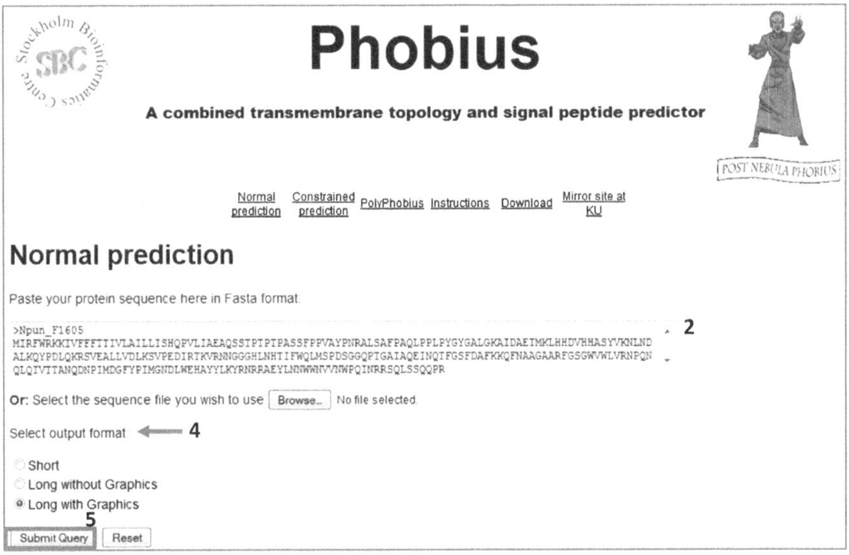


Fig. 4.19 Phobius interface at Stockholm Bioinformatics Centre

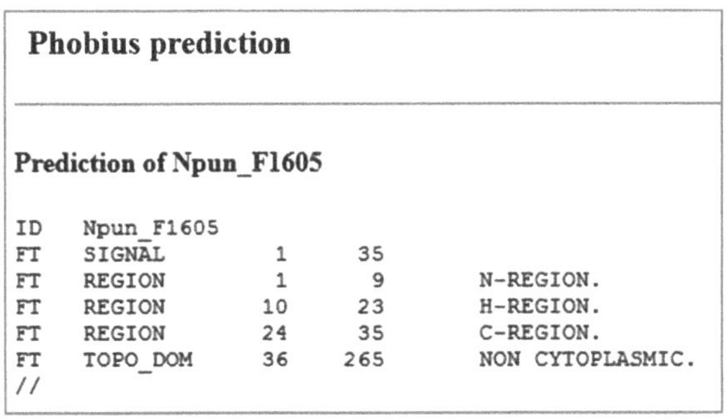


Fig. 4.20 The long format result

6. The second part of the result is a graphical output displaying the **Phobius posterior probabilities plot** (Fig. 4.21).
7. The plot shows the posterior probabilities of cytoplasmic/non-cytoplasmic/TM helix/signal peptide which is obtained by calculating the total probability that a residue belongs to a helix, cytoplasmic, or non-cytoplasmic summed over all possible paths through the model. In short, the plot gives an idea of the certainty of each segment in the prediction. Possible weak TM helices are not predicted.
8. In the **short output format**, one line is produced for each protein with no graphics (Fig. 4.22).

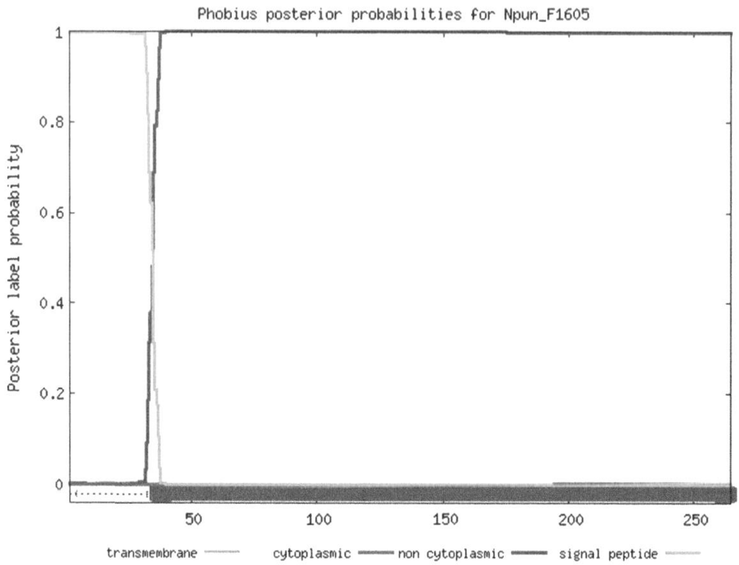


**Fig. 4.21** The posterior probabilities plot

**Fig. 4.22** The short format result

**Phobius prediction**

| SEQENCE ID | TM | SP | PREDICTION |
|---|---|---|---|
| Npun_F1605 | 0 | Y | n10-23c35/36o |

9. Each line starts with the sequence identifier (Käll et al. 2007) and then with these fields:

| | |
|---|---|
| **'TM'** | The number of predicted transmembrane segments. |
| **'SP'** | Y/N indicator if a signal peptide was predicted or not. |
| **'PREDICTION'** | Predicted topology of the protein. |

## Other Results

Since the program also predicts the transmembrane regions, the following examples are presented so as to give a clear picture how other predictions look like:

Transmembrane proteins without signal peptides (Fig. 4.23).

Non-transmembrane proteins with signal peptides (Fig. 4.24).

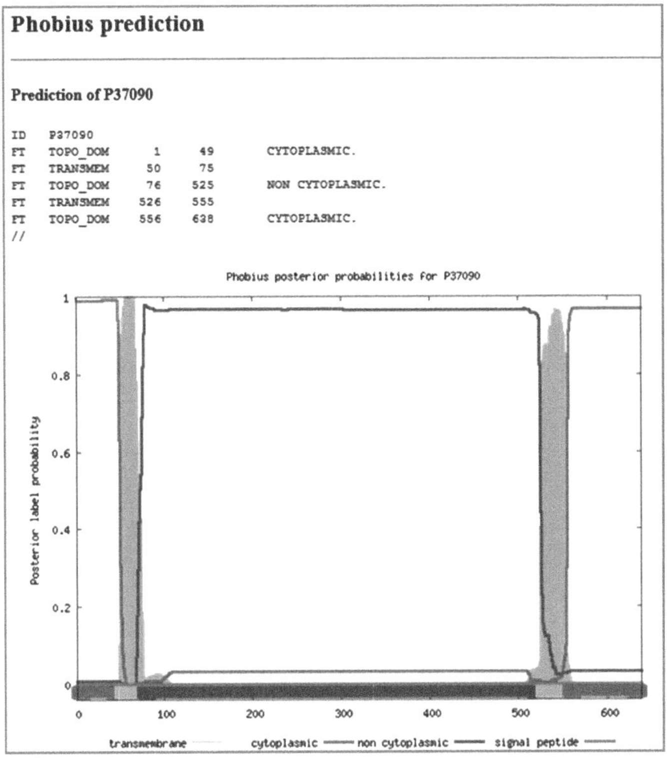


**Fig. 4.23** Amiloride-sensitive sodium channel subunit beta (P37090) from rat

Non-transmembrane proteins without signal peptides (Fig. 4.25).
Transmembrane proteins with signal peptides (Fig. 4.26).

## Exercise

1. For the following proteins identify the source, whether they have transmembrane segments and if so specify how many are present. Prepare the results as a table.

   a. **Q6IWH7**
   b. **P0C9J4**
   c. **P14009**
   d. **Q43495**
   e. **P50927**

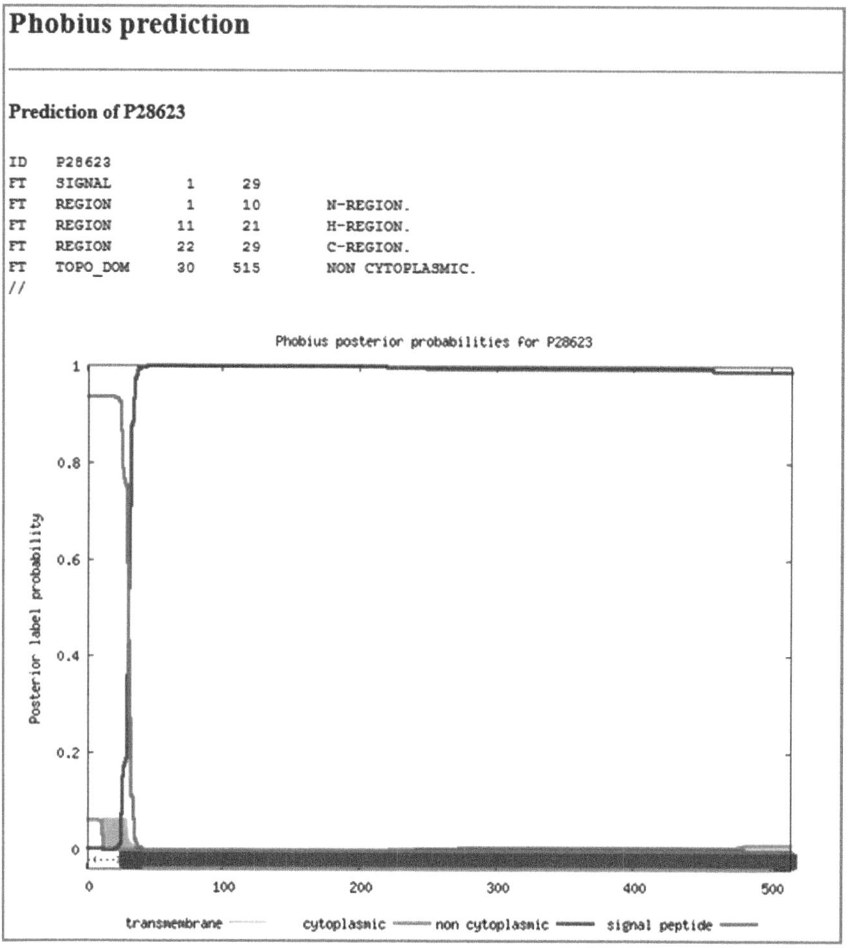


**Fig. 4.24** Endoglucanase D (P28623) from *Clostridium cellulovorans* (strain ATCC 35296)

## 4.5 Predicting Subcellular Location Using TargetP

### Introduction

TargetP 1.1 server predicts the subcellular location of eukaryotic proteins. The protein location is detected by checking the presence of any of the N-terminal presequences like mitochondrial targeting peptide (**mTP**), chloroplast transit peptide (**cTP**) or secretory pathway signal peptide (**SP**) (Emanuelsson et al. 2000). If the server predicts any N-terminal presequence, then a potential cleavage site can also be predicted. TargetP predictions are mainly based on the 130 N-terminal residues of each input sequence. Hence, it is strongly recommended to have the N-terminus for all sequences to be checked. Prediction becomes more difficult and less reliable, when N-terminal residues are missed. There are other programs

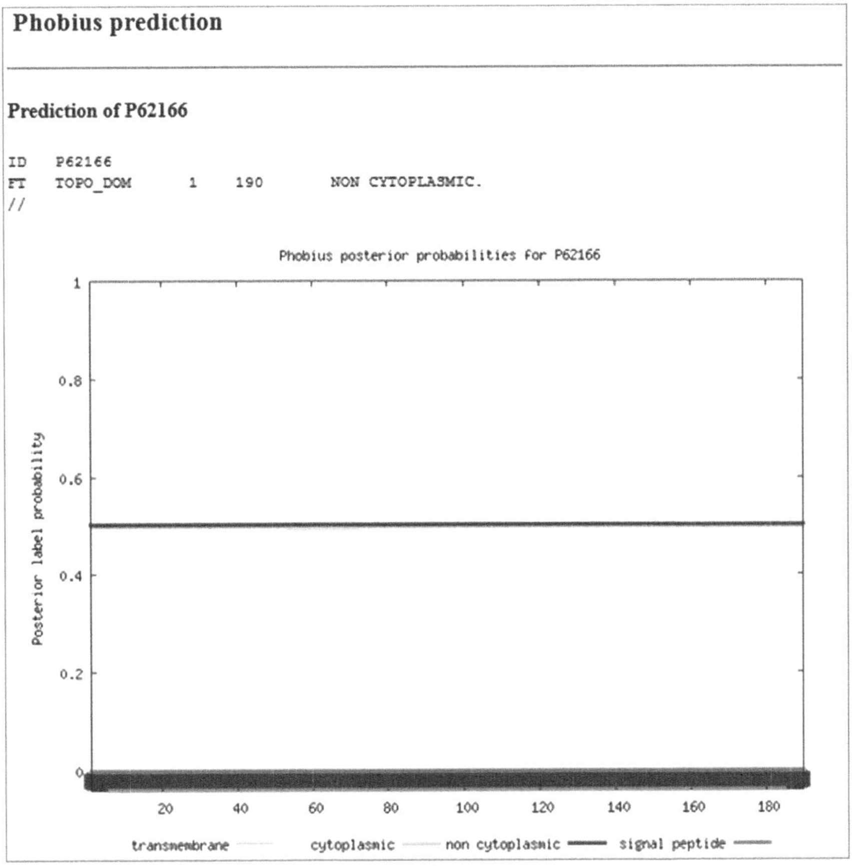


**Fig. 4.25** Neuronal calcium sensor 1 (P62166) from *Homo sapiens* (Human)

available that can predict subcellular location of proteins like PROT-PROP J-tool (Senthilkumar et al. 2012) based on the amino acid sequence.

**Exploration Activity**

A-kinase anchor protein 7 isoforms alpha (**O43687**) from Homo sapiens (Human) is used as an example here. This enzyme targets the cAMP-dependent protein kinase (PKA) to the plasma membrane, and permits functional coupling to the L-type calcium channel (Fraser et al. 1998).

1. The program can be accessed from the site http://www.cbs.dtu.dk/services/TargetP/.
2. You can provide a single or even multiple sequences in FASTA format into the upper window of the main server page (Fig. 4.27).
3. Alternatively, you can also upload a FASTA file from your local disk by browsing the disk.

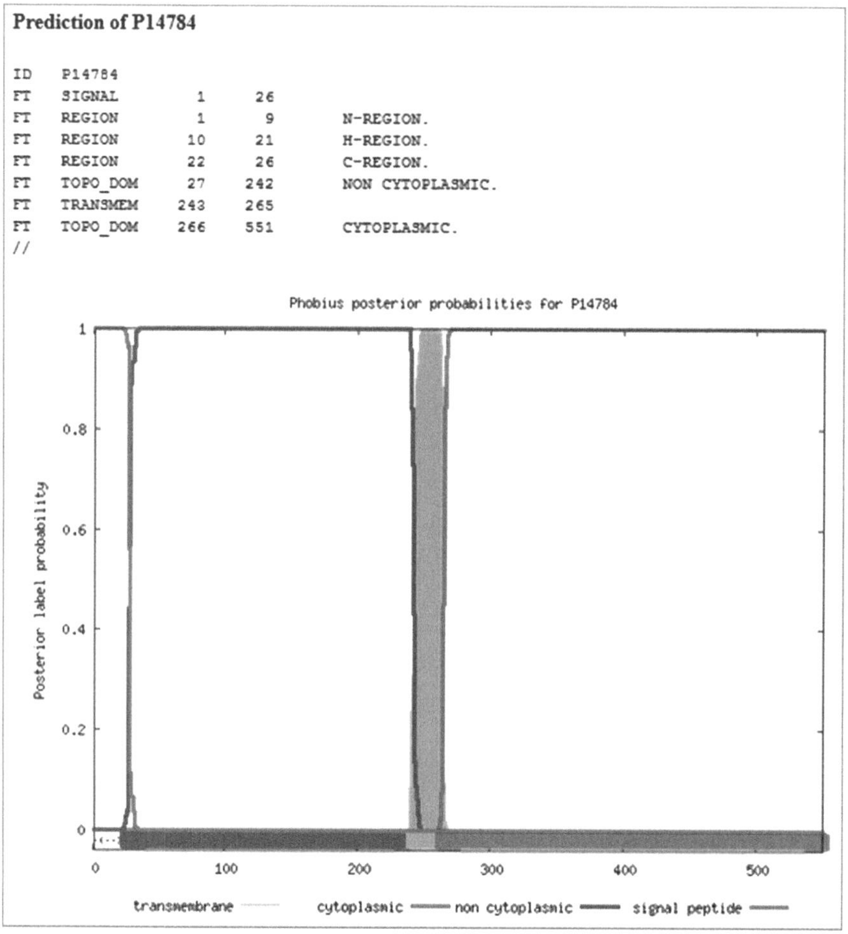


**Fig. 4.26** Interleukin-2 receptor subunit beta (P14784) from *Homo sapiens* (Human)

4. In a submission, 2000 sequences and 200,000 amino acids with each sequence containing not more than 4000 amino acids have been set as the limit by the server.
5. Choose the appropriate organism group. **Non-plant** version predicts the locations secretory pathway, mitochondrion, etc. **Plant** version can also predict chloroplast location.
6. If the sequences are predicted to contain any N-terminal presequence, a potential cleavage site can also be predicted by choosing the option '**Perform cleavage site predictions**' under '**Prediction scope**'. Programs ChloroP and SignalP will be used for predicting **cTP** and **SP** cleavage sites, respectively.

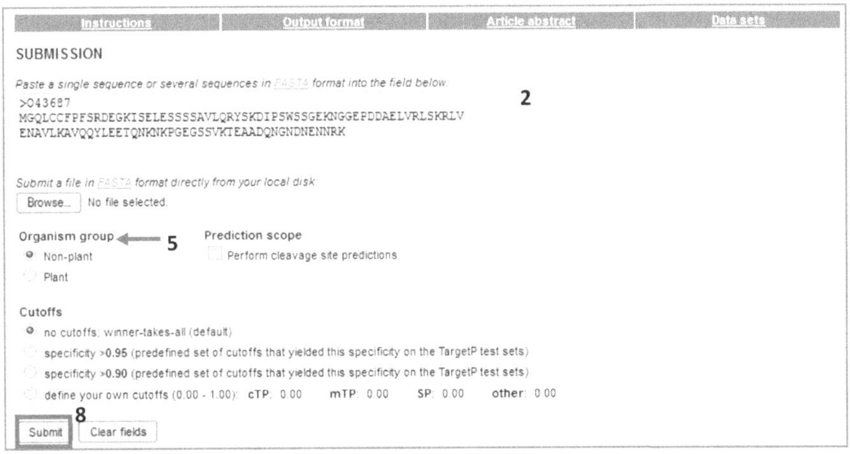


**Fig. 4.27** TargetP interface of Centre for Biological Sequence Analysis at Technical University of Denmark

**Fig. 4.28** The TargetP result

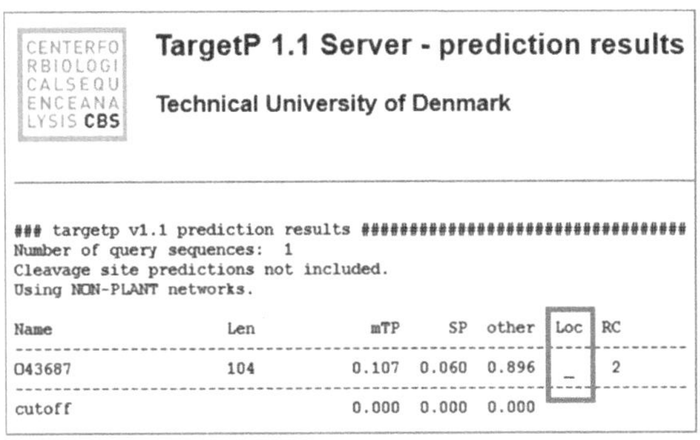


7. Leave '**Cutoffs**' as default.
8. Click '**Submit**' button.

### Interpreting Results

The output is a table in plain text (Fig. 4.28).

The descriptions of the columns are as follows:

**Name**: Sequence name (only first 20 characters will be displayed).

**Len**: The length of the sequence.

**cTP**, **mTP**, **SP**, **other**: Displays the final scores of prediction. According to TargetP the location that has highest score is the most probable location and it also indicates how certain the prediction is.

**Loc**: localization prediction is based on the above scores. The other possible values are as follows:

**'C'** Chloroplast, i.e. the sequence has **cTP**, which is a chloroplast transit peptide
**'M'** Mitochondrion, i.e. the sequence has **mTP**, which is a mitochondrial targeting peptide;
**'S'** Secretory pathway, i.e. the sequence has **SP**, which is a signal peptide;
'_' Refers to any other location;
'*' "don't know"; shows that the cut-off parameter needs to be revised

For our sequence the prediction is '**Any other location**'.

In the Reliability class (RC), the score can range from 1 to 5, where 1 indicates the strongest prediction. RC measures the size of the difference ('diff') between the highest (winning) and the second highest output scores. Based on the difference there are five reliability classes, which is defined as follows:

1. diff over 0.800
2. diff between 0.600 and 0.800
3. diff between 0.400 and 0.600
4. diff between 0.200 and 0.400
5. diff below 0.200.

Hence, lower RC indicates safer prediction. In our case it is 2, which indicates near strong prediction.

**TPlen**: Predicted presequence length; any value appears here only when TargetP was asked to perform cleavage site predictions.

**Other Examples**
Location: Secretory Pathway (Fig. 4.29).
Location: Chloroplast (Fig. 4.30).
Location: Mitochondrion (Fig. 4.31).

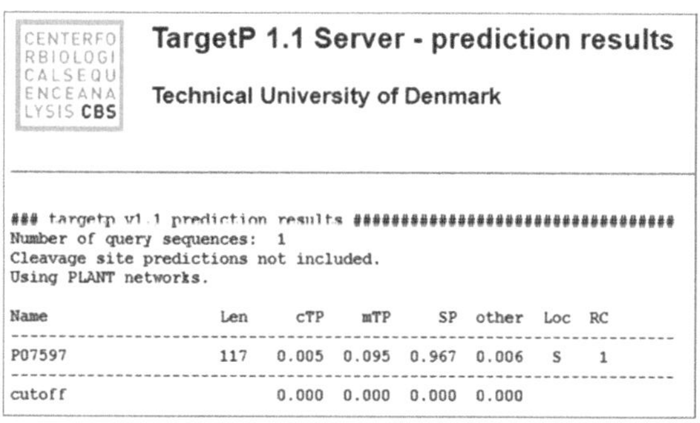

TargetP 1.1 Server - prediction results

Technical University of Denmark

```
### targetp v1.1 prediction results ##################################
Number of query sequences:  1
Cleavage site predictions not included.
Using PLANT networks.

Name                  Len    cTP    mTP     SP  other  Loc  RC
----------------------------------------------------------------------
P07597                117  0.005  0.095  0.967  0.006   S    1
----------------------------------------------------------------------
cutoff                     0.000  0.000  0.000  0.000
```

**Fig. 4.29** Non-specific lipid-transfer protein 1 (P07597) from *Hordeum vulgare* (Barley)

**Fig. 4.30** Major basic nuclear protein 2 (Q01238) from *Crypthecodinium cohnii* (Dinoflagellate)

**TargetP 1.1 Server - prediction results**

**Technical University of Denmark**

```
### targetp v1.1 prediction results ##################################
Number of query sequences:  1
Cleavage site predictions not included.
Using PLANT networks.

Name                  Len     cTP     mTP     SP   other  Loc  RC
----------------------------------------------------------------------
Q01238                102   0.420   0.277  0.033   0.164   C    5
----------------------------------------------------------------------
cutoff                      0.000   0.000  0.000   0.000
```

**Fig. 4.31** Photosystem I reaction centre subunit IV, chloroplastic (P12352) from *Chlamydomonas reinhardtii* (*Chlamydomonas smithii*)

**TargetP 1.1 Server - prediction results**

**Technical University of Denmark**

```
### targetp v1.1 prediction results ##################################
Number of query sequences:  1
Cleavage site predictions not included.
Using PLANT networks.

Name                  Len     cTP     mTP     SP   other  Loc  RC
----------------------------------------------------------------------
P12352                 97   0.397   0.555  0.014   0.150   M    5
----------------------------------------------------------------------
cutoff                      0.000   0.000  0.000   0.000
```

### Exercise

1. For the following proteins identify the source, subcellular location and their probable function and prepare the result as a table.

   a. **P28321**
   b. **P11602**
   c. **P0A9M0**
   d. **P19236**
   e. **Q8W471**.

## 4.6 Protein BLAST (blastp)

### Introduction

Standard protein–protein BLAST (**blastp**) helps in identifying a query amino acid sequence and for finding similar sequences in protein databases by finding local regions of similarity. When the program detects sequence similarity spanning whole sequence, it also reports a global alignment which is preferable in case of protein identification (Altschul et al. 1990).

**Table 4.2** Algorithm parameters

| Algorithm parameters | Default | Special cases | |
|---|---|---|---|
| | | Short query | Large sequence family |
| SEG filter | Off | Off | On |
| Scoring matrix | PAM30 | PAM30 | BLOSUM62 |
| Word size | 3 | 2 | 3 |
| E-value | 10 | 20,000 | 10 |
| Gap costs | 9,1 | 9,1 | 11,1 |
| Max target sequences | 100 | 100 | 20,000 |

**Table 4.3** Choosing substitution matrix based on query length

| Query length | Substitution matrix | Gap costs |
|---|---|---|
| Lesser than 35 | PAM-30 | (9,1) |
| Between 35 and 50 | PAM-70 | (10,1) |
| Between 50 and 85 | BLOSUM-80 | (10,1) |
| Over 85 | BLOSUM-62 | (11,1) |

The algorithm parameters for protein BLAST are given in Tables 4.2 and 4.3.

**Exploration Activity**
**Doing Protein BLAST Search**

1. Prepare the **amino acid query sequence** (sequence to be BLAST) you want to use as raw or in FASTA format.
2. Go to NCBI home page available at http://www.ncbi.nlm.nih.gov/ and choose BLAST program (Fig. 4.32).
3. Choose the **protein BLAST program** (Fig. 4.33).

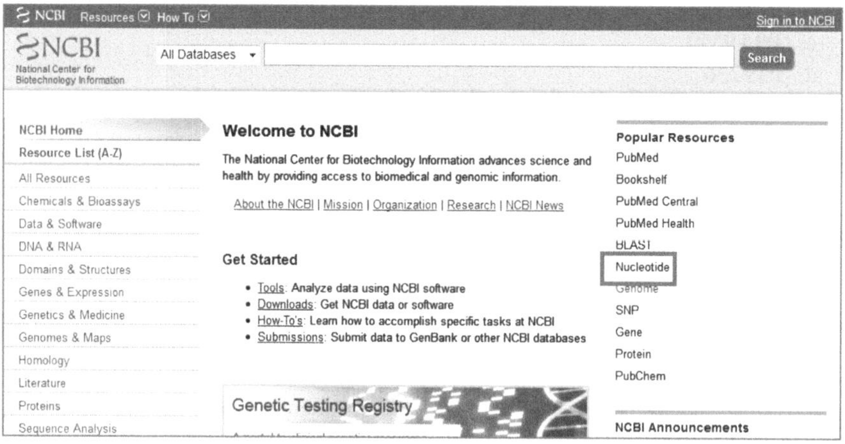


**Fig. 4.32** NCBI home page

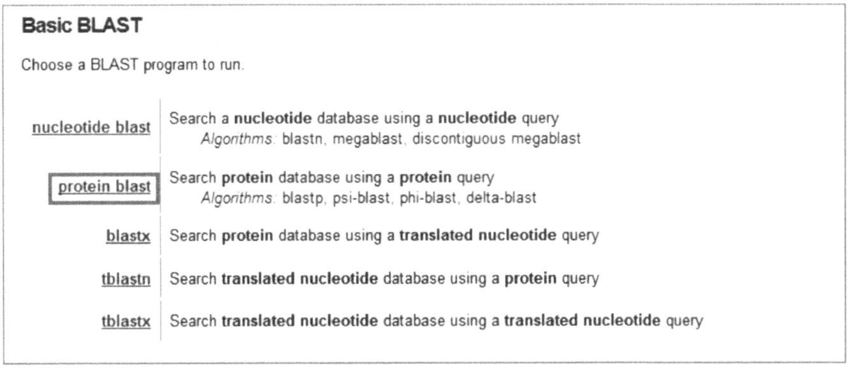


**Fig. 4.33** Basic BLAST programs in NCBI

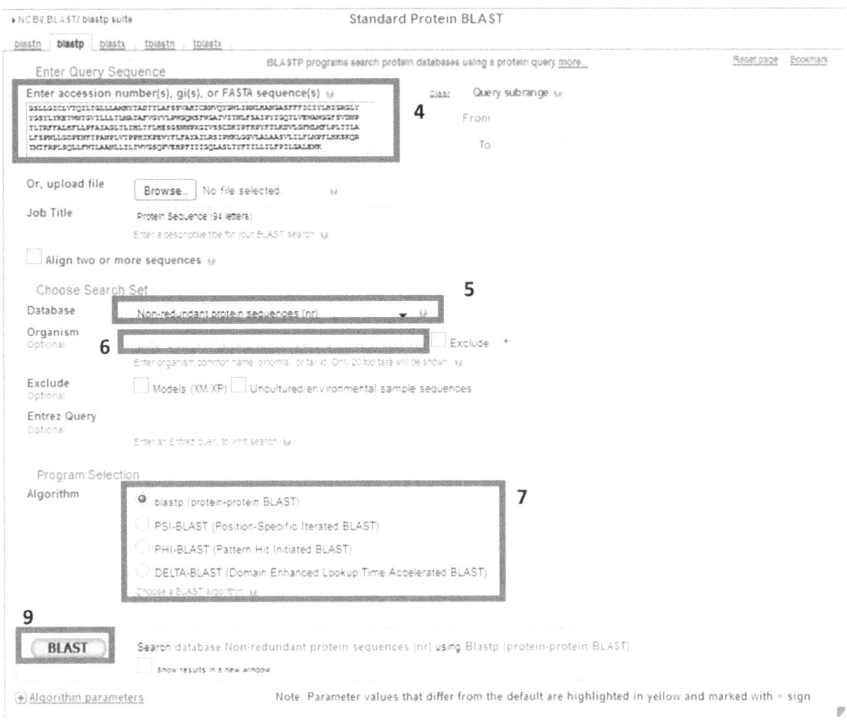


**Fig. 4.34** Protein BLAST interface in NCBI

4. Copy and paste the sequence in the sequence box or alternatively, you can also upload the sequence in FASTA format from your computer (Fig. 4.34).
5. Select the **database** you want to search against, preferably ‘Others’ (nr—non redundant).

**Fig. 4.35** BLASTP algorithm default parameters

6. If you wish to search against any specific organism, under 'Organism', enter first few characters of the organism's name and select the organism for the dropdown list.
7. Choose the **BLASTP** under 'Program Selection'. It compares a protein query to a protein database (Fig. 4.34).
   a. DELTA-BLAST builds a position-specific substitution matrix (PSSM) using the results from Conserved Domain Database search and searches the sequence database.
8. Adjust the parameters, if required. Any parameter changed is highlighted in yellow (Fig. 4.35).
9. Click '**BLAST**' button at the end of the page (Fig. 4.34).

**Interpretation of Protein BLAST Results**
**Significance of Algorithm Parameters**:

Protein BLAST is more sensitive to the algorithm parameters. It is necessary to change the parameters depending upon the query length as mentioned in Tables 4.1 and 4.2.

Consider the following example to test the importance of algorithm parameters. The cytochrome b from *Cathartes aura* (Sequence ID: emb|CAA60423.1|Length: 342) is used for protein BLAST search with default parameters (Fig. 4.35). The results are shown in Fig. 4.36.

**Descriptions**

**Sequences producing significant alignments:**

Select: All None Selected:0

Alignments

| | Description | | Max score | Total score | Query cover | E value | Max ident | Accession |
|---|---|---|---|---|---|---|---|---|
| ☐ | cytochrome b [Cathartes aura] >gb\|AAR19059.1\| cytochrome b [Cathartes aura] | | 1153 | 1153 | 100% | 0.0 | 99% | YP_423947.1 |
| ☐ | cytochrome b [Cathartes aura] >gb\|ACA28884.1\| cytochrome b [Cathartes aura] | → | 1153 | 1153 | 100% | 0.0 | 100% | CAA60423.1 |
| ☐ | cytochrome b [Cathartes melambrotus] >gb\|AAN77010.1\| cytochrome b [Cathartes burrovianu | | 1149 | 1149 | 100% | 0.0 | 99% | AAN77009.1 |
| ☐ | cytochrome b [Cathartes melambrotus] | | 1145 | 1145 | 100% | 0.0 | 99% | AAN77008.1 |
| ☐ | cytochrome b [Vultur gryphus] | | 1118 | 1118 | 100% | 0.0 | 97% | CAA60443.1 |
| ☐ | cytochrome b [Sarcoramphus papa] | | 1114 | 1114 | 100% | 0.0 | 97% | CAA60440.1 |
| ☐ | cytochrome b, partial (mitochondrion) [Vultur gryphus] | → | 1105 | 1105 | 97% | 0.0 | 98% | AAA19479.1 |
| ☐ | cytochrome b [Puffinus huttoni] | | 1102 | 1102 | 100% | 0.0 | 96% | AAC68641.1 |
| ☐ | cytochrome b [Pterodroma phaeopygia] >gb\|ADV17984.1\| cytochrome b [Pterodroma phaeop | | 1102 | 1102 | 100% | 0.0 | 96% | ADV17982.1 |
| ☐ | cytochrome b [Pterodroma phaeopygia] | | 1102 | 1102 | 100% | 0.0 | 96% | ADV17977.1 |

**Fig. 4.36** Protein BLAST results of cytochrome b from *Cathartes aura* with default parameters

**Descriptions**

**Sequences producing significant alignments:**

Select: All None Selected:0

Alignments

| | Description | Max score | Total score | Query cover | E value | Max ident | Accession |
|---|---|---|---|---|---|---|---|
| ☐ | cytochrome b [Cathartes aura] >gb\|ACA28884.1\| cytochrome b [Cathartes aura] | 648 | 648 | 100% | 0.0 | 100% | CAA60423.1 |
| ☐ | cytochrome b [Cathartes aura] >gb\|AAR19059.1\| cytochrome b [Cathartes aura] | 647 | 647 | 100% | 0.0 | 99% | YP_423947.1 |
| ☐ | cytochrome b [Cathartes melambrotus] >gb\|AAN77010.1\| cytochrome b [Cathartes burr | 646 | 646 | 100% | 0.0 | 99% | AAN77009.1 |
| ☐ | cytochrome b [Cathartes melambrotus] | 644 | 644 | 100% | 0.0 | 99% | AAN77008.1 |
| ☐ | cytochrome b [Vultur gryphus] | 634 | 634 | 100% | 0.0 | 97% | CAA60443.1 |
| ☐ | cytochrome b [Sarcoramphus papa] | 632 | 632 | 100% | 0.0 | 97% | CAA60440.1 |
| ☐ | cytochrome b [Calonectris borealis] | 630 | 630 | 100% | 0.0 | 96% | CAC07322.1 |
| ☐ | cytochrome b [Calonectris diomedea] >gb\|AAN71699.1\| cytochrome b [Calonectris dior | 630 | 630 | 100% | 0.0 | 96% | AAC79888.1 |
| ☐ | cytochrome b [Puffinus huttoni] | 630 | 630 | 100% | 0.0 | 96% | AAC68641.1 |
| ☐ | RecName: Full=Cytochrome b; AltName: Full=Complex III subunit 3; AltName: Full=Cor | 629 | 629 | 100% | 0.0 | 96% | O79215.1 |
| ☐ | cytochrome b [Eupodotis senegalensis] | 628 | 628 | 100% | 0.0 | 94% | AAP43980.1 |
| ☐ | cytochrome b [Fulmarus glacialis] | 628 | 628 | 100% | 0.0 | 95% | CAC07232.1 |
| ☐ | cytochrome b [Pterodroma phaeopygia] >gb\|ADV17984.1\| cytochrome b [Pterodroma pl | 629 | 629 | 100% | 0.0 | 96% | ADV17982.1 |
| ☐ | cytochrome b [Calonectris diomedea diomedea] >emb\|CAC07333.1\| cytochrome b [Ca | 627 | 627 | 100% | 0.0 | 95% | CAC07330.1 |
| ☐ | cytochrome b [Calonectris borealis] >emb\|CAC07321.1\| cytochrome b [Calonectris bor | 627 | 627 | 100% | 0.0 | 95% | CAC07320.1 |

**Fig. 4.37** Change in ranking due to altered algorithm parameter conditions

When the same sequence is searched with changed parameters: Matrix: BLOSUM 62, Word Size: 3, Gap Cost: 11, 1, Filter: ON, Max Target Sequence: 20,000, the BLAST results are quiet different and more informative (Fig. 4.37). In comparison with default parameters, there is a phenomenal change in ranking due to altered algorithm parameter conditions.

**Exercise**

1. What matrix do you use when query length is below 35?
2. When do you change the default algorithm parameters in protein BLAST?
3. How will you change the display of results in protein BLAST?
4. What is the use of protein BLAST?
5. Do a protein BLAST search for the accession YP_003844692 and identify the family it belongs to and its close match from a different organism.

## 4.7 Position-Specific Iterated (PSI)-BLAST

**Introduction**
A more sensitive version of BLAST program is position-specific iterated (PSI)-BLAST that can find distantly related proteins or novel members of a protein family (Altschul et al. 1997). Unlike BLAST, PSI-BLAST uses a statistically driven searching method and position-specific scoring matrices (PSSMs) for scoring matches between query and database sequences, being more sensitive than BLAST.

**The Algorithm**
PSI-BLAST makes a standard protein–protein BLAST search in the first round and generates a multiple alignment of the highest scoring pairs (HSP) of the BLASTp run above a e-value cut-off (default = 0.005). It then calculates a profile or a position-specific score matrix (PSSM) from the multiple alignment of related proteins and comes out with an amino acid substitution pattern that occur at a specific sequence position for a specific protein family. This makes highly conserved positions to receive higher scores and weakly conserved positions to near zero scores. This newly generated profile will be substituted with the original substitution matrix and searching will be done. If new sequences are detected above the mentioned e-value cut-off (threshold), it will be again added to alignment and the profile is reconstructed for another round of searching. This process continues iteratively until desired or no new sequences are detected above the defined threshold (which is called convergence). The process makes PSI-BLAST far more effective in detecting distantly related proteins than BLASTP which performs a single query alone (Altschul et al. 1997).

**Significance of PSI-BLAST**
PSI-BLAST performs more sensitive protein–protein similarity searches is helpful in cases where standard protein–protein BLAST search returns no significant hits or comes out with hypothetical proteins. PSI-BLAST can be used for Superfamily and Functional analysis.

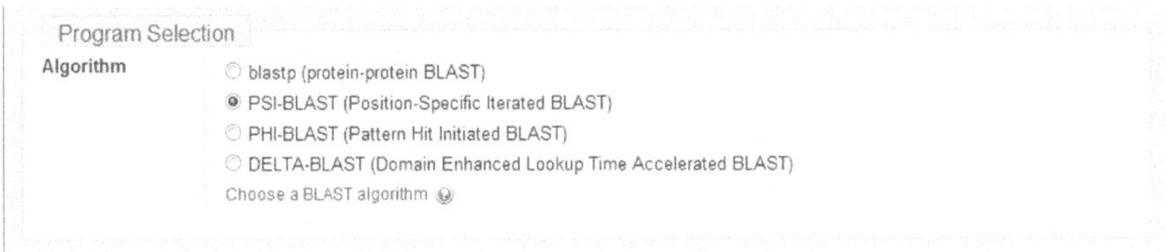


**Fig. 4.38** Program selection in protein BLAST

## Exploration Activity

1. Follow the steps 1–6 as mentioned in protein BLAST. Hypothetical protein from *Dickeya dadantii* 3937 (Accession number: ADM97672) was used as a query.
2. Choose the '**PSI-BLAST**' under '**Program Selection**' (Fig. 4.38). PSI-BLAST allows the user to build a PSSM (position-specific scoring matrix) from the previous BLASTP results.
3. Adjust algorithm parameters according to query length as mentioned in protein BLAST (Fig. 4.39). In addition, PSI-BLAST includes parameters like Upload PSSM, PSI-BLAST Threshold and Pseudocount.
4. Set the threshold value of 0.005 and Pseudocount to '0'.
5. You can also bypass the initial BLASTP step either by providing a PSSM saved from a previous search or by specifying a set of aligned sequences that can be used for generating the initial PSSM.
6. Click '**BLAST**' button at the end of the page (Fig. 4.39).

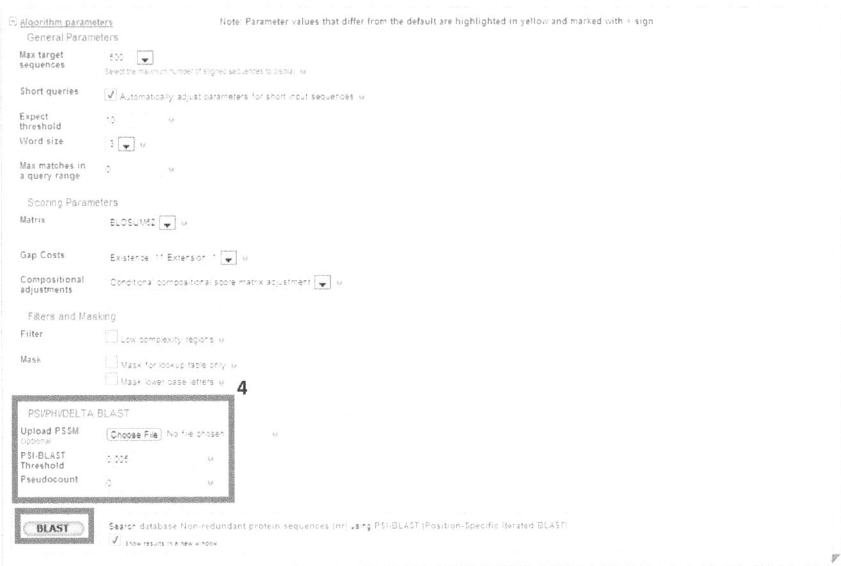


**Fig. 4.39** PSI-BLAST algorithm parameters

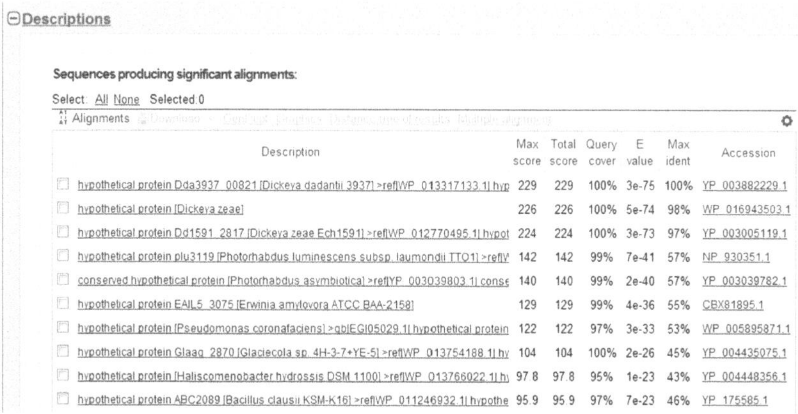


| | Description | Max score | Total score | Query cover | E value | Max ident | Accession |
|---|---|---|---|---|---|---|---|
| ☐ | hypothetical protein Dda3937 00821 [Dickeya dadantii 3937] >ref\|WP 013317133.1\| hyp | 229 | 229 | 100% | 3e-75 | 100% | YP 003882229.1 |
| ☐ | hypothetical protein [Dickeya zeae] | 226 | 226 | 100% | 5e-74 | 98% | WP 016943503.1 |
| ☐ | hypothetical protein Dd1591 2817 [Dickeya zeae Ech1591] >ref\|WP 012770495.1\| hypot | 224 | 224 | 100% | 3e-73 | 97% | YP 003005119.1 |
| ☐ | hypothetical protein plu3119 [Photorhabdus luminescens subsp. laumondii TTO1] >ref\|V | 142 | 142 | 99% | 7e-41 | 57% | NP 930351.1 |
| ☐ | conserved hypothetical protein [Photorhabdus asymbiotica] >ref\|YP 003039803.1\| conse | 140 | 140 | 99% | 2e-40 | 57% | YP 003039782.1 |
| ☐ | hypothetical protein EAIL5 3075 [Erwinia amylovora ATCC BAA-2158] | 129 | 129 | 99% | 4e-36 | 55% | CBX81895.1 |
| ☐ | hypothetical protein [Pseudomonas coronafaciens] >gb\|EGI05029.1\| hypothetical protein | 122 | 122 | 97% | 3e-33 | 53% | WP 005895871.1 |
| ☐ | hypothetical protein Glaag 2870 [Glaciecola sp. 4H-3-7+YE-5] >ref\|WP 013754188.1\| hy | 104 | 104 | 100% | 2e-26 | 45% | YP 004435075.1 |
| ☐ | hypothetical protein [Haliscomenobacter hydrossis DSM 1100] >ref\|WP 013766022.1\| hy | 97.8 | 97.8 | 95% | 1e-23 | 43% | YP 004448356.1 |
| ☐ | hypothetical protein ABC2089 [Bacillus clausii KSM-K16] >ref\|WP 011246932.1\| hypothe | 95.9 | 95.9 | 97% | 7e-23 | 46% | YP 175585.1 |

**Fig. 4.40** BLASTP results of hypothetical protein from *Dickeya dadantii* 3937

## Interpretation of Result

1. A conventional search with the protein BLAST (BLASTP) retrieve records with similar hypothetical proteins only, which is not useful in determining the function (annotation) of the protein (Fig. 4.40). On the other hand, PSI-BLAST is more informative in annotating this hypothetical protein.
2. In PSI-BLAST, the first iteration presents two tables (Fig. 4.41), namely,
   - Sequences producing significant alignments with E-value BETTER than threshold and
   - Sequences with E-value WORSE than threshold.
3. PSSM is query specific and it is possible to save a PSSM created during a PSI-BLAST search and use it to search a different database with the same query also. For doing this change "Alignment" to "PSSM" in the pull-down menu found in the Format section of a "Formatting BLAST" page after the first iteration. Then do the search, by coping the resulting ASCII encoded PSSM and pasting it into the PSSM window of a new PSI-BLAST search page.
4. Select all the sequences under '**Select for PSI blast**' to include them for the second round of iteration.
5. You can add database hits falling outside the inclusion threshold to your PSSM for the next round by checking the box next to the hit or by unchecking the checkbox you can remove the previously selected hits.
6. Click button '**GO**' to 'Run PSI-Blast iteration 2 with max = 500'.
7. Search result includes both LuxR transcriptional factor protein and Lipase protein from the second table (Fig. 4.42).
8. Select both the proteins for next round of iteration 3.
9. Click button 'GO' to 'Run PSI-Blast iteration 3 with max = 500'.

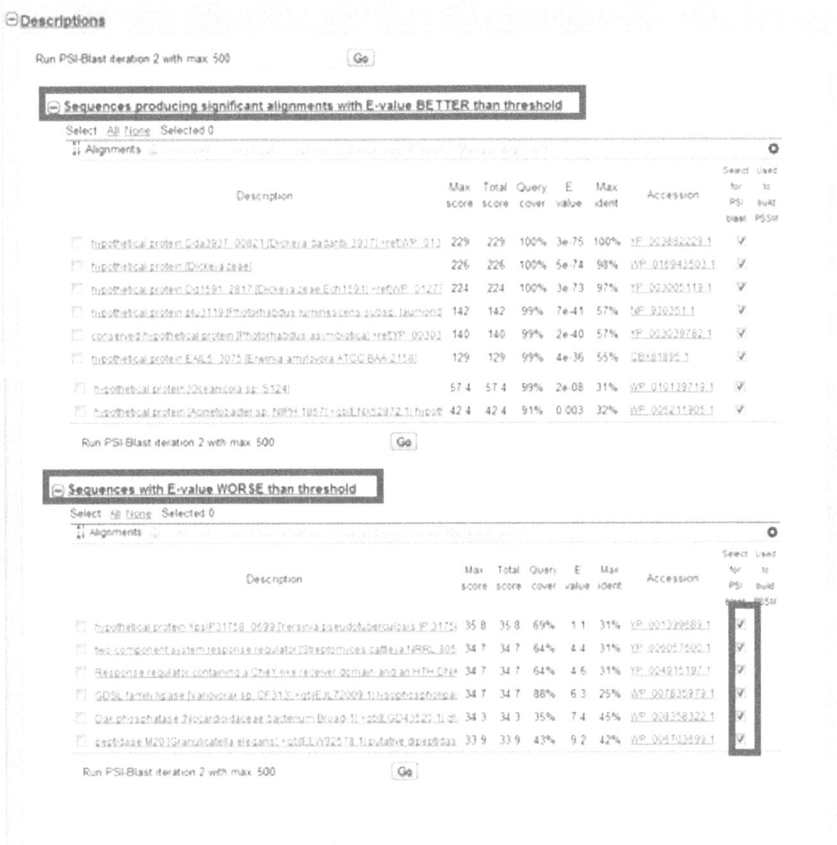


**Fig. 4.41** PSI-BLAST results of hypothetical protein from *Dickeya dadantii* 3937

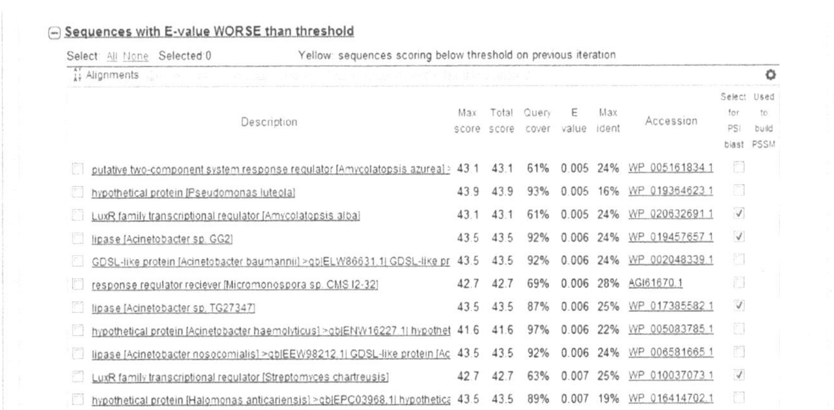


**Fig. 4.42** Second iteration showing LuxR and lipase proteins

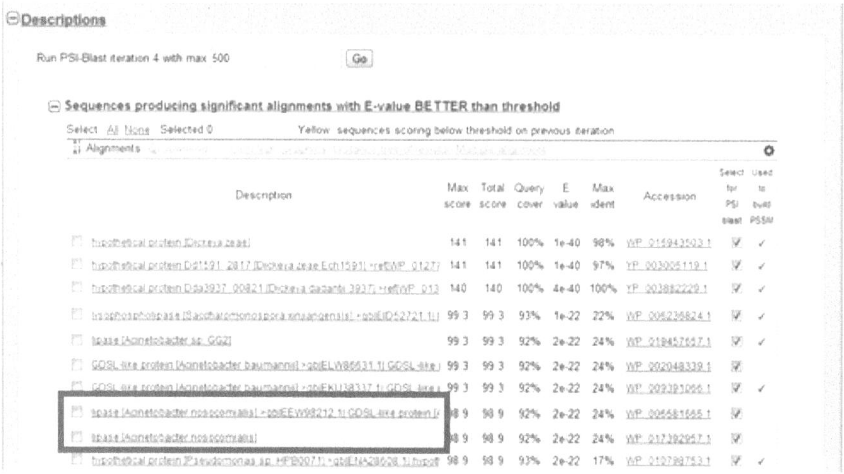

Fig. 4.43 Third iteration of PSI-BLAST showing convergence. *Yellow* sequences refer to scoring below threshold on previous iteration (colour figure online)

10. PSI-BLAST on iteration 3 produces only 'Sequences producing significant alignments with E-value BETTER than threshold' table, which indicates convergence (Fig. 4.43).
11. When no more matches are found in subsequent iterations, it can be said that the PSI-BLAST search has converged.
12. The PSI-BLAST result shows hypothetical protein and 'Lipase'.
13. This indicates that this hypothetical protein could be attributed to the function of 'Lipase' and not as transcriptional factor.

**Exercise**

1. How does PSI-BLAST differ from BLASTP?
2. Mention the application of PSI-BLAST?
3. Define the following: Iteration, Pseudocount, PSI Threshold value, convergence.
4. What is PSSM?

## 4.8 Creating Pattern from Alignment

**Introduction**

The patterns are regular expressions that are accepted by PHI-BLAST. They indicate allowed and disallowed amino acids in specific positions. The general pattern syntax in PHI-BLAST follows the PROSITE conventions.

```
CXXC1_BOVIN      RPLPTQPQPQPSQKLGRLREDEGAVASAAVKEPPEATATPEPLSDEDLPLDSELYQDFCA
CXXC1_HUMAN      RPLPTQQQPQPSQKLGRIREDEGAVASSTVKEPPEATATPEPLSDEDLPLDPDLYQDFCA
CXXC1_MOUSE      RPPPTQQQPQQSQKLGRIREDEGTVLSSVVKEPPEATATPEPLSDEDLALDPDLYQDFCA
                 ** *** *** ******:*****:* *:.*******************.**.:*******
```

**Fig. 4.44** Result from a multiple sequence alignment

```
CXXC1_BOVIN      RPLPTQPQPQPSQKLGRLREDEGAVASAAVKEPPEATATPEPLSDEDLPLDSELYQDFCA
CXXC1_HUMAN      RPLPTQQQPQPSQKLGRIREDEGAVASSTVKEPPEATATPEPLSDEDLPLDPDLYQDFCA
CXXC1_MOUSE      RPPPTQQQPQQSQKLGRIREDEGTVLSSVVKEPPEATATPEPLSDEDLALDPDLYQDFCA
                 ** *** *** ******:*****:* *:.*******************.**.:*******
```

**Fig. 4.45** Region selected for creating a pattern

A protein characters supported by PHI-BLAST patterns include **'ABCDEFGHIKLMNPQRSTUVWXYZ' and for** DNA, **'ACGT'**.

Other Delimiters:

[] refers to one of the characters within in the brackets, e.g. [ABCD] means one occurrence of A or B or C or D

'-' means nothing (this is a spacer character used by PROSITE)

**x(3)** means 3 positions in which any residue is allowed

**x(2,6)** means 2–6 positions where any residue is allowed

### Exploration Activity

1. Select few sequences to which you want to construct the pattern. Do a multiple sequence alignment using MUSCLE (Fig. 4.44).
2. Pick a small region attributing to any functional importance and see which amino acid residues are used (Fig. 4.45).
3. Create a pattern using the appropriate syntax. Here at the third position, A or T can be there and in fifth position it could be A or L. Thus for the above-selected region, the generated pattern could be as follows (Fig. 4.46).

### Exercise

1. What is a pattern?
2. Explain the steps involved in creating a pattern.
3. What is the meaning of in the syntax: x, x(10), [AGHT]

**Fig. 4.46** A pattern created following the syntax

```
E G A V A S A A V K E
E G A V A S S T V K E
E G T V L S S V V K E

E G A V A S A A V K E
    T   L   S T
              V
```

**EG[AT]V[AL]S[AS][ATV]VKE**

## 4.9 Pattern-Hit Initiated (PHI)-BLAST

**Introduction**

PHI-BLAST is yet another version of BLAST that not only looks for similar sequences but also a signature. This can detect statistically significant similar sequences that a traditional single-pass database search method misses out.

Protein sequence analyses have often observed that a group of related proteins always had 'elements' which are conserved patterns of amino acids known as 'motifs' or 'signatures' corresponding to some important functional or structural characteristics. Analysing these signatures often helped in characterizing a protein family, its function or even structural domain. If the pattern is too weak, then there is a possibility of getting many false positives. In order to prevent this, PHI-BLAST considers only patterns that are expected to match less frequently than once per 5000 database residues. PHI-BLAST takes only patterns of domains and protein families that have been studied and characterized by Prosite (http://us.expasy.org/prosite/) and the pattern must at least four completely specified residues.

**Significance of PHI-BLAST**

PHI-BLAST combines pattern-searching techniques with significant sequence similarity searching thereby making it more effective than the regular protein–protein BLAST.

**A Word of Caution**

Though PHI-BLAST is good at to identifying weakly related proteins, it cannot be used to delineate all members of a protein family. The program can identify several false positives, when weak patterns are used and there is also a possibility for losing some sequences, as it is too specific. To overcome this situation, PHI–BLAST has been developed to make its output to be used as input for PSI-BLAST, another BLAST variant that uses PSSM and multi-pass searching, making it more sensitive.

**Exploration Activity**

In this exercise the sequences homologous to GATA-binding factor 3 and its conserved pattern is to be analysed.

The GATA families of transcription factors are proteins that bind to DNA sites having consensus sequence (A/T)GATA(A/G) that is found within the regulatory region of a number of genes. GATA-3 is a transcriptional activator that binds to the enhancer of the T-cell receptor α and delta genes (Ko and Engel 1993).

For running PHI-BLAST, enter your query into the 'Search' box and pattern into the '**PHI pattern**' box in the "Options" section of the page. You can provide only one pattern for searching and it must follow the **PROSITE syntax conventions.**

1. Follow the steps 1–4 as mentioned in protein BLAST. GATA-binding factor 3 from *Gallus gallus* (P23825) was used as a query.
2. Under **database**, select the '**Protein Data Bank protein (pdb)**' (Fig. 4.47).

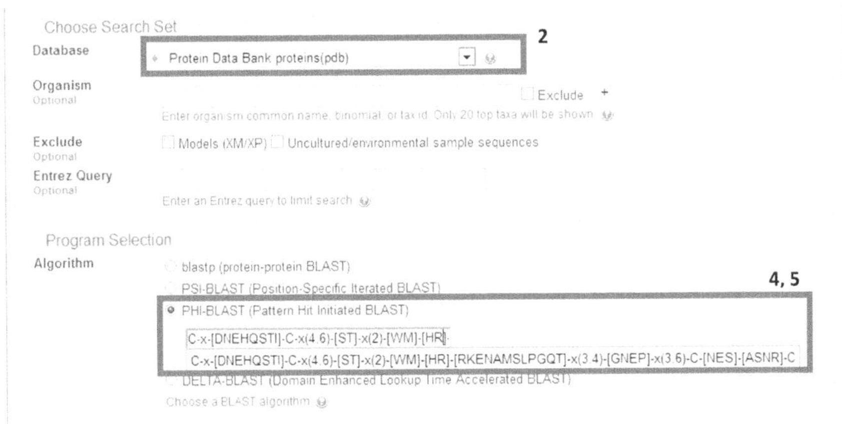


**Fig. 4.47** Choosing PDB and entering pattern for PHI-BLAST search

3. If you wish to search against any specific organism, under '**Organism**', enter first few characters of the organism's name and select the organism for the dropdown list.
4. Choose the **PHI-BLAST** under '**Program Selection**' (Fig. 4.47). PHI-BLAST limits alignments to those that match a pattern in the query and performs the search.
5. Enter the pattern (Fig. 4.47) for GATA-type zinc finger-domain profile: C-x-[DNEHQSTI]-C-x(4,6)-[ST]-x(2)-[WM]-[HR]-[RKENAMSLPGQT]-x(3,4)-[GNEP]-x(3,6)-C-[NES]-[ASNR]-C. Patterns are available at Prosite (http://us.expasy.org/prosite/). The user can search for patterns for their own sequence of interest.
6. Adjust algorithm parameters according to query length as mentioned in protein BLAST (Fig. 4.48). In addition PHI-BLAST includes parameters, like Upload PSSM, PSI-BLAST Threshold and Pseudocount.
7. Set the threshold value of 0.005 and Pseudocount to '0'.
8. You can also bypass the initial BLASTP step either by providing a PSSM saved from a previous search or by specifying a set of aligned sequences that can be used for generating the initial PSSM.
9. Click '**BLAST**' button at the end of the page (Fig. 4.48).

**Interpretation of Results**

In the first iteration PHI-BLAST result shows the sequences homologous to GATA-binding factors containing the specified pattern whose structures are solved from PDB (Fig. 4.49).

From the second iteration onwards and up to convergence, PHI-BLAST tries to add (shown in yellow) the sequences homologous to GATA-binding factors containing the specified pattern whose structures are solved from PDB (Fig. 4.50).

Up to four iterations, 25 sequences are added containing the specified pattern.

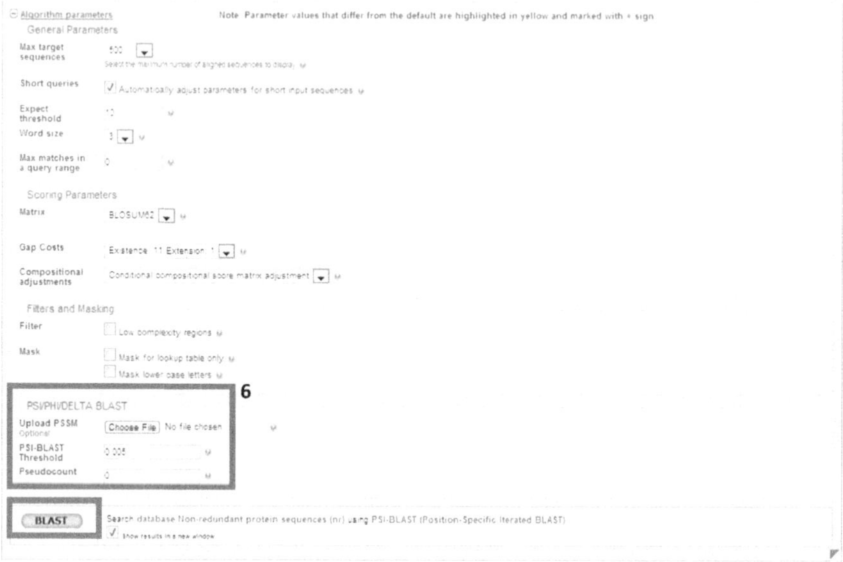

Fig. 4.48 PHI-BLAST algorithm parameters

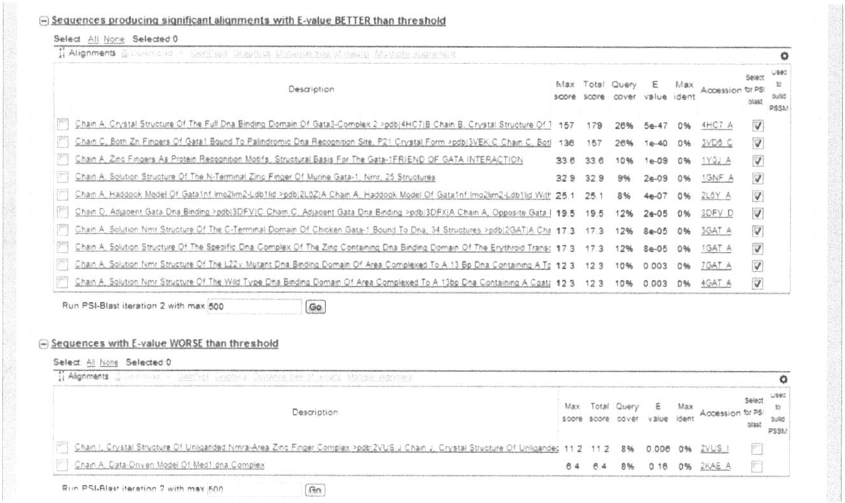

Fig. 4.49 First iteration results of PHI-BLAST

PHI-BLAST is more preferable than just searching for pattern occurrences because it eliminates those cases where the pattern occurrence might be random and because of homology.

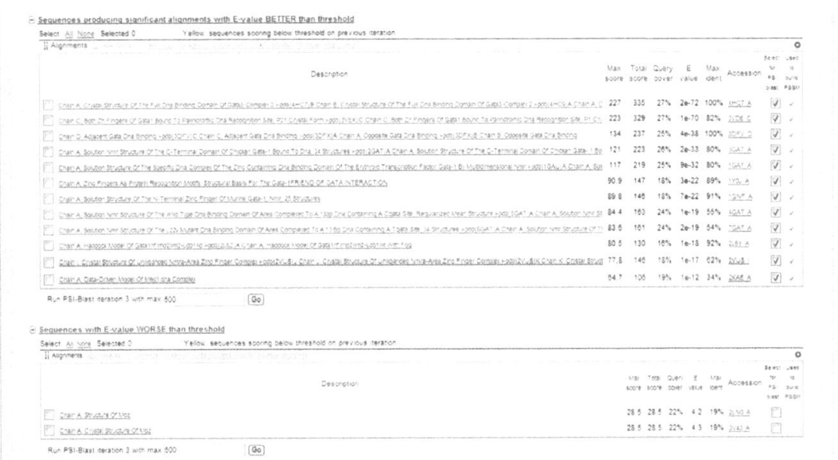


**Fig. 4.50** In second iteration of PHI-BLAST, *yellow* sequences refer to scoring below threshold on previous iteration (colour figure online)

## Exercise

1. What is the difference between PSI-BLAST and PHI-BLAST?
2. What is a motif?
3. How can we get the pattern for PHI-BLAST analysis?
4. What is the application of PHI-BLAST?
5. When false positive results are obtained in PHI-BLAST?

# 4.10 Domain Enhanced Lookup Time Accelerated-BLAST (DELTA-BLAST)

## Introduction

In Position-Specific Iterated BLAST (PSI-BLAST), the program automatically generates multiple sequence alignments and their related PSSMs from the results of database searches, iteratively.

In contrast, DELTA-BLAST searches a database containing pre-constructed PSSMs generated from Conserved Domain Database (CDD), which is an NCBI resource that identifies conserved domains within protein sequences, before making a search in a protein sequence database (Fig. 4.51). DELTA-BLAST initially uses RPS-BLAST (Reverse PSI-BLAST) to align a query sequence to conserved domains in CDD, and later performs a sequence database search using a PSSM derived from the aligned domains. The construction of PSSM is similar to that of

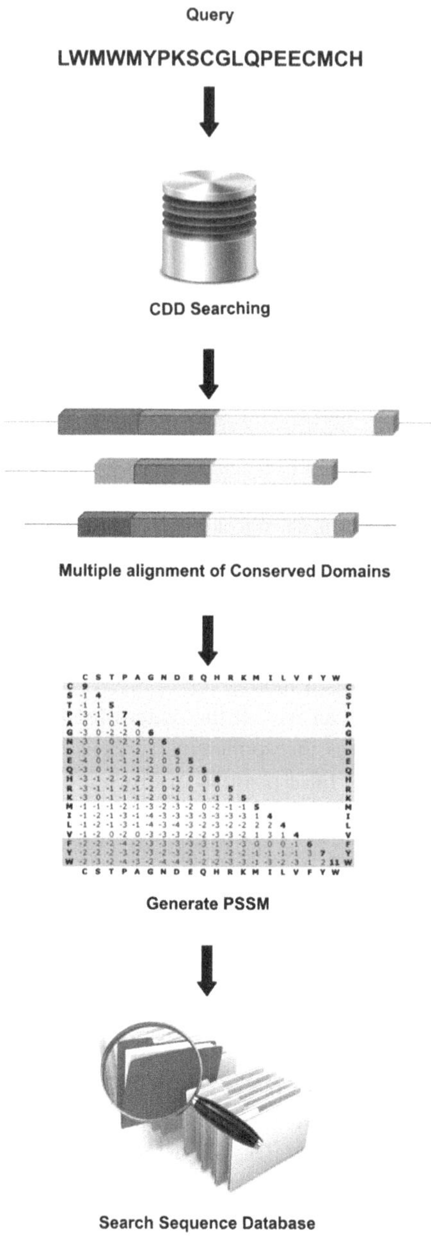


**Fig. 4.51** DELTA-BLAST methodology (Boratyn et al. 2012)

PSI-BLAST, but it begins by aligning the query to CDs instead of the individual sequences to minimize the time spent in the initial BLASTP search. When used along with CDD, DELTA-BLAST becomes more sensitive than the PSI-BLAST (Boratyn et al. 2012).

**Exploration Activity**

This exercise tries to test the presence of cytochrome b like domain of *Cathartes aura* in *Arabidopsis and Nostoc.*

1. Follow the steps 1–4 as mentioned in protein BLAST. Cytochrome b from *Cathartes aura* (Sequence ID: emb|CAA60423.1|Length: 342) was used as a query.
2. Under **database**, select the '**Reference proteins (refseq_protein)**' (Fig. 4.52).
3. Under '**Organism**', enter first few characters of the organism's name and select the organism for the dropdown list. Type *Arabidopisis* and select *Arabidopsis thaliana* (taxid:3702) and click '+' button to add another organism. Now type *Nostoc* and select *Nostoc punctiforme* PCC 73102 (taxid: 63737). Note: *The user can include as many organisms as they want to look for.*
4. Choose the **DELTA-BLAST** under '**Program Selection**' (Fig. 4.52). DELTA-BLAST constructs a PSSM from the results of a Conserved Domain Database search and then searches the sequence database.
5. Adjust algorithm parameters according to query length as mentioned in protein BLAST (Fig. 4.53).
6. Set the PSI-BLAST threshold value as 0.005, DELTA-BLAST threshold to 0.05 and Pseudocount to '0'.
7. Click '**BLAST**' button at the end of the page (Fig. 4.53).

**Interpretation of Results**

8. In the conventional protein BLAST (BLASTP) the program gives preferences to complete length of the sequence and tries to fetch the more identical one and then homologous ones. Thus the number of hits generated by this BLAST will be high (Fig. 4.54).

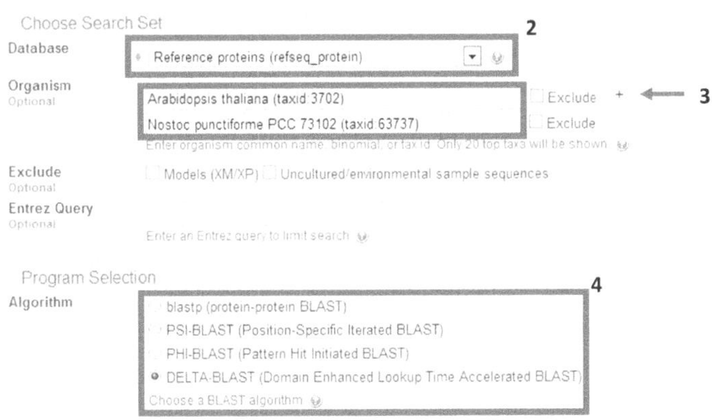


**Fig. 4.52** DELTA-BLAST algorithm parameters

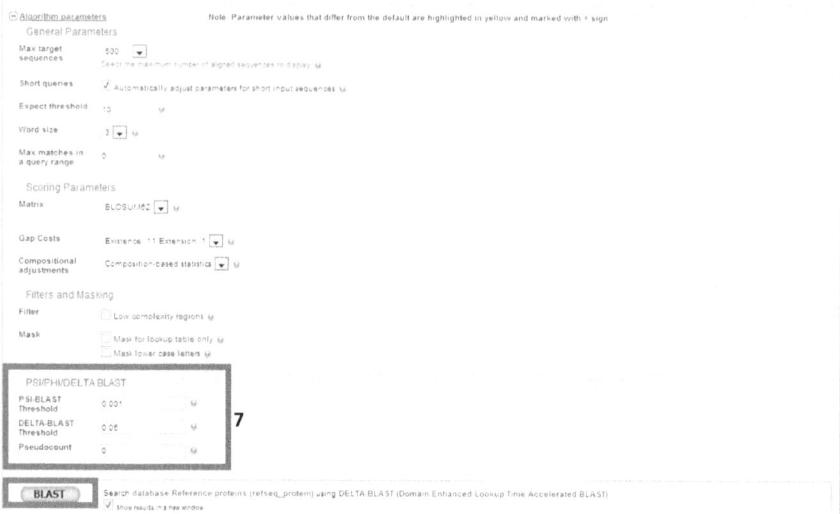


**Fig. 4.53** Algorithm parameters for DELTA-BLAST

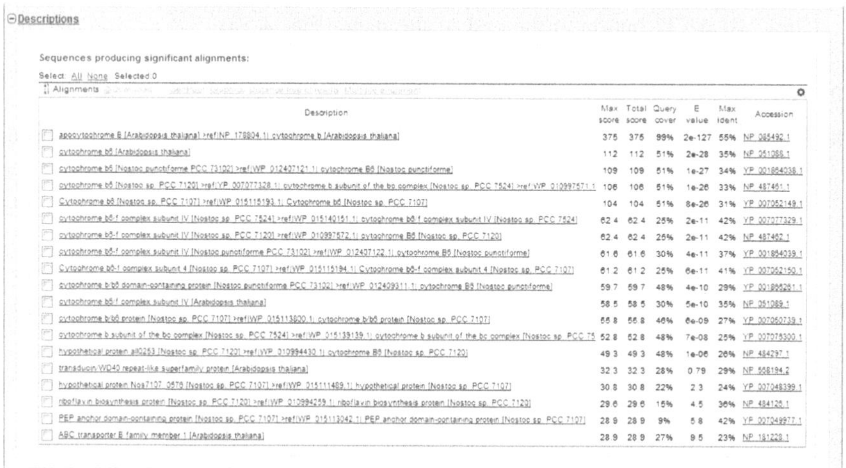


**Fig. 4.54** Result for protein BLAST

9. On the other hand, DELTA-BLAST focuses more on the domain alone and thus tries to bring more distantly related one, thereby limiting the number of hits (Fig. 4.55).
10. When the programs cannot add more sequences during iteration with the specified threshold, it displays the following message (Fig. 4.56). It indicates the '**convergence**' stage and if the user wishes, then he can reduce the threshold value and can search again.

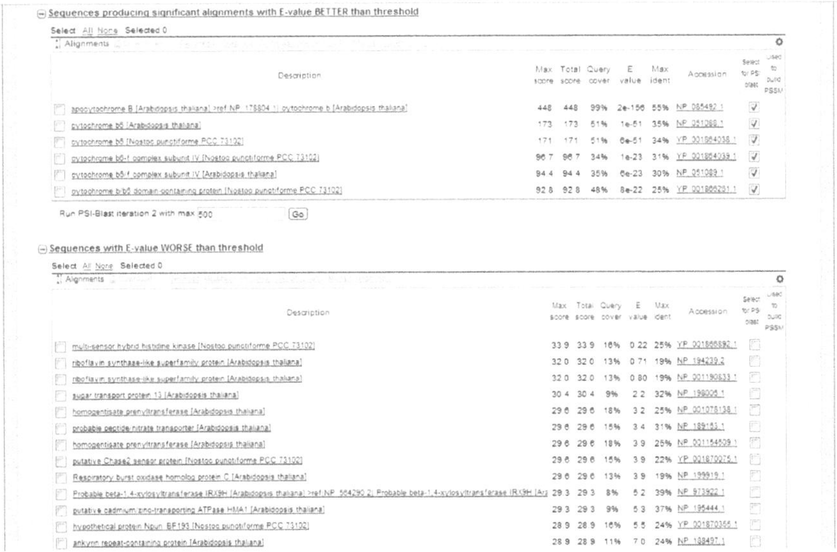

Fig. 4.55 Result for DELTA-BLAST

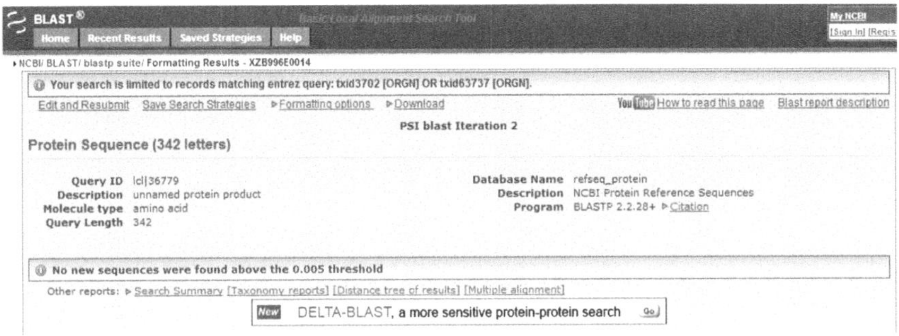

Fig. 4.56 Convergence reported for a particular threshold value

## Exercise

1. What is the difference between PSI-BLAST and DELTA-BLAST?
2. How DELTA-BLAST differs from BLASTP?
3. Why two threshold values, PSI-BLAST threshold and DELTA-BLAST threshold, are used in DELTA-BLAST?

# Chapter 5
# Protein Structure Analysis

## 5.1 Protein Primary Structure Analysis—ProtParam

### Introduction

Amino acids have been grouped into different categories like polar, nonpolar, acidic, basic, large, small, aliphatic and aromatic, and each group has its own functional features (Table 5.1). ProtParam from ExPASy (Swiss Institute of Bioinformatics) is one of the programs that computes various physicochemical properties for a Swiss-Prot/TrEMBL accession number or ID, or user-defined protein sequence. Any white space or numbers that are provided with the sequences are ignored (Gasteiger et al. 2005). There are other programs available that can calculate the properties of proteins like PROT-PROP J-tool (Senthilkumar et al. 2012) based on the amino acid sequence.

### Amino Acids and Their Properties

### Using ProtParam

The physicochemical parameters of two proteins, *Homo sapiens* glutathione peroxidise (**P07203**) and *Bacillus polymyxa* intracellular serine protease (**P29139**) are being compared. The results are tabulated for easy comparison.

1. ProtParam program can be accessed from http://web.expasy.org/protparam/.
2. Put in your protein sequence retrieved from UniProtKB (ExPASy). You can cut and paste either plain protein sequence in the text box provided or as a Swiss-Prot/TrEMBL accession number or ID, to the text above it (Fig. 5.1).
3. Posttranslational modifications of the proteins cannot be specified or the program cannot identify if the protein is dimer or multimer. If you do know that your protein forms a dimer, you may just append a second copy of the sequence to the first in the text box (Fig. 5.2).
4. Click '**Compute parameters**' button.



K. Syed Ibrahim et al., *Bioinformatics—A Student's Companion*,
DOI 10.1007/978-981-10-1857-2_5

**Table 5.1** Amino acids grouped by their common side chain features

| Amino acid group | Amino acid name | Functional features |
|---|---|---|
| Small and nonpolar | Glycine<br>Alanine<br>Proline | Non-reactive in chemical reactions, Proline and Glycine disrupt secondary structures |
| Small and polar | Cysteine<br>Serine<br>Threonine | Serve as posttranslational modification sites and participate in active site and metal binding sites |
| Large and polar | Glutamine<br>Asparagine | Participate in hydrogen bonding or in enzyme active sites |
| Large and polar (basic) | Arginine<br>Lysine<br>Histidine | Found in the surface of globular proteins providing salt bridges; Histidine participate in enzyme active site or metal binding sites |
| Large and polar (acidic) | Glutamate<br>Aspartate | Found in the surface of globular proteins providing salt bridges |
| Large and nonpolar (aliphatic) | Leucine<br>Isoleucine | Non-reactive in chemical reactions, participate in hydrophobic interactions |
| Large and nonpolar (aromatic) | Phenylalanine<br>Tyrosine<br>Tryptophan | Providing sites for aromatic packing interactions, Tyrosine and Tryptophan being weakly polar can serve as a site for phosphorylation and hydrogen bonding |

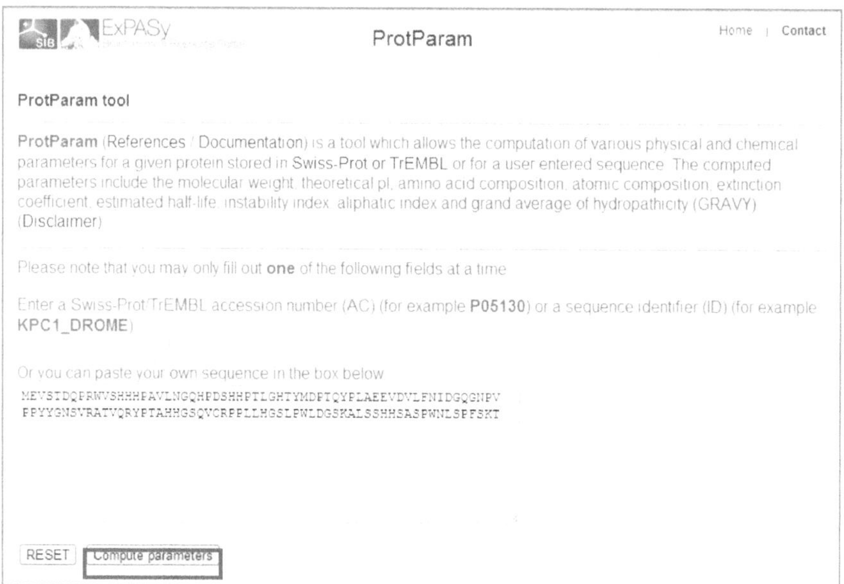


**Fig. 5.1** ProtParam interface at ExPASy

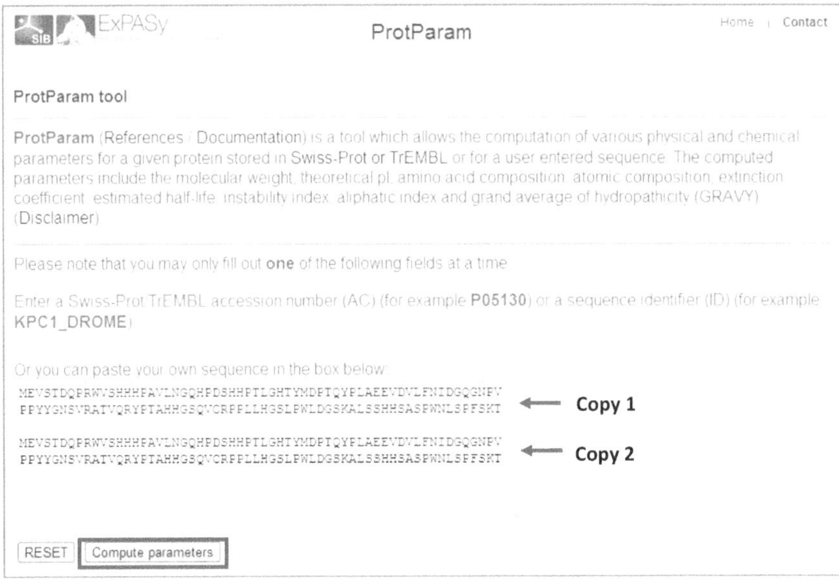


**Fig. 5.2** Entering sequences in case of a dimer

**Table 5.2** Comparison of *Homo sapiens* glutathione peroxidise and *Bacillus polymyxa* intracellular serine protease

| | *Homo sapiens* glutathione peroxidase (P07203) | *Bacillus polymyxa* intracellular serine protease (P29139) |
|---|---|---|
| Number of amino acids | 203 | 326 |
| Molecular weight | 22088.1 | 35173.6 |
| Theoretical pI | 6.15 | 4.73 |
| Total number of negatively charged residues (Asp + Glu) | 21 | 47 |
| Total number of positively charged residues (Arg + Lys) | 20 | 28 |
| Ext. coefficient | 17,210 | 26,025 |
| In vivo half-life | 30 | 30 |
| The instability index (II) | 47.96 (unstable) | 29.84 (stable) |
| Aliphatic index | 86.11 | 93.04 |
| Grand average of hydropathicity (GRAVY) | −0.070 | −0.214 |

**Interpreting Results**

The results are tabulated for easy comparison (Table 5.2).

The parameters computed by ProtParam include the molecular weight, theoretical pI, amino acid composition and atomic composition, as basic information about a protein. Also information about extinction coefficient, estimated half-life, instability index, aliphatic index and grand average of hydropathicity (GRAVY) is also listed in the result page.

**Isoelectric Point (pI)**

Isoelectric point (pI) is a pH at which an amino acid does not move in an electric field. This means that its net charge becomes zero (zwitterion). The pH of the surrounding aqueous environment determines the net charge of the amino acid.

pI < 7 indicates that the protein is acidic and pI > 7 indicates that the protein is basic. pI of 7 indicates neutral. Theoretical isoelectric point (pI) is useful in developing buffer systems for protein purification. Here, the protease is found to be more acidic than the glutathione peroxidise (Table 5.2).

**Extinction Coefficients**

The extinction coefficient of a protein indicates the amount of light it absorbs when subjected to a particular wavelength. The extinction coefficient of a native protein in water can be computed from the molar extinction coefficient of tyrosine, tryptophan and cystine at 280 nm wavelength. The extinction coefficient helps in accurate determination of protein concentration when experimental measurement is difficult or impossible.

Higher extinction coefficient indicates presence of Cys, Trp and Tyr at high concentrations. When ProtParam returns no value, it shows that the protein is devoid of Tyr, Cys or Trp which indicates that these proteins cannot be analysed using UV spectroscopy. The results here indicate that these proteins can be analysed using UV spectroscopy.

**In Vivo Half-Life**

The time taken by a protein to become half in a cell after its synthesis is called half-life of the protein. As ProtParam uses 'N-end rule' that relates the half-life of a protein to the identity of its N-terminal residue, it is not applicable for proteins that are N-terminally modified.

Higher the half-life value higher the stability of the protein in the cell, whereas lower values indicate proteins may degrade soon.

**Instability Index (II)**

The instability index is determined by the order of certain amino acids in its sequence and measures the stability of the protein in vitro.

Protein with instability index lesser than 40 is predicted to be stable and in vivo half-life is more than 16 h, whereas protein may be unstable if it is above 40 and half-life is less than 5 h. In case of unstable proteins, amino acids methionine, glutamine, proline, glutamic acid and serine are found to occur with a relatively high frequency. In our case, glutathione peroxidise is unstable, whereas the protease is found to stable (Table 5.2).

**Aliphatic Index**

The aliphatic index of a protein (volume of a protein filled by aliphatic side chains like isoleucine, alanine, valine and leucine) denotes the thermostability of globular proteins. Increase in thermostability of proteins is due the substitutions of amino

**Table 5.3** Hydropathy values of respective amino acids (Kyte and Doolite 1982)

| | | | |
|---|---|---|---|
| Alanine | 1.8 | Arginine | −4.5 |
| Leucine | 3.8 | Asparagine | −3.5 |
| Methionine | 1.9 | Lysine | −3.9 |
| Phenylalanine | 2.8 | Aspartic acid | −3.5 |
| Cysteine | 2.5 | Proline | −1.6 |
| Isoleucine | 4.5 | Glutamine | −3.5 |
| Valine | 4.2 | Serine | −0.8 |
| | | Glutamic acid | −3.5 |
| | | Threonine | −0.7 |
| | | Glycine | −0.4 |
| | | Tryptophan | −0.9 |
| | | Histidine | −3.2 |
| | | Tyrosine | −1.3 |

acids: Gly to Ala, Ser to Thr, Lys to Arg, Asp to Glu, Met to Ala/Leu, Cys to Ile/Ala/Val and Trp to Tyr (Gromiha et al. 1999). Here, both show nearly same values but still the protease appears to be more stable than the glutathione peroxidise (Table 5.2).

**GRAVY (Grand Average of Hydropathy)**

It denotes the hydrophobic nature of a protein and calculated by Kyte–Doolittle scale (Kyte and Doolite 1982). GRAVY is the sum of hydrophobic values of all the amino acids. The values of different amino acids are given in Table 5.3.

Increasing positive score indicates greater hydrophobicity. Proteins with very low GRAVY index show better interaction with water.

From the analysis, it is evident that protease from Bacillus is in a better interaction with water compared to the human's glutathione peroxidise (Table 5.2).

**Exercise**

1. What will be MW of a protein in case of a trimer?
2. What is the importance of hydrophobicity?
3. What do you infer from a protein having a lower GRAVY index?
4. What is the importance of having knowledge on the pI of a protein?
5. What is the significance of extinction coefficient of a protein?
6. For the proteins **P10477** and **C7ASI9**, perform a ProtParam analysis and answer the questions.
   a. What is its theoretical pI?
   b. What is the Val percent of composition?
   c. Do they have high concentration of Cys, Trp and Tyr?
   d. What is the nature of hydrophobicity?

## 5.2 Protein Secondary Structure Prediction

**Introduction**

Proteins are polypeptide chains made up of amino acids. The amino acids are connected by peptide bonds, rigid in nature, between the carboxylic acid group (COOH) of one amino acid and the amino group ($NH_3$) of the next. Limitations to the folding of a protein chain are due to rigidity which prevents rotation around a peptide bond. The typical bond length between C–N, C=N and C=O are 1.47, 1.27 and 1.33 Å, respectively. The shape of the peptide chain is determined by type of bond, rotation and torsional angles (Table 5.4).

**Protein Secondary Structure**

Local structures of a protein with regular conformations are known as secondary structures. They are stabilized by hydrogen bonds formed between carbonyl oxygen and amino hydrogen of different amino acids. There are two types of protein secondary structures, alpha helix and beta pleated sheet (Pauling and Cory 1951).

**α-Helices**

α-Helices are right handed, exhibiting a rightward spiral form and contains 3.6 amino acids per helical turn. The structure is stabilized by hydrogen bonds formed between the main chain atoms of residues i and i + 4. The mean phi angle is −62° and the mean psi angle is −41° (Ramachandran et al. 1963). The hydrogen bonds are nearly parallel with the helical axis. Hydrophobic residues of the helix tend to face inside and hydrophilic residues of the helix face outside. Thus, every third residue along the helix tends to be a hydrophobic residue. Ala, Gln, Leu and Met are commonly found in an α-helix, but not Pro, Gly and Tyr. These rules are useful in guiding the prediction of protein secondary structures.

**β-Sheets**

The β-strand conformation is pleated with main chain backbone zigzagging and side chains positioned alternately on opposite sides of the sheet. β-Strands are stabilized by hydrogen bonds between residues of adjacent strands. β-strands near the surface of the protein tend to show an alternating pattern of hydrophobic and hydrophilic regions, whereas strands buried at the core of a protein are nearly all hydrophobic. The β-strands can run in the same direction to form a parallel sheet or can run every other chain in reverse orientation to form an antiparallel sheet, or a mixture of both. The hydrogen bonding patterns are different in each configuration.

**Table 5.4** Bonds and their torsional angles

| Bond | Rotation | Torsion angle defined |
|---|---|---|
| NH to Cα | Free | *phi* |
| Cα to C=O | Free | *psi* |
| C=O to NH (peptide bond) | Rigid planar due to double bond character | *omega* |

Because of the long-range nature of residues involved in this type of conformation, it is more difficult to predict β-sheets than α-helices (Jin 2006).

**Coils and Loops**

There are also local structures that do not belong to regular secondary structures (α-helices and β-strands). The irregular structures are coils and loops. The loops are often characterized by sharp turns or hairpin-like structures. If the connecting regions are completely irregular, they belong to random coils. Residues in the loop or coil regions tend to be charged, polar and are located on the surface of the protein structure. They can be functionally significant because these locations are often the active sites of proteins.

Coiled coils are a special type of supersecondary structure characterized by a bundle of two or more α-helices wrapping around each other. The helices forming coiled coils have a unique pattern of hydrophobicity, which repeats every seven residues (five hydrophobic and two hydrophilic) (Bujnicki 2008).

Alpha-helices and beta-sheets are positioned at the core of the protein, whereas loops are present in outer regions. Random coils, loops or turns don't have a stable secondary structure.

**Amino Acids Contributing to Secondary Structure**

The main criterion for alpha helix preference is that the amino acid side chain should cover and protect the backbone H-bonds in the core of the helix. Amino acids like Ala, Leu, Met, Phe, Glu, Gln, His, Lys and Arg are preferred in alpha-helix. The extended structure leaves the maximum space free for the amino acid side chains and as a result, those amino acids with large bulky side chains prefer to form beta sheet structures like just plain large (Tyr, Trp, Phe, Met), bulky and awkward due to branched beta carbon (Ile, Val, Thr) and large S atom on beta carbon (Cys). The remaining amino acids have side chains which disturb secondary structure are known as secondary structure breakers, e.g. Gly, Pro, Asp, Asn and Ser. Clusters of breakers give rise to regions known as loops or turns which mark the boundaries of regular secondary structure, and serve to link up secondary structure segments (Geourjon and Deleage 1994).

## 5.2.1 Secondary Structure Prediction Using SOPMA

**SOPMA** (Self-Optimized Prediction Method with Alignment) is a modified version of **SOPM** (Self-Optimized Prediction Method) predicting the secondary structure of a protein based on **homologue method** (Fig. 5.3). The SOPMA derived account

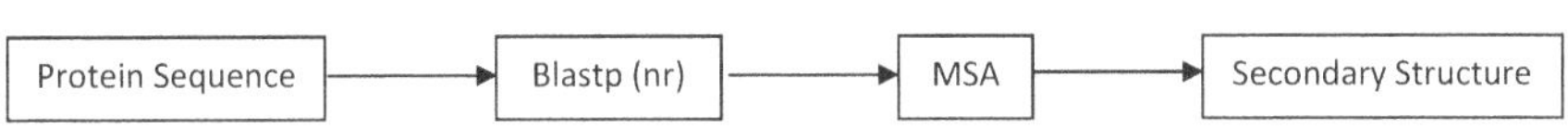


**Fig. 5.3** Steps involved in secondary structure prediction using SOPMA

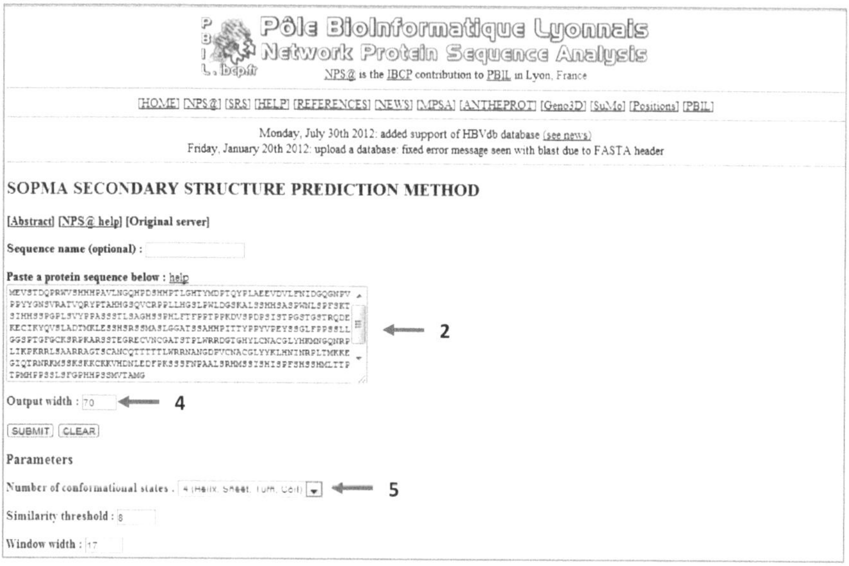


**Fig. 5.4** The SOPMA interface

information from an alignment of sequences belonging to the same family. If in any case there are no homologous sequences, then the SOPMA prediction is the SOPM one (Geourjon and Deléage 1995).

**Exploration Activity**

The chick GATA-binding factor 3 (**P23825**, Length: 444) is used for secondary structure analysis.

1. SOPMA program can be accessed from http://npsa-pbil.ibcp.fr/cgi-bin/npsa_automat.pl?page=/NPSA/npsa_sopma.html.
2. **Put in your protein sequence**. You can cut and paste the protein sequence in the box (Fig. 5.4). Here, the chick GATA-binding factor 3 is used.
3. **If required an optional title can be provided**.
4. **Output width indicates the number of amino acid sequences to be displayed in a single line in the output**.
5. You can set the **number of conformational states** to predict: 3 (Helix, Sheet, Coil) or 4 (Helix, Sheet, Turn, Coil).
6. The **Similarity threshold** parameter is the threshold below which a subject peptide is rejected when it is compared with a query peptide of the sequence (default value = 8).
7. The **Window width** parameter sets the length of the peptides to use (default value = 17).
8. Click on the '**Submit**' button to compute.

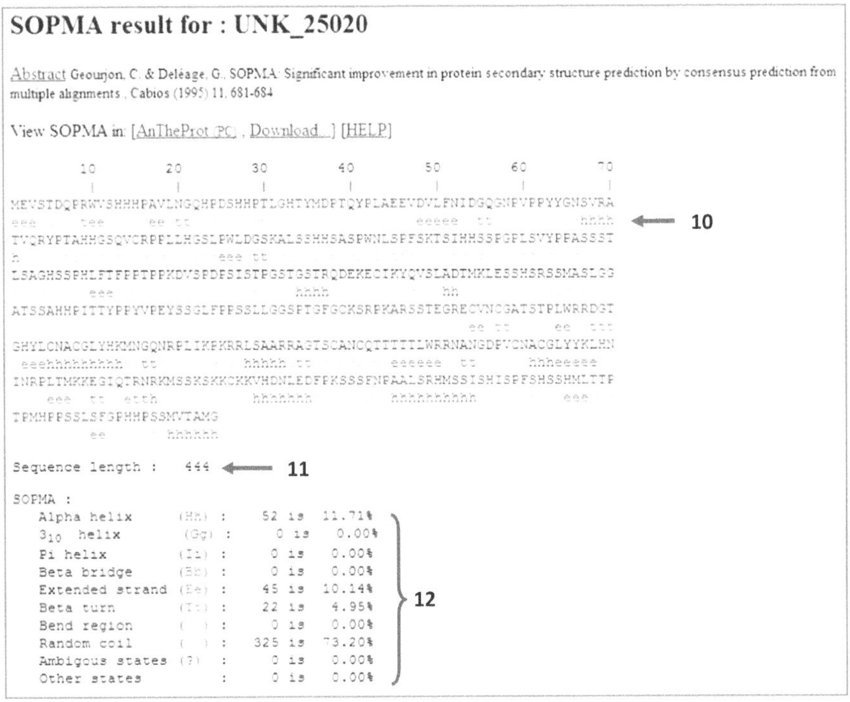


**Fig. 5.5** The SOPMA results

## Interpreting Results

9. The first line provides link to view the prediction in **ANTHEPROT**, a Windows-based software for protein analysis and a download link for downloading **ANTHEPROT** software (Fig. 5.5).
10. A **colour-coded prediction** below every sequence line indicates the corresponding predicted states like Helix, Sheet, Turns and Coils.
11. The **sequence length** of the user-provided sequence is 444.
12. The **percentage of each secondary structure element** in the sequence under SOPMA.
13. **Two graphical outputs**. The first is the linear representation of secondary structure for better visualization of the prediction (Fig. 5.6a), and the second indicates the score curves for each predicted state (Fig. 5.6b).
14. The graphical output is followed by the list of parameters used for predicting the secondary structure (Fig. 5.6b).
15. It is then followed by the links for prediction result as **text file** and **intermediate result text files**.
16. A more detailed result can be obtained from the **Prediction result in text file**.

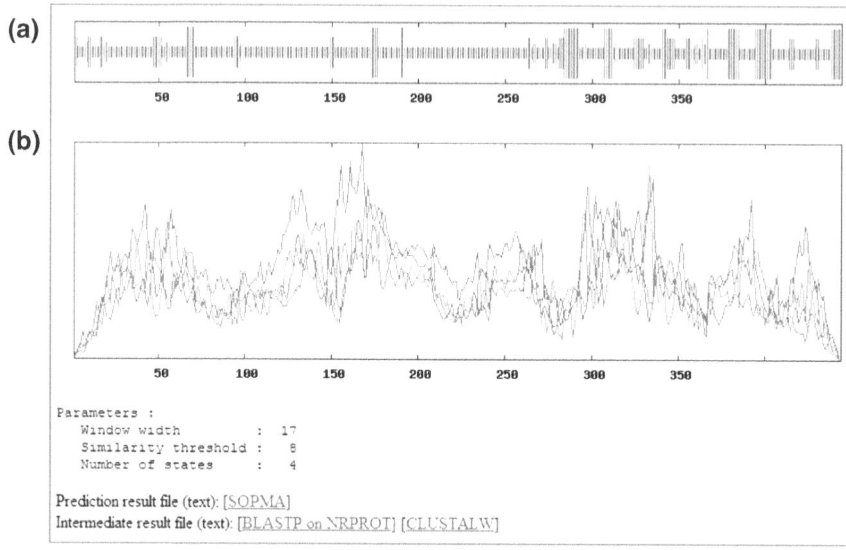


**Fig. 5.6** The SOPMA graphical results. **a** Linear representation, **b** curve score

Secondary state rates:

| | | |
|---|---|---|
| Alpha helix | 55 | 12.39% |
| Beta sheet | 47 | 10.59% |
| Turn | 35 | 7.88% |
| Coil | 307 | 69.14% |

The protein thus contains both helical and beta sheet structures in approximately equal proportions. Coil structural content of a protein is very high due to the presence of more flexible glycine and hydrophobic proline amino acids. Proline creates bends in polypeptide chains and disorders the secondary structure.

## Exercise

1. What amino acids contribute to alpha helix structure?
2. What are the secondary structure breakers?
3. What are the different forms of beta sheets?
4. What do you mean by $3_{10}$ and Pi helix?
5. How SOPMA predicts the secondary structure of a protein?
6. For the protein with UniProt id **O81360** perform a secondary structure prediction and tabulate the secondary structure states.

## 5.2.2 *Secondary Structure Prediction Using PSIPRED*

**Introduction**

PSIPRED, a web-based program, predicts protein secondary structures using a combination of evolutionary information and neural networks. PSIPRED predicts secondary structure by incorporating two feed-forward neural networks which perform an analysis on output obtained from PSI-BLAST. A profile is extracted from the multiple sequence alignment which is generated from three rounds of automated PSI-BLAST (Position-Specific Iterated—BLAST). This profile is then used as input for a neural network prediction. A unique filtering algorithm is implemented to filter out unrelated PSI-BLAST hits during profile construction so as to achieve higher accuracy. PSIPRED 3.2 achieves an average $Q_3$ score of 81.6% by using a very stringent cross-validation method for evaluating the method's performance (Buchan et al. 2013).

**Exploration Activity**

The chick GATA-binding factor 3 (**P23825**, Length: 444) is used for secondary structure analysis.

1. PSIPRED can be accessed from http://bioinf.cs.ucl.ac.uk/psipred/.
2. Under Input tab, first choose the prediction method from the list of methods provided. By default '**PSIPRED**' will be selected (Fig. 5.7).

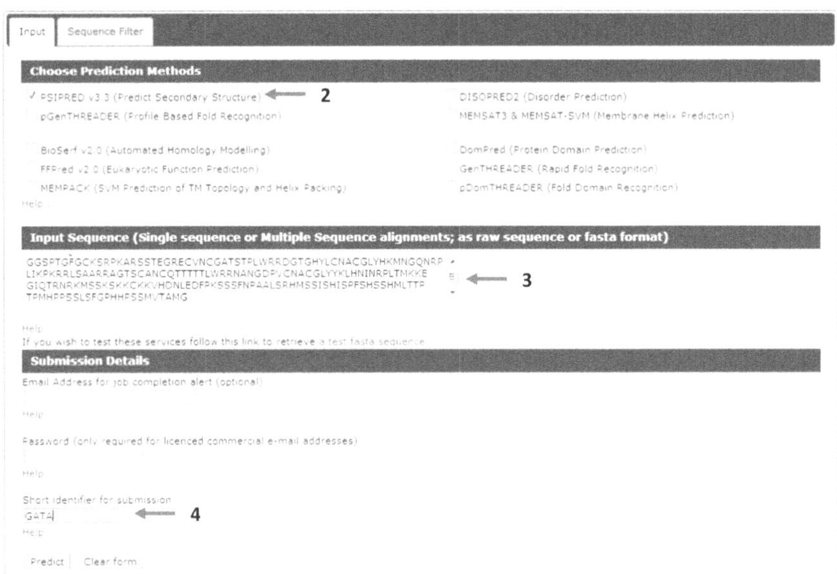


**Fig. 5.7** PSIPRED interface

3. Enter valid amino acid sequence as single letter code. Chick GATA-binding factor 3 sequence in plain format is pasted. It also supports FASTA format and FASTA formatted multiple sequence alignments as input. The upper limit to the length of the input sequence is **1500** residues.
4. Provide a short identifier for the submission under '**Submission details**' section (compulsory). Here GATA is given as identifier.
5. Other entries in the '**Submission details**' section are optional.
6. Click '**Predict**' button at the bottom.

**Interpreting Results**

7. The results summary page gives a brief summary of the results of the submitted sequence (Fig. 5.8).
8. The top of the page shows Job ID details including a short identifier provided by the user for the job and a unique private ID assigned by the server.
9. The Summary tab annotates the residues as per the predicted secondary structure.
10. The PSIPRED diagrammatic output annotates the query sequence with secondary structure in the form of cartoons and confidence value at each position in the alignment showed as blue bar graphs (Figs. 5.9 and 5.10).
11. The third download tab allows the user to download the PSIPRED results and raw scores as plain text. It also provides links to download PSIPRED diagram in Postscript or PDF format (Fig. 5.11).

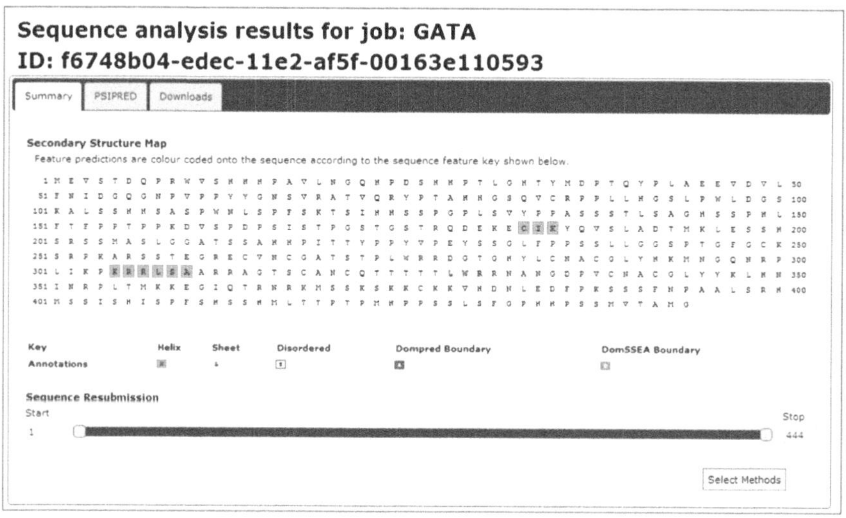


**Fig. 5.8** The PSIPRED result summary

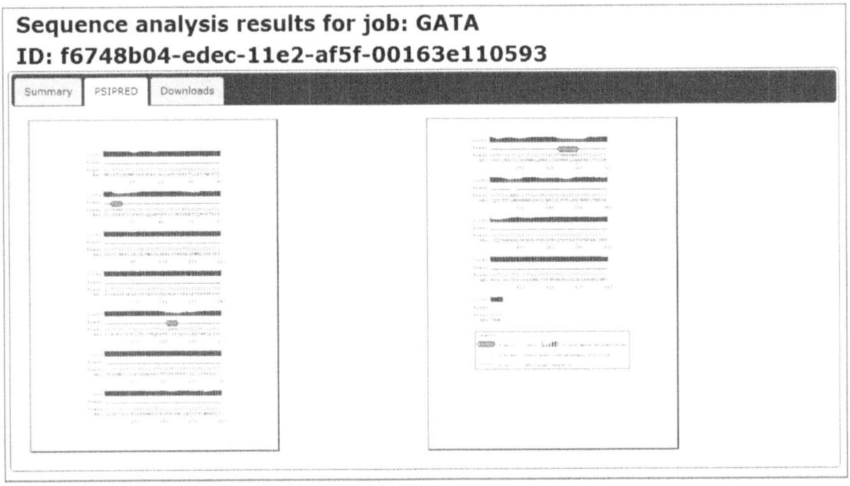


**Fig. 5.9** The graphical representation of secondary structure

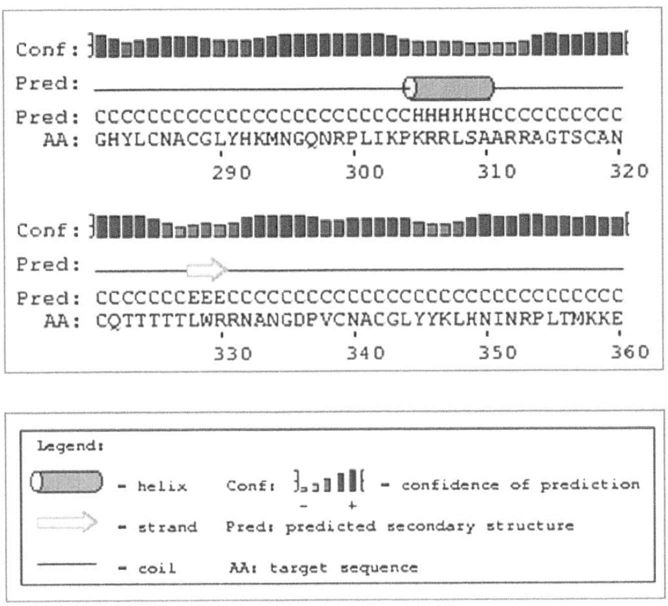


**Fig. 5.10** Enlarged view of the PSIPRED secondary structure assignment with legend

The PSIPRED results indicate that the protein contains 10–12% of helix and beta sheet and the rest comprising coiled regions. This result correlates with our earlier prediction by SOPMA.

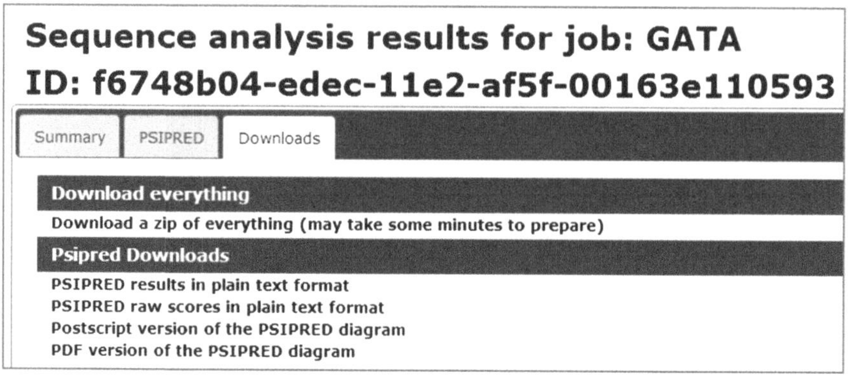


**Fig. 5.11** Downloading the PSIPRED results

**Exercise**

1. What are the major secondary structure elements?
2. What are *phi*, *psi* and *omega* angles?
3. What is the importance of predicting secondary structure?
4. How PSIPRED predicts the secondary structure of a protein?
5. Which portion of a peptide is rigid and flexible?
6. Perform a secondary structure analysis for the protein with UniProt id O81360 (used in last exercise) and compare the results.

## 5.3 Protein Tertiary Structure Prediction by Homology Modelling

**Introduction**
Even though there is a huge growth of training data, quality of prediction has not improved much. Also secondary structures are not completely determined by local forces and long-distance interactions do not appear in sliding window. Empirical studies have also shown that same amino acid sequences can assume multiple secondary structures. It is for this reason secondary structure prediction does not work always. Hence, predicting the tertiary structure becomes impossible.

**Structure Prediction**
The conventional techniques include X-ray crystallography and NMR. These experimental methods are time-consuming and have limitations in their approach. Currently, it takes 1–3 years to solve a protein structure. Moreover, certain proteins like membrane proteins and soluble proteins are extremely difficult to solve by X-ray or NMR techniques.

**The Need**

There are many important proteins for which the sequence information is available, but their three-dimensional structures remain unknown. The full understanding of the biological roles of these proteins requires knowledge of their structures. Hence, the lack of such information hinders many aspects of the analysis, ranging from protein function to ligand binding mechanisms in enzyme catalysis. It may be useful for the rational design of biochemical experiments like site-directed mutagenesis, protein stability or functional analysis. In addition to serving as a theoretical guide to design experiments for protein characterization, the model can help to rationalize the experimental results obtained with the protein of interest. In short, the modelling study helps to advance our understanding of protein functions.

**Computational Methods**

There are two approaches for predicting three-dimensional and structural modelling of protein namely knowledge-based and ab initio methods. The former predicts protein structures based on information of protein structural information that exists in databases. It is of two types: The **Homology Modelling** which builds models based on an experimentally determined structure that is closely related at the sequence level and **Threading** which identifies proteins that are structurally similar, with or without detectable sequence similarities. The later, **Ab initio** method is simulation-based and predicts structures based on physicochemical principles governing protein folding without the use of structural templates (Jin 2006).

**Homology Modelling**

Homology modelling, also known as comparative modelling, predicts protein structures of user-defined sequence (**Target**) based on sequence homology with known structures (**Templates**). If two proteins are expected to share a high enough sequence similarity, then they are likely to have very similar three-dimensional structures. If one of the protein sequences has a known structure, then the structural information can be copied to the unknown protein with a high degree of confidence.

**Steps in Homology Modelling**

The overall homology modelling procedure consists of six steps (Fig. 5.12).

The first step is template selection. This involves recognition of homologous sequences in the protein structure database to be used as templates for modelling. As a rule of thumb, the template protein should have at a minimum of 30% sequence identity with the query sequence. When multiple database structures with significant similarity are found as a result of the search, it is recommended to use structure(s) with the highest percentage (sequence) identity and highest resolution as a template. If there is a situation in which no highly similar sequences are found in the structure database, template selection becomes difficult and one should use either a more sensitive profile-based PSI-BLAST method or threading (fold recognition method) to identify distant homologs.

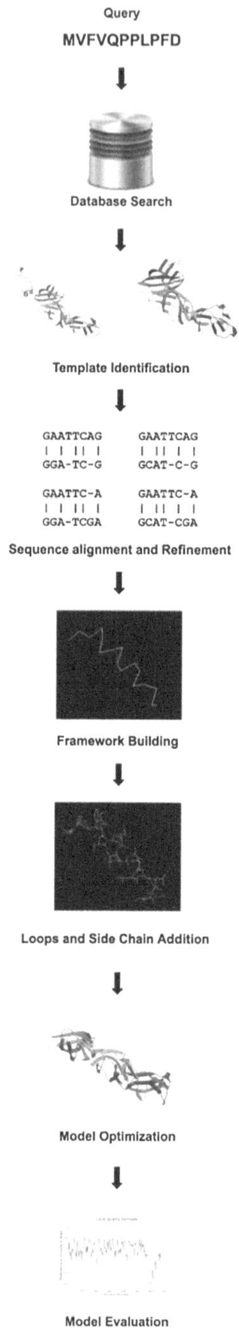


**Fig. 5.12** Steps involved in homology modelling

The second step is the alignment of the target and template sequences. This is the most critical step in homology modelling, which directly affects the quality of the final model. Hence, one should use the best alignment program (that may not be error free) and visually inspect and refine manually the alignment to ensure that conserved key residues are correctly aligned.

The third step is building a framework structure for the target protein with main chain atoms. If aligned residues are identical, the side chain atoms' coordinates can also be copied along with the main chain atoms. If they differ, only the backbone atoms are copied. It is the simplest way to use only one template structure with the best quality and highest resolution. If multiple template structures are available for modelling, the template structures have to be optimally aligned and superimposed which allows the user to choose either average coordinate values of the templates or the best parts from each of the templates to model.

The fourth step is the addition and optimization of side chain atoms and loops. Recent programs use the concept of 'Rotamers' for side chain predicting. These are preferred side chain torsion angles extracted from other previously solved protein crystal structures. A rotamer library is a collection of preferred side chain conformations where the rotamers are ranked based on their frequency of occurrence. Thus, for predicting side chain conformation, possible rotamers having lowest interaction energy with nearby atoms are alone selected. While aligning sequence for modelling, insertions and deletions produce gaps. Closing these gaps requires loop modelling, which is also difficult and a major source of error in homology modelling. With no mature methods available, currently two approaches, the database searching method and ab initio methods, are in use. This fourth step is also a crucial step in homology modelling and poses error in modelling procedure.

The fifth step is refining and optimizing the entire model based on the energy criteria. Any structural irregularities like bond lengths, unfavourable bond angles, or even close atomic contacts are corrected by applying energy minimization procedure on the entire model. This actually moves the atoms so as to make the overall conformation with the lowest energy potential. The ultimate goal of 'Energy Minimization' is to ease steric collisions and strains without altering the overall structure significantly. However, using energy minimization requires utmost care as excessive energy minimization often disturbs the overall structure by moving residues away from their correct positions.

The final step involves evaluating the overall quality of the obtained model. This makes sure that the structural features of the model obey the physicochemical rules. Evaluation involves checking any anomalies in close contacts, $\phi$–$\psi$ angles, bond lengths, etc. If any structural irregularities are detected, the region is said to have errors and requires further refinement. Usually model building is iterated until satisfactory results are obtained.

## 5.3.1 Homology Modelling Using SwissModel

**SwissModel** offers three different types of modelling requests (*automated mode, alignment mode, project mode*) that differ in the amount of user intervention (Schwede et al. 2003). Two methods the automated and alignment are discussed below.

**Automated Mode**
The "automated mode" in **SwissModel is** the simplest homology modelling procedure that requires either UniProt accession code or just amino acid sequence as input and selects suitable templates automatically based on the BLAST E-value. This method requires **high target–template similarity and when they share** more than 50% of sequence identity, the models generated are highly reliable.

**Using SwissModel in Automated Mode**

1. SwissModel Automatic Modelling Mode can be accessed from http://swissmodel.expasy.org/.
2. Select '**Automated Mode**' from the menu (Fig. 5.13).
3. Provide an **email ID** so that the link to the result can be mailed to you once completed. This is optional when you wish not to close the page after submitting.
4. Provide an optional **title for the job**.
5. Enter **valid amino acid sequence** as single letter code or a UniProt accession code.
6. '**Advanced options**' allows users to choose a different template other than the ranked one in the automated process specifying the PDB-ID along with the

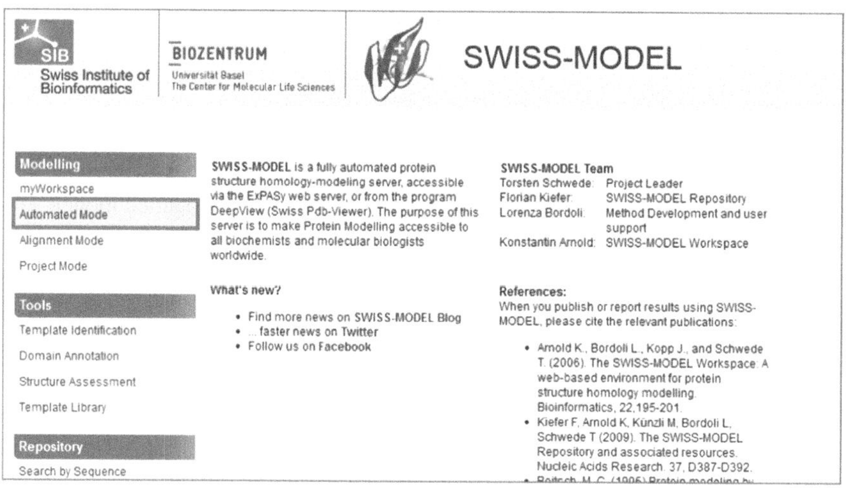


**Fig. 5.13** SwissModel server at ExPAsy

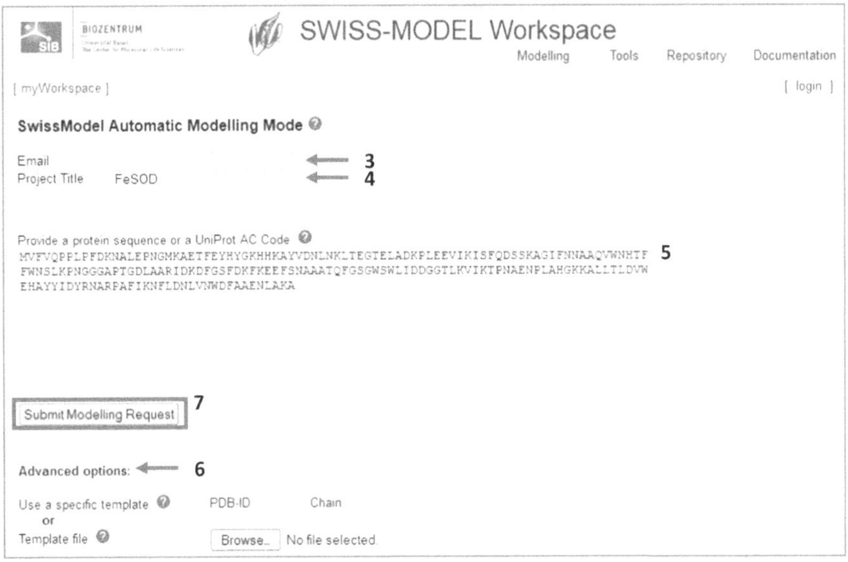


**Fig. 5.14** SwissModel automatic modelling mode interface at ExPAsy

ChainID or by uploading a file in PDB format. But one should make sure that the files do not have any ligands or modified amino acids, other than a single protein chain.

7. Click '**Submit Modeling Request**' button (Fig. 5.14) to initiate modelling.

**Using SwissModel Alignment Mode**

The "**alignment mode**" on the other hand allows the user to use alternative alignments so as to get an optimal result.

1. SwissModel Alignment mode can be accessed from http://swissmodel.expasy.org/.
2. Click on the '**Alignment Mode**' to do modelling under alignment mode (Fig. 5.15).
3. Provide an email ID so that the link to the result can be mailed to you once completed. This is optional when you wished not to close the page after submitting.
4. Provide an optional title for the job.
5. Choose the appropriate alignment format from the list (FASTA, MSF, CLUSTALW, PFAM and SELEX).
6. Prepare a multiple sequence alignment, preferably with T-COFFEE, with the target and template sequences.
7. Copy and paste the target–template alignment in the box.
8. Alternatively you can also upload the alignment file by selecting the alignment file by clicking '**Browse**' button (Fig. 5.16).

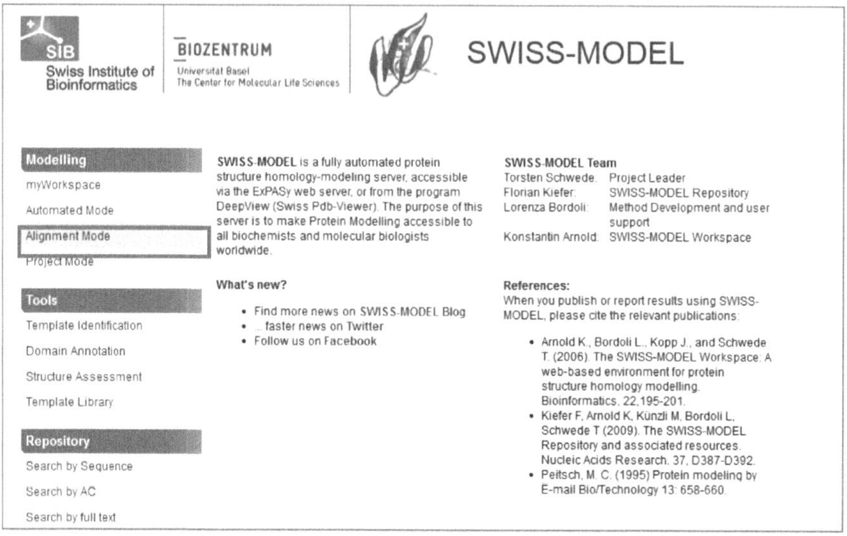


**Fig. 5.15** SwissModel server at ExPAsy

9. The alignment made by the server is displayed at the bottom part of the page. Select the target sequence name, the template sequence and specify the correct chain ID.
10. The alignment now represents the correct mapping of the template structure on the target sequence. Make sure to check before submission.
11. Click '**Submit Alignment**' button (Fig. 5.16) to initiate modelling.

**Interpreting Results**

The results containing the coordinates of the model, the alignment, modelling log files and evaluations made on the quality can be accessed and downloaded via web browser from the workspace (Fig. 5.17).

**Model Summary**

1. The result page provides information on template(s) used, its resolution and residue range that is being modelled, sequence identity with the target are provided.
2. The hyperlinked PDB ID (**1my6B**) provides cross references to structural information databases.
3. The modelled structure can be downloaded as PDB format for visualization with other molecular visualization tool. There is also a provision to download the DeepView project files for further manipulation.

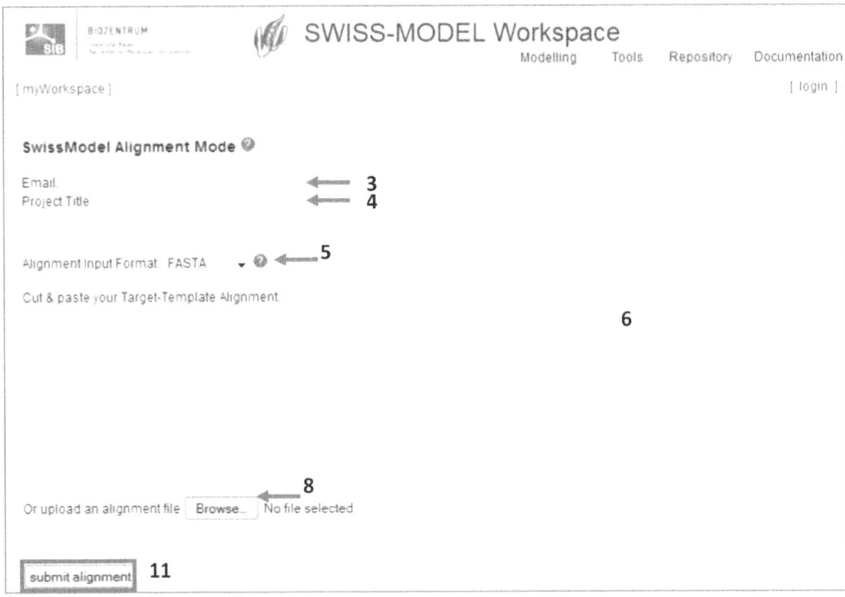


**Fig. 5.16** SwissModel alignment mode interface at ExPAsy

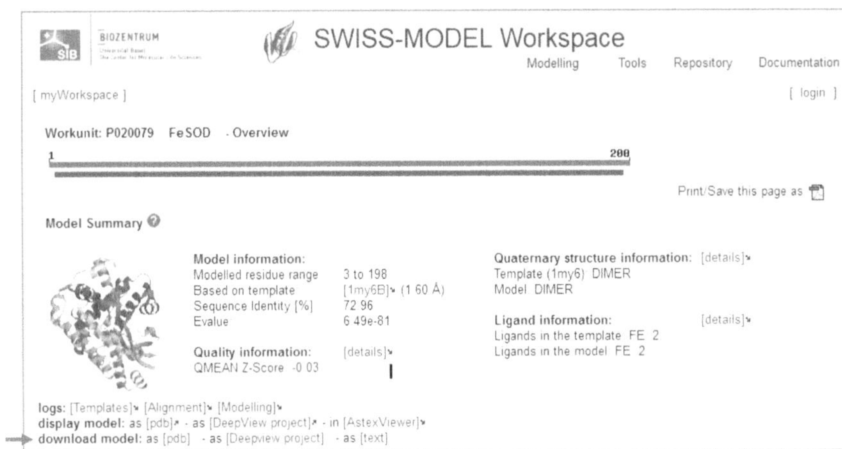


**Fig. 5.17** Result page displaying model information summary

## Protein Structure and Model Assessment—Global and Local

4. Swiss model represents both global and local quality information for the modelled protein (Figs. 5.18, 5.19 and 5.20). Automated SWISS-MODEL pipeline evaluates by the EVA project continuously. The local quality of the

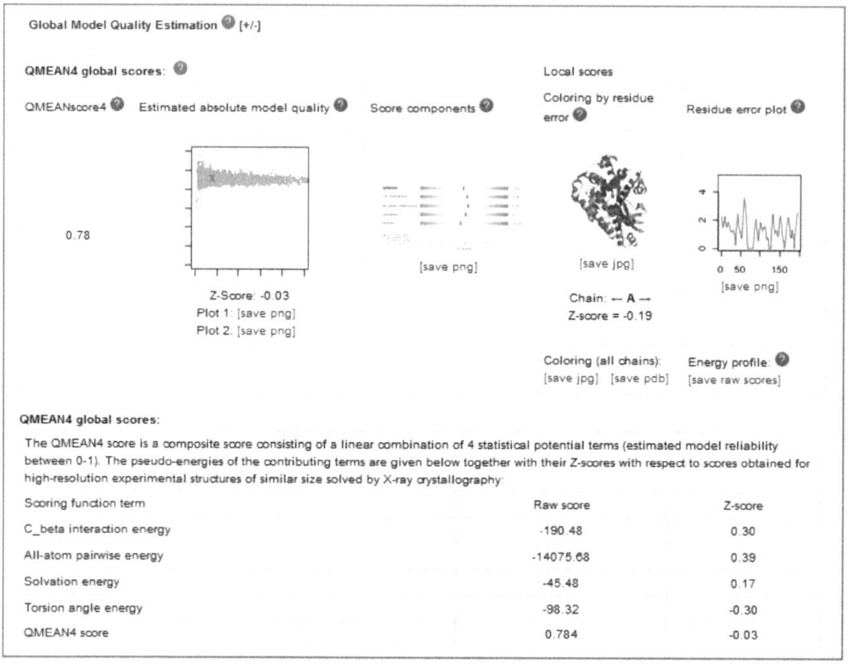


**Fig. 5.18** Global model quality estimation

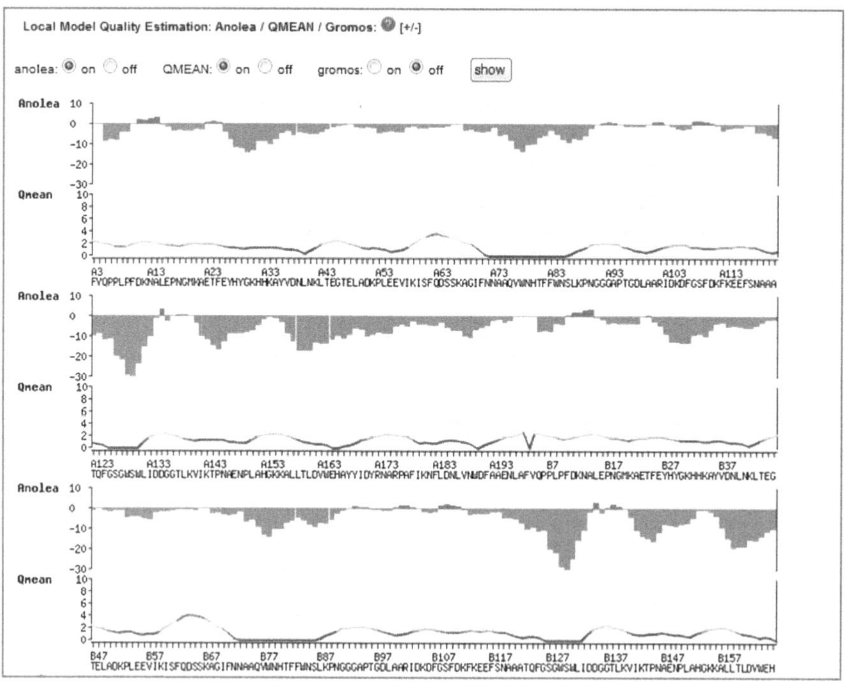


**Fig. 5.19** Local model quality estimation

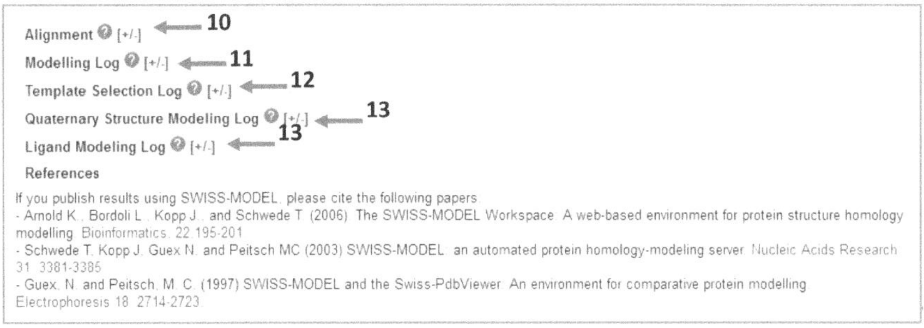


**Fig. 5.20** Other information list as tabs

predicted structure is depicted by means of graphical plots of ANOLEA mean force potential and GROMOS empirical force field energy.

### Anolea

5. Packing quality of the models is assessed by atomic empirical mean force potential ANOLEA that performs energy calculations on a protein chain and evaluates the "Non-Local Environment" (NLE) of each heavy atom in the molecule (Melo and Feytmans 1998).
6. The energy for each amino acid of the protein chain is represented in the plot's Y-axis.
7. Green bar symbolizes favourable energy environment (negative energy) and red bar represents unfavourable energy (positive values) for a given amino acid.

### QMEAN

8. QMEAN estimates the global model quality.
9. **It describes t**he local geometry that is analysed by a torsion angle potential over three consecutive amino acids, a solvation potential describing the burial status of the residues, a distance-dependent pairwise residue-level potential for assessing long-range interactions and two simple terms describing the agreement of predicted and calculated secondary structure and solvent accessibility (Benkert et al. 2008).
10. Besides that, QMEAN also provides a pseudo energy of the whole model. This can be used to compare and rank alternative models of the same target. When the predicted energy is lower the, the model is better.

### GROMOS

11. GROMOS is a molecular dynamics computer simulation package used for the analysing conformations obtained either by experiment or by computer simulation (van Gunsteren et al. 1996). The y-axis of the plot represents the energy

for each amino acid of the protein chain. Green bar represent favourable energy environment (negative energy) and red bar represent unfavourable energy (positive values) for a given amino acid.

### Other Information

12. Other information like alignment, modelling log, template selection log, etc. are listed at the bottom of the page. The users can click the [+/−] symbol to view the details.

### Alignment Output

13. The sequence alignment between target and template made in the modelling procedure and the assigned secondary structure is displayed under 'Alignment' (Fig. 5.21).

### Modeling Log

14. The log depicts every individual step like which parts of the model have been built by ab initio during the modelling procedure using SWISS-MODEL server pipeline.

```
Alignment [+/-]

TARGET    1       FVQPPLP FDKNALEPNG MKAETFEYHY GKHHKAYVDN LNKLTEGTEL
1my6_1    1    a--fvqeplp fdpgalepyg msaktlefhy gkhhkgyvdn lnkltqdtel

TARGET                  h          hhhhhhhh hhhhhhhhhh hhhh      h
1my6_1                  h          hhhhhhhh hhhhhhhhhh hhhh      h


TARGET    48   ADKPLEEVIK ISFQDSSKAG IFNNAAQVWN HTFFWNSLKP NGGGAPTGDL
1my6_1    49   adksledvir ttygdaakvg ifnnaaqvwn htffwnslkp ggggvptgdv

TARGET         h   hhhhhh hh       hh hhhhhhhhhh hhhhhhsss         hh
1my6_1         h   hhhhhh hh       hh hhhhhhhhh    hhhhsss         hh


TARGET    98   AARIDKDFGS FDKFKEEFSN AAATQFGSGW SWLIDDGGTL KVIKTPNAEN
1my6_1    99   aarinsafgs ydefkaqfkn aaatqfgsgw awlvleagtl kvtktanaen

TARGET         hhhhhhh    hhhhhhhhhh hhhh    sss ssssss  ss sssssss
1my6_1         hhhhhhh    hhhhhhhhhh hhhh    sss ssssss  ss sssssss


TARGET    148  PLAHGKKALL TLDVWEHAYY IDYRNARPAF IKNFLDNLVN WDFAAENLA
1my6_1    149  plvhgqvpll tidvwehayy ldyqnrrpdf idnflnqlvn wdfvaknla-

TARGET              ssss ssss             hhhh hhhhhh sss  hhhhhhhh
1my6_1              ssss ssss             hhhh hhhhhh sss  hhhhhhhh


TARGET    197  --- FVQPPL PFDKNALEPN GMKAETFEYH YGKHHKAYVD NLNKLTEGTE
1my6_1    198  aa--fvqepl pfdpgalepy gmsaktlefh ygkhhkgyvd nlnkltqdte
                 *
TARGET                  h         hhhhhhh hhhhhhhhhh hhhhh
1my6_1                  h         hhhhhhh hhhhhhhhhh hhhhh
```

**Fig. 5.21** Target–template sequence alignment and the assigned secondary structure

**Template Selection Log**

12. Steps involved in searching the SWISS-MODEL Template library for suitable templates are provided by Template selection log.
13. If the modelled protein has on Quaternary Structure and Ligand information, the **Quaternary Structure Modeling** Log and **Ligand Modeling Log** will be also be listed (Fig. 5.20).

Homology modelling was performed using the templates identified on the basis of significant sequence similarity. PDB ID **1my6** chain B (1.60 Å) (Kerfeld et al. 2003) with a sequence similarity of 72.96% was used as the template by SWISS-MODEL server. The predicted structure consisted of 13 α-helices and 3 β-strands. This can be determined by visualizing the downloaded pdb in structure visualization tools like RasMol. The modelled FeSOD structure resembles the previously described eubacterial FeSOD structures in overall fold (Perry et al. 1010). The structure can be then further analysed using Ramachandran plot.

**Exercise**

1. What is homology modelling?
2. What are the steps involved in homology modelling?
3. What is meant by 'Target' and 'Template'?
4. On what condition one can go for homology modelling?
5. What is meant by 'Structure analysis' and why it is important?
6. What is 'Ramachandarn Plot'?
7. Model the structure for the proteins with Uniprot id **O81360** and **Q40412** and compare the structural elements.

### *5.3.2 Protein Tertiary Structure Prediction by Threading (Fold Recognition)*

**Introduction**

Threading predicts structure of an unknown protein sequence by selecting the best-fitting fold in a structural database. Though there are millions of protein sequences, there are only lees than thousand protein folds available which show that protein structures tend to be more conserved than protein sequences. Hence there is high probability that many proteins can still share a similar fold even if they lack sequence similarities. It is this that made computational methods to predict protein even without sequence similarities. Threading becomes handy in a situation where no highly similar sequences are found in the structure database and template selection becomes difficult to identify distant homologs. The method takes protein sequence for which folds are predicted in a fold library which is then used for

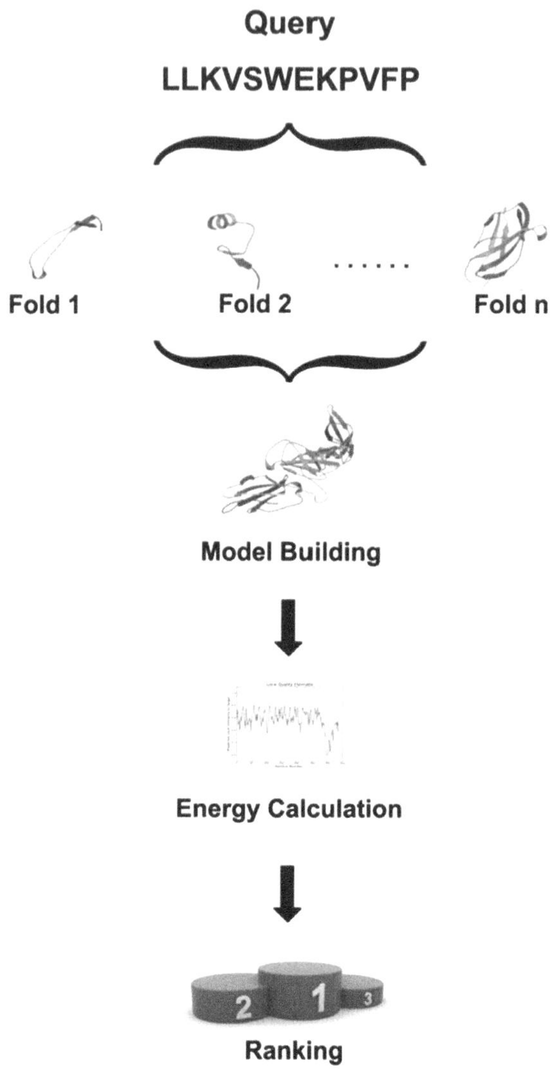


**Fig. 5.22** Threading workflow

building model. The models built are then energy minimized and scored (Fig. 5.22).

The threading algorithms can be pairwise energy-based or profile-based, where the former is originally referred to as threading and latter as fold recognition.

**Using Phyre2**

Phyre2 a new system that is designed to predict three-dimensional (3D) structure for the protein sequence, uses advanced remote homology detection methods for building 3D models and predicting ligand binding sites. It also analyses the effect of amino acid variants [e.g. nonsynonymous SNPs (nsSNPs)] for a user provided protein sequence. The older version, Phyre, used a profile-profile alignment algorithm, whereas Phyre2 uses the alignment of hidden Markov models via HH search to significantly improve accuracy of alignment and detection rate (Kelley et al. 2015).

Phyre2 detects sequence homologues with PSI-BLAST and predicts secondary structure and disorder with PSIPRED and DISOPRED and constructs a hidden Markov model based on the homologues detected. This is scanned against a library of HMMs of proteins and 3D models are constructed based on the alignments. Insertions and deletions are modelled using a loop library and a set of empirical energy terms and the amino acid side chains using a rotamer library from Roland Dunbrack's laboratory and in-house approach to avoid steric clashes. Binding site prediction is done by 3DLigandSite and Transmembrane helix and topology prediction is made by memsat-svm (Kelley et al. 2015).

As Phyre2 uses remote homology detection techniques such as profile–profile and HMM–HMM matching, it can detect homology even when the sequence identity between template and target is **as low as 15%** with high confidence.

**Exploration Activity**

1. Phyre2 can accessed http://www.sbg.bio.ic.ac.uk/phyre2/html/page.cgi?id=index.
2. You can sign in for Phyre2 account if wish to model more proteins.
3. Click '**Login**' link and provide the ID and password details to login.
4. Or else, simply provide a valid email and an optional job title.
5. Copy and paste protein sequence for the protein to be modelled.
6. Let the '**Modeling mode**' be Normal (default).
7. Click '**Phyre Search**' button (Fig. 5.23).

**Interpreting Results**

8. The link to the will be emailed to the email address provided while modelling along with the modelled structure in pdb format. Clicking the link displays the results.
9. The first part is the header section that provides the information like email ID, job ID, date and time of submission. (Fig. 5.24).
10. It also provides a link to download the complete results as zipped (compressed) file.
11. It is then followed by the summary that talks about the template used, template information, confidence and coverage information, snap shot of the modelled structure along with the facility to view the modelled structure in JMol via link. Clicking the snapshot will allow us to download the modelled structure (Fig. 5.25).

**Secondary Structure and Disorder Prediction**

12. On average Phyre2 makes 78–80% accurate secondary structure and disorder prediction if substantial number of diverse sequence homologues detectable

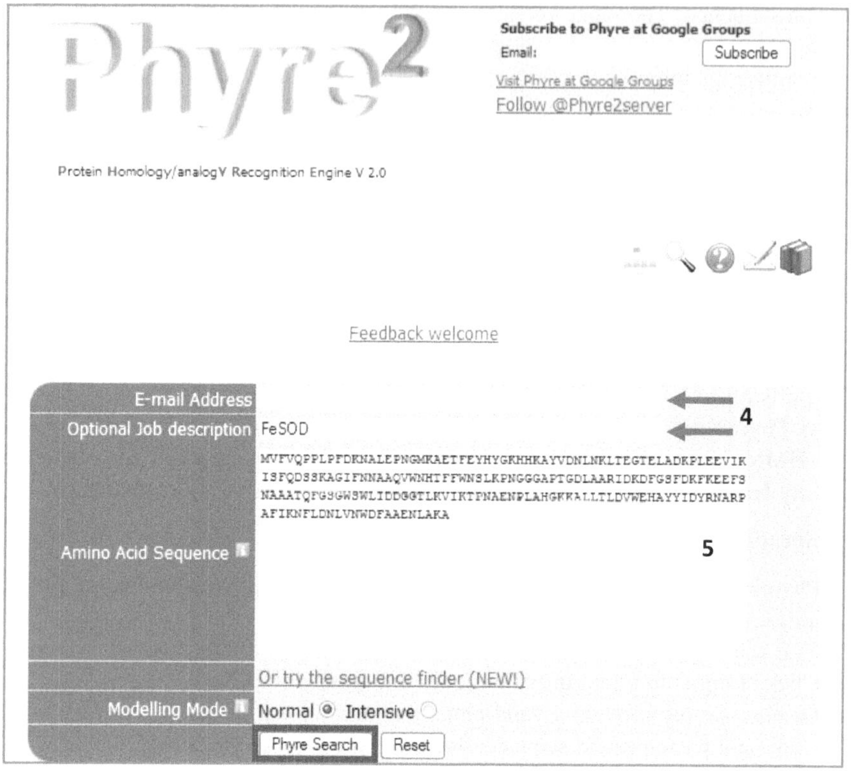


**Fig. 5.23** Phyre2 interface at Structural Bioinformatics Group, Imperial College, London

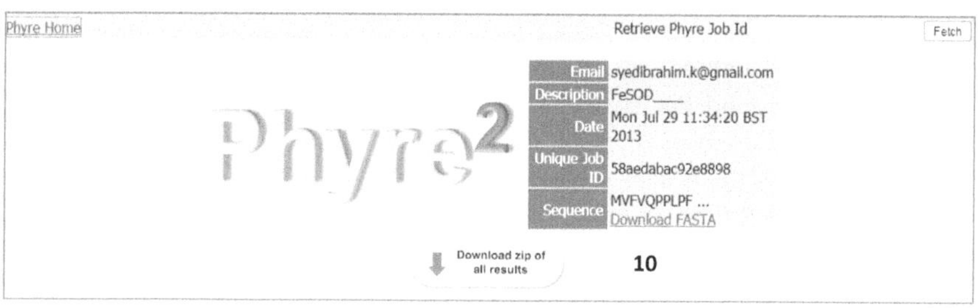


**Fig. 5.24** The header section of the Phyre2 results

from the sequence database. If very few homologues are detected, accuracy may fall to approximately 65%.

13. In the secondary structure and disorder prediction the residues coloured according is represented in '**sequence**', followed by '**secondary structure**' that shows either α-helix, β-strand or coil. **Green helices** shows α-helices, **blue arrows** shows β-strands and **faint lines** represents coil. The '**SS confidence**'

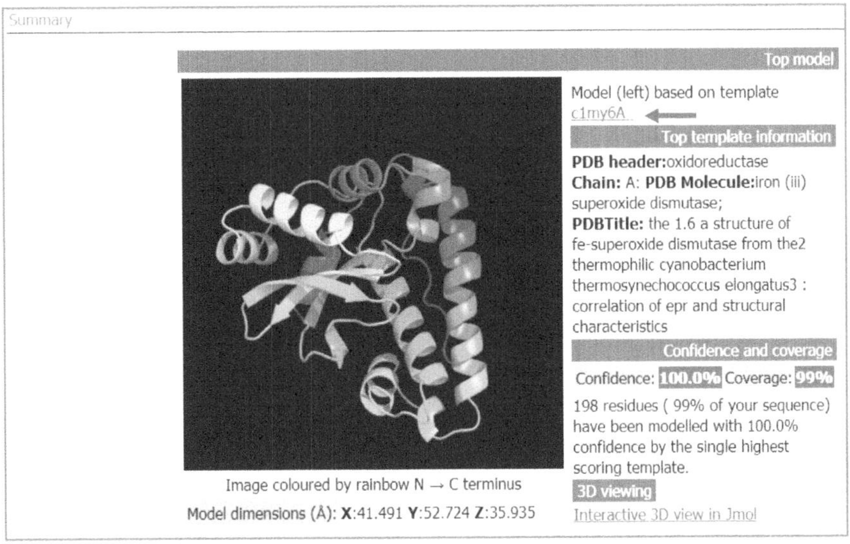


**Fig. 5.25** The result summary for *N. punctiforme* FeSO

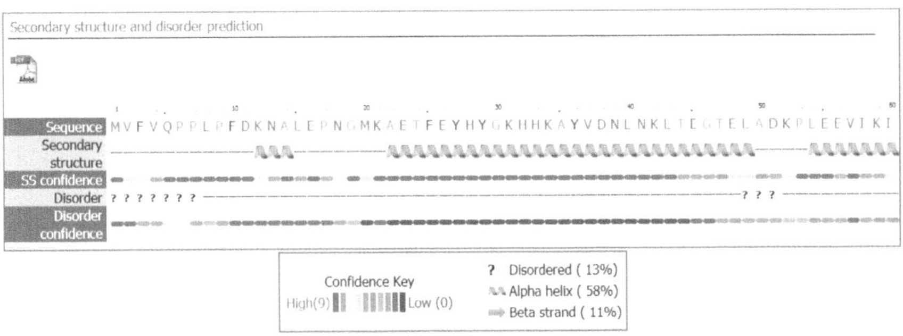


**Fig. 5.26** Secondary structure and disorder prediction

line depicts the confidence in prediction, the **red** being high confidence and **blue** being low. O**range**, **yellow or green** indicates a weaker structure prediction.

14. As the name suggests the '**Disorder**' line predicts disordered regions in the protein and are indicated by question marks (?). It can be noted that weakly predicted region normally coincides where there is relatively a strong prediction of disorder (Fig. 5.26).
15. Phyre2 does not individually predict β-turns, β-bends π-helices, or $3_{10}$-helices. Instead β-turns and β-bends are treated as coil and π-helices and $3_{10}$-helices are considered α-helices.

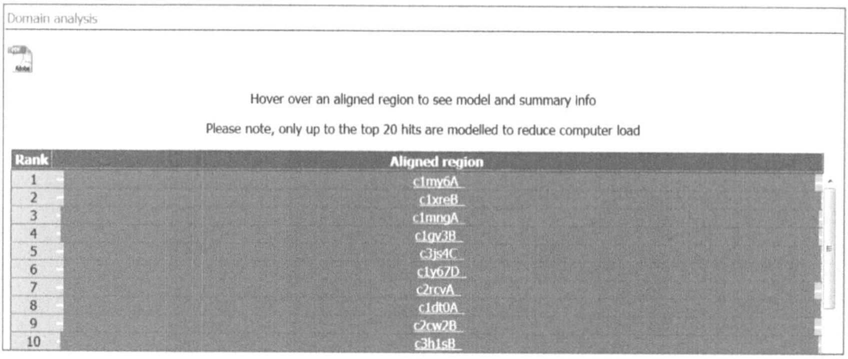


**Fig. 5.27** Domain analysis

## Domain Analysis

16. This section illustrates where sequence matches are found. The red indicates a higher confidence in match to the entire length of the protein to fold library having entry c1my6A. The 'c' in the entry indicates that this protein is taken as a whole chain from the PDB, with PDB bearing the identifier 1my6 having chain A (Fig. 5.27).
17. The colour changes like oranges and green (not shown) indicate a drop in confidence with the match. Phyre2 has detected these matches but not used for modelling.

## Detailed Template Information (and Models)

18. This section displays a table containing information on the template including alignment coverage, model snapshot, confidence score, percent of identity and template information (Fig. 5.28).

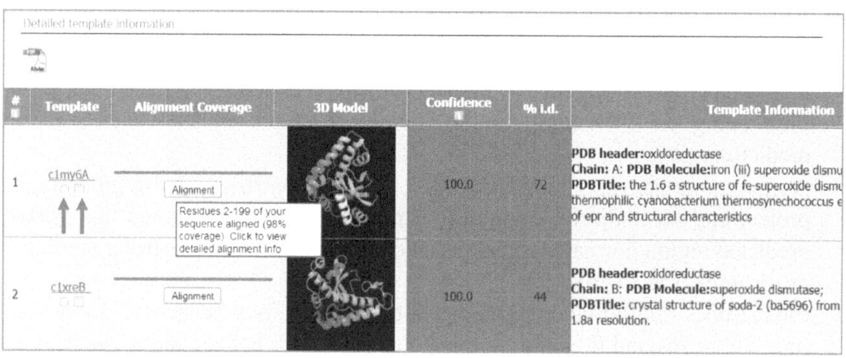


**Fig. 5.28** Detailed template information

19. The matches are ranked by a raw alignment score which in turn is based on the similarity, distribution probability, secondary structure similarity and insertion and deletions' presence or absence.
20. Below the template ID is a radio button and a check box. Selecting the radio button allow us to select the model as the master in the multiple alignment and selecting the check box indicate the selected model to be aligned with the master.
21. Hovering over '**Alignment**' button provides information about the position and percentage of the alignment. Clicking it to see the detailed alignment information.
22. To download the coordinates of the model in PDB format click the picture.
23. The '**Confidence**' DOES NOT represent the expected accuracy of the model rather it represents the probability (from 0 to 100) that the match between your sequence and this template is a true homology. Over 90% confidence indicates that the protein is modelled at high accuracy (2–4 Å **RMSD** from native, true structure).
24. The last column provides **information like from where the template** was taken, say SCOP or the PDB and along with information on the potential function of your protein.
25. Superposition of structures can be obtained by clicking button "**Generate superposition of selected models**" (Fig. 5.29) by selecting the tick bock present below the ID.
26. Superimposed structures are displayed in JMOL window (requires Java) and also permits to download the superposed structures by clicking the link '**Download superposition**' (Fig. 5.30).
27. This is helpful in identifying which regions of the models agree and disagree and which regions of the model are reliable and which regions require cautious handling.

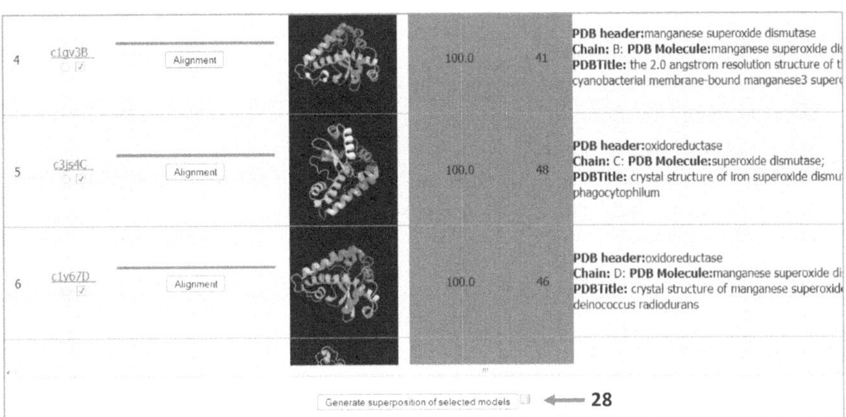


Fig. 5.29 Superposing selected structures

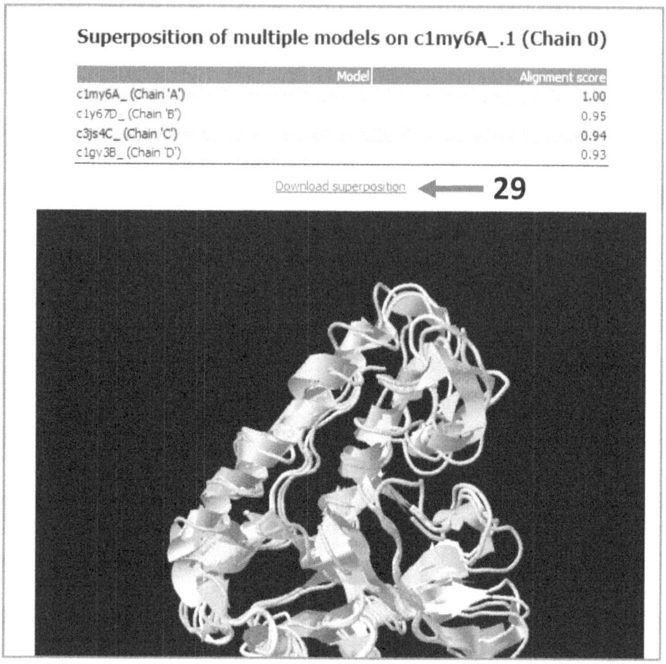


**Fig. 5.30** JMol window showing superposed structures

## Ligand and Transmembrane Helix Prediction

28. If the submitted sequence has ligand binding sites or transmembrane regions phyre2 submits to **3DLigandsite server** and **memsat-svm server** respectively to identify the potential ligand binding sites and transmembrane regions.

## Exercise

1. What is fold recognition?
2. On what condition one can go for fold recognition?
3. What is meant by domain?
4. What is meant by coverage?
5. Model the structure for the proteins with Uniprot id **O81360** and **Q40412** (used in last exercise) and compare the results.

## 5.4 Protein Tertiary Structure Analysis

### 5.4.1 *RAMPAGE*

**Introduction**

Geometrical structure validation generally revolves around three major components: backbone conformation, side chain conformation and Cα geometry. The Ramachandran plot is the representation of the phi–psi torsion angles for all residues in the structure but not those at the end of the chain which show either empirical scatter plots of the conformations observed in the database of known 3D structures, or contours of calculated energies, or steric criteria as a function of and for a dipeptide (Ramachandran et al. 1963). Glycine residues are plotted as triangles as they not restricted to the regions of the plot appropriate to the other side chain types. Since refinement process does not optimize the phi–psi values, phi–psi plots for individual proteins have also become central for structure validation in recent years.

**Exploration Activity**

1. RAMPAGE can be accessed from http://mordred.bioc.cam.ac.uk/~rapper/rampage.php.
2. The structure to be analysed is uploaded as in PDB format (FeSOD structure earlier modelled with Phyre2). The file size is limited to 20 MB (Fig. 5.31).
3. Click '**Submit to RAMPAGE**' button for analysis.

**Interpreting Results**

4. The result page displays the file name, links to download the images in PDF and postscript formats. This is followed by the graphical output displaying the Ramachandran plot (Fig. 5.32).

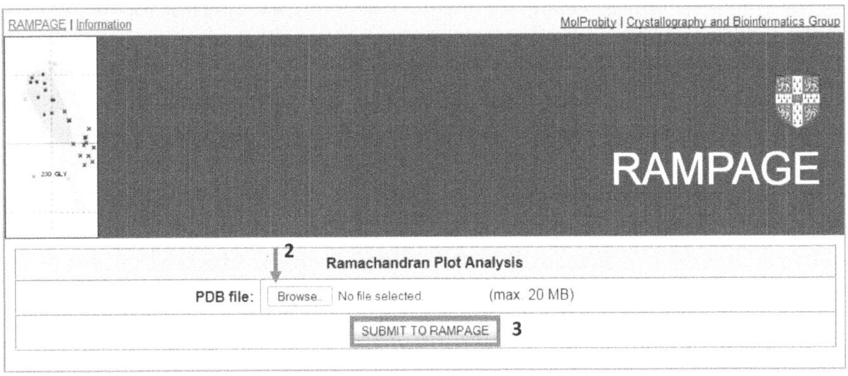


**Fig. 5.31** RAMPAGE interface by Crystallography and Biocomputing Group at Cambridge University

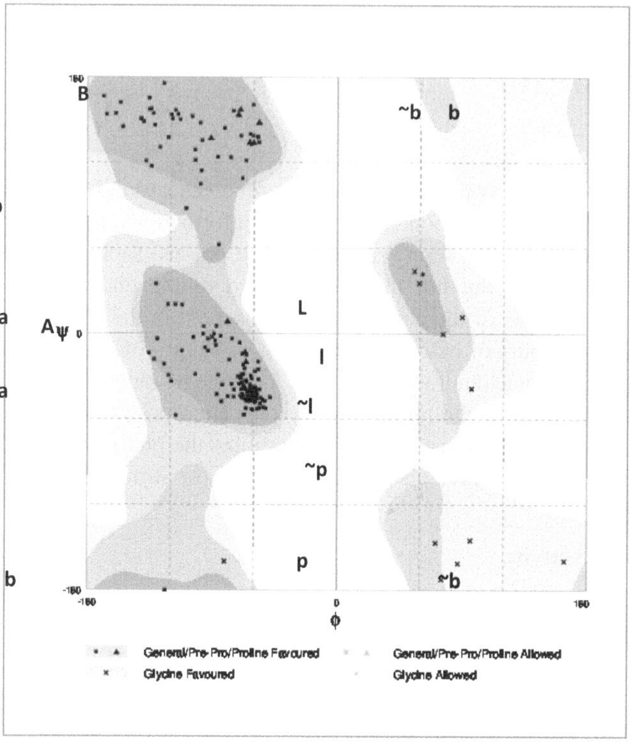


**Fig. 5.32** Ramachandran plot for the submitted protein structure

5. It is followed by the information on Glycine, Proline and Pre-Proline falling regions (Fig. 5.33).
6. It is followed by the information like the number of residues in favoured, allowed and outlier regions (Fig. 5.34).
7. The colouring or the shading on the plot represents the different regions like the **darkest** areas corresponding to core regions and it also represents the most favourable combinations of phi–psi values.
8. Having over 90% of the residues in the core regions is always preferred. This percentage serves as a better guide to stereochemical quality.

**Ramachandran Plot Regions**

The different regions (Table 5.5) on the Ramachandran plot are as described by Morris et al. (1992). These regions were taken from the observed phi-psi distribution for 121,870 residues from 463 known X-ray protein structures.

The two most favoured regions, 'core' and 'allowed' regions correspond to $10° \times 10°$ pixels with more than 100 and 8 residues in them, respectively. Morris et al. (1992) defined 'generous' regions by extending out by 20° (two pixels) all round the 'allowed' regions.

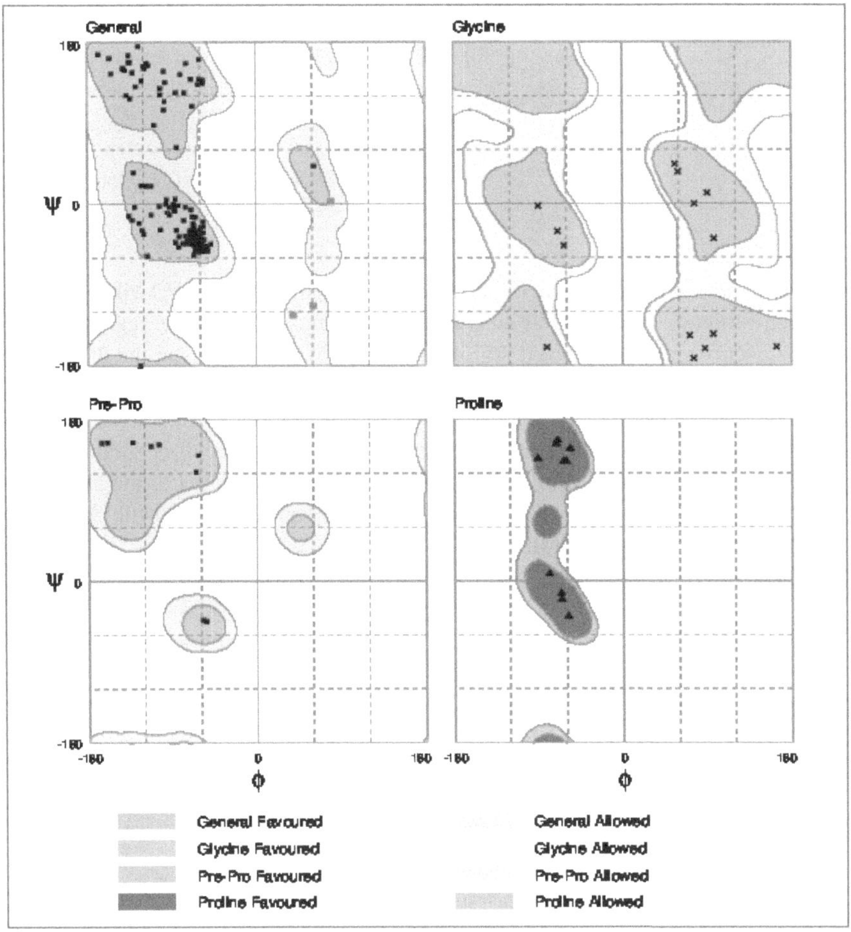


**Fig. 5.33** Glycine, Proline and Pre-Proline falling regions in Ramachandran plot

## Evaluation of residues

```
Residue [   63 :GLN] (  79.82,    2.53) in Allowed region
Residue [  146 :ASN] (  60.85,-113.49) in Allowed region
Residue [  173 :ARG] (  39.42,-123.99) in Allowed region
Number of residues in favoured region    (~98.0% expected)  :  193 ( 98.5%)
Number of residues in allowed region     ( ~2.0% expected)  :    3 (  1.5%)
Number of residues in outlier region                        :    0 (  0.0%)
```

**Fig. 5.34** Residues in favoured and allowed regions in Ramachandran plot

Table 5.5 The regions in Ramachandran plot

| | | | |
|---|---|---|---|
| A | Core alpha | L | Core left-handed alpha |
| a | Allowed alpha | l | Allowed left-handed alpha |
| ~a | Generous alpha | ~l | Generous left-handed alpha |
| B | Core beta | p | Allowed epsilon |
| b | Allowed beta | ~p | Generous epsilon |
| ~b | Generous beta | | |

The predicted models from Phyre2 were assessed using the RAMPAGE. The modelled structures had ~98% residues in the most favourable regions, 2.0% residues occurring in the allowed regions and none of the residues in the disallowed regions (Fig. 5.34).

In the model, all residues except Asn146 and Arg173 fall within allowed regions of the Ramachandran plot due to the deviation from the allowed Φ–Ψ angles. These residues form parts of tight turns. Similar distorted conformations of the analogous residues are observed in the *E. coli* FeSOD (Lah et al. 1995) and the Anabaena MnSOD (Atzenhofer et al. 2002).

**Exercise**

1. For the models you generated in last two exercises (for proteins **O81360** and **Q40412**) using homology modelling and Phyre2 and compare the structure analysis results.

## *5.4.2 Protein Structure Analysis Using SAVeS*

**Introduction**

SAVeS (Structure Analysis and Verification Server) is a metaserver for checking and validating protein structures during and after model refinement. It contains the following structure analysis tools.

**PROCHECK**

This analyses overall geometry residue-by-residue and generates stereochemical quality of a protein as postscript files (Laskowski et al. 1993).

**WHAT_CHECK**

Does extensive inspection of stereochemical parameters of the residues in the model (Hooft et al. 1996) and is basically derived from protein verification tools available in the WHATIF program (Vriend 1990).

**ERRAT**

Non-bonded interactions between different atom types are statistically analyzed and the value of the error function versus position of a 9-residue sliding window is plotted by comparing highly refined structures (Colovos and Yeates 1993).

**VERIFY_3D**

From the amino acid sequence it predicts the compatibility of its 3D model by assigning a structural class based on its location and environment which is then compared to results of good structures (Bowie et al. 1991).

**PROVE**

The program PROVE (PROtein Volume Evaluation) calculates the volumes of atoms in macromolecules from PDB-deposited structures having better than 2.0 Å resolution and with R-factor of 0.2 or better (Pontius et al. 1996).

**Exploration Activity**

1. SAVeS server can be accessed from http://nihserver.mbi.ucla.edu/SAVES/.
2. The structure to be analysed is uploaded as in PDB format (FeSOD structure earlier modelled with Phyre2) by clicking '**Browse**' button to locate the file in your computer and 'Upload Files' to submit for evaluation (Fig. 5.35).
3. After uploading the file, the program asks to choose the programs to be added for analysis. A user can either run all programs or select individual programs.
4. Click '**Run all programs**' after selecting the desired programs (Fig. 5.36).

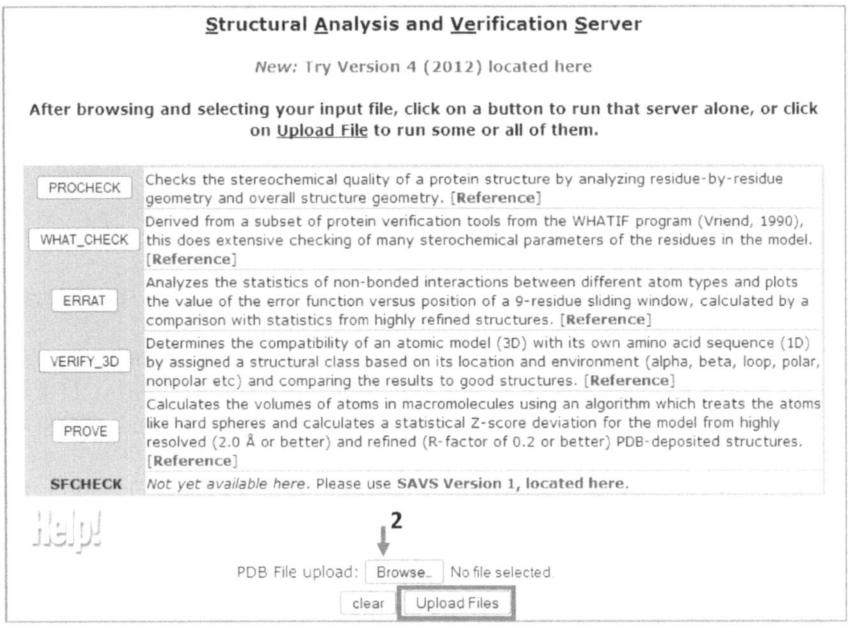


**Fig. 5.35** SAVeS interface of NIH MBI Laboratory for Structural Genomics and Proteomics at UCLA

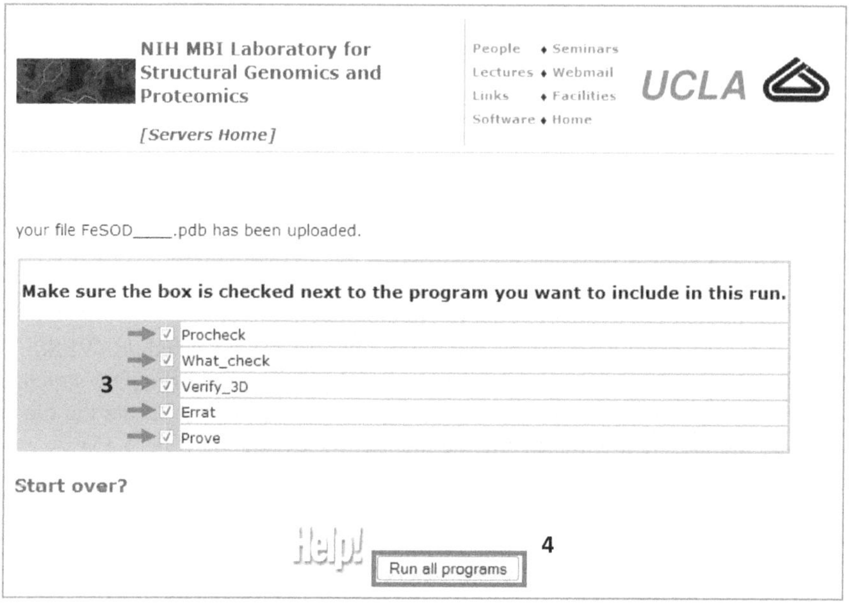


**Fig. 5.36** Programs available in SAVeS for analysis

### Interpretation of the Results

5. The program displays the results in table format. Links to download PROCHECK results in postscript, PDF or JPG is provided. Also provided are links to view results in more detailed form.
6. The first part deals with the PROCHECK results. This displays the Ramachandran plot analysis (Fig. 5.37).
7. The second part displays the results from What_Check, Verify_3D, Errat and Prove (Fig. 5.38).
8. After evaluation, the server shows the outputs of each program, which are labelled by a simple three-colour scheme Green (GOOD: SATISFACTORY), Yellow (WARNING: INSPECTION SUGGESTED) and Red (ERROR: INSPECTION RECOMMENDED).
9. For Errat (the data plot produced for each chain in the input structure), the average 3D-1D score of the residues should be >95%.
10. Verify 3D uses the averaged data points produced for each amino acid in the sequence, the number of times the value is greater than 0.2 is converted into the percentage of the sequence that has positions with values >0.2. If this percentage is <20%, it indicates Good. If it is 20–45%, it indicates Warning and if $\geq 45\%$, it indicates Error (Colovos and Yeates 1993).

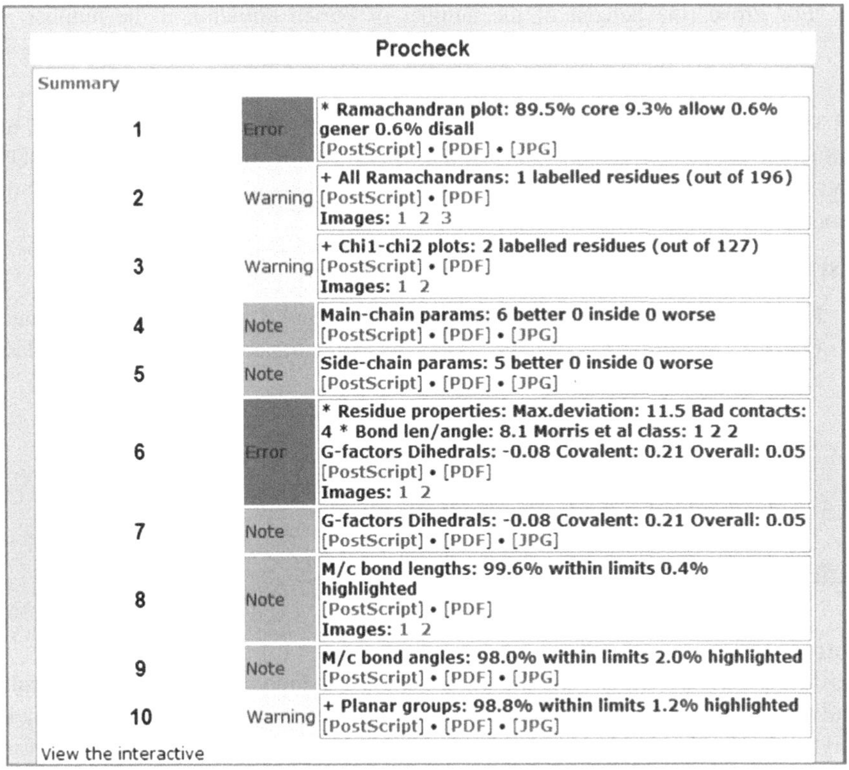


**Fig. 5.37** Procheck results

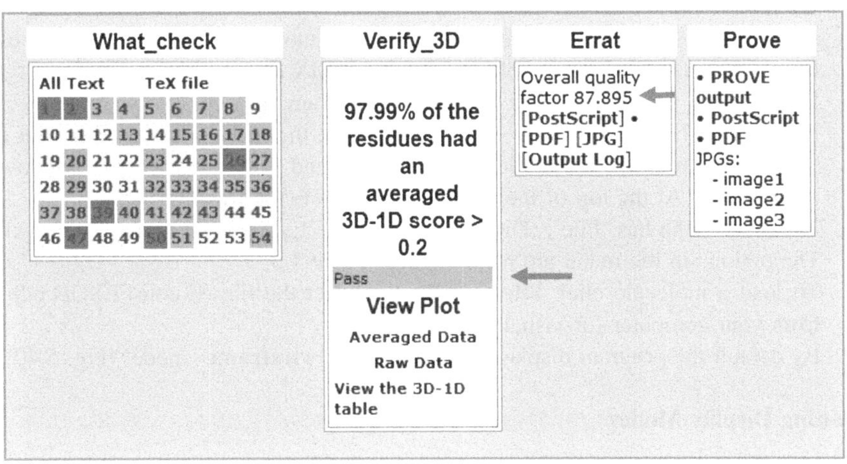


**Fig. 5.38** Results from What_Check, Verify_3D, Errat and Prove

11. For Prove (the percent of the number of buried outliers), if the number is <1.0%, it indicates Good. If it is 1–5%, it indicates Warning and if >5%, it indicates Error.

The modelled structures when validated by VERIFY3D showed that 97.99% of the amino acids have scored $\geq 0.2$ in the 3D/1D profile, the Errat score of 87.895 and Prove score of 0.116 (buried atoms 22) which also indicated a good quality of modelled structure.

**Exercise**

1. For the models you generated in earlier exercises (for proteins **O81360** and **Q40412**) perform the structure analysis and compare results with the last exercise.

## 5.5 Protein Structure Visualization

### *5.5.1 RasMol*

**Introduction**

RasMol is an important scientific tool for the visualisation of proteins, nucleic acids and small molecules and to prepare publication-quality images was created by Roger Sayle in 1992. This program can be operated by point and click method using mouse or by command line interface. But the former has limited options. Hence more controlled operations can be done only with the command line interface.

**Exploration Activity**

1. RasMol is available to download from http://rasmol.org for different platforms like Microsoft Windows, Apple Macintosh, UNIX etc.
2. Double click program icon to activate the program.
3. RasMol displays two windows on all platforms, the main graphics area with a black background (Fig. 5.39a) and a command line or terminal window (Fig. 5.39b). At the top of the graphics window is the RasMol menu bar.
4. The main menu has 'File', 'Display', 'Colours', 'Export', 'Options', 'Settings'. The options in the menu are mentioned in Table 5.6.
5. To load a molecule click **File > Open**.... Select the file (E-coli-FESOD.pdb) from your computer for visualization.
6. By default the program displays the content as '**wireframe**' model (Fig. 5.40).

**Setting Display Modes**

7. The display modes of a given object can be altered clicking '**Display**' option in the menu bar and selecting the options like wireframe or ball and stick (Fig. 5.41).

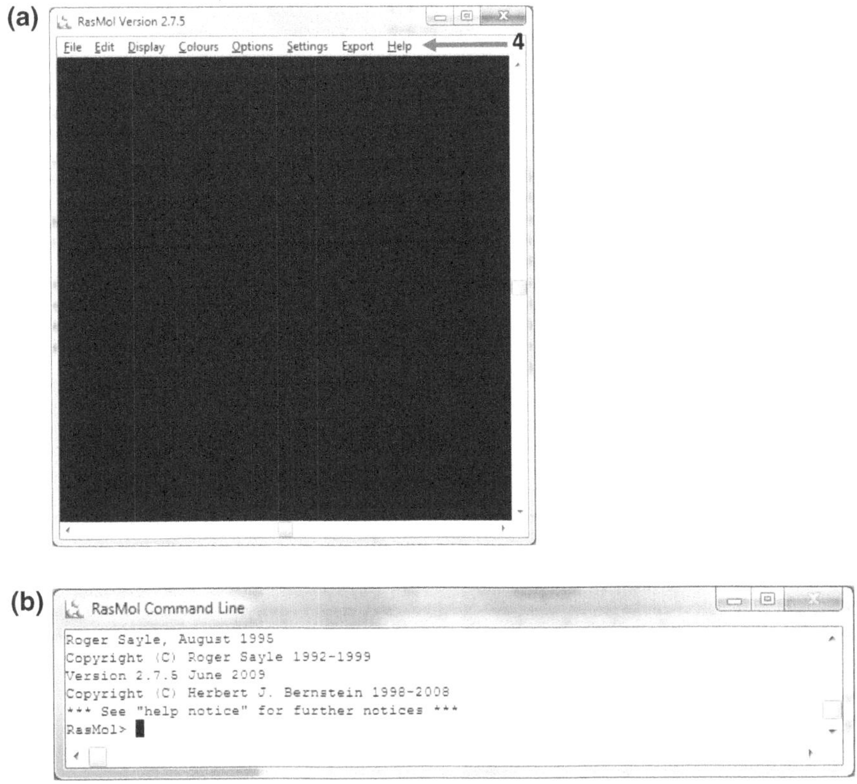


**Fig. 5.39** The RasMol program. **a**) RasMol terminal window. **b**) RasMol command line

**Table 5.6** The options in the menu bar of the RasMol program

| File | Edit | Display | Colours | Options | Export |
|---|---|---|---|---|---|
| Open… | Select | Wireframe | Monochrome | Slab mode | BMP… |
| Information | all | Backbone | CPK | Hydrogens | GIF… |
| Close | Cut | Sticks | Shapely | Heteroatoms | PostScript… |
| Print | Copy | Spacefill | Group | Specular | PPM… |
| Print setup | Paste | Ball and | Chain | Shadows | IRIS RGB… |
| Exit | Delete | stick | Temperature | Stereo | Sun Raster… |
| | | Ribbons | Structure | Labels | BMP… |
| | | Strands | User | | PICT… |
| | | Cartoons | Model | | Vector PS… |
| | | Molecular | Alt | | Molscript… |
| | | surface | | | Kinemage… |
| | | | | | POV Ray3… |
| | | | | | VRML… |
| | | | | | Ramachandran… |
| | | | | | Raster 3D… |
| | | | | | Rasmol Script… |

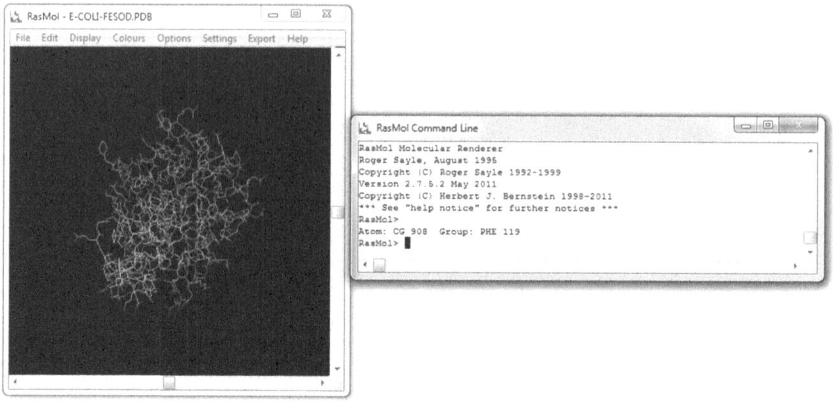


**Fig. 5.40** The wireframe representation by RasMol

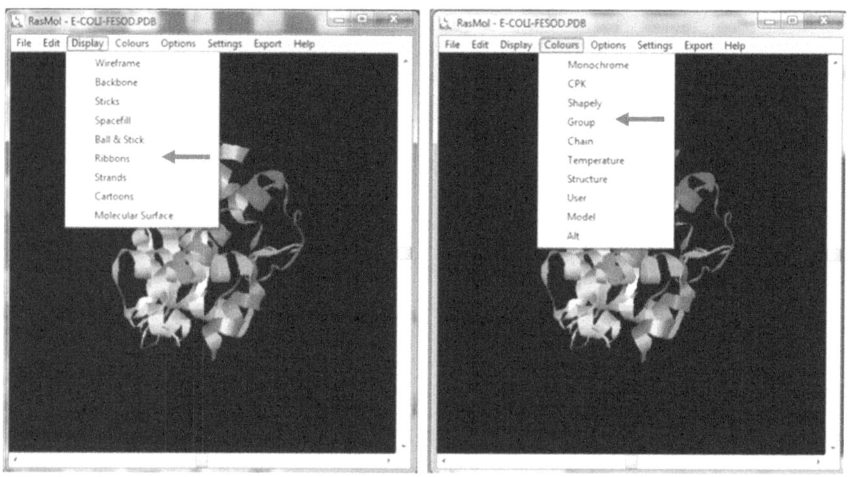


**Fig. 5.41** A ribbon representation with colour based on group

8. The display can be represented in different colours by selecting the '**Colours**' options in the menu bar by selecting the appropriate option (Fig. 5.41).

### Rotating Molecules

9. The two scroll bars one situated the right is used to rotate the molecule about the y-axis and the other at the bottom may be used to rotate the molecule about the x-axis.

10. The scroll bar is initially positioned in the centre.
11. To specify the **centre of rotation**, under '**Settings**' in the menu bar select '**Pick Centre**' and select the portion of the molecule which you want to set as centre of rotation.
12. To specify a **bond as the axis of rotation**, under '**Settings**' in the menu bar select '**Rotate Bond**' and select the portion of the molecule which you want to set.

### Selecting Parts of Molecules

13. Under '**Setting**' in the menu bar select '**Pick Ident**' (default).
14. Moving the mouse pointer over the terminal window displays the cross-hair cursor (+).
15. Clicking on mouse button when placed over appropriate item, the nearest atom to the point is identified and the information is reflected in the command line window also (Fig. 5.42).
16. Clicking the molecule by holding the shift key or ctrl key displays above information along with the atom coordinates (Fig. 5.42).
17. This can also be done selecting '**Setting**' in the menu bar and choosing '**Pick Cord**' and click an atom.

### Displaying Distances Between Two Atoms

18. To display the distance between two atoms, under '**Setting**' in the menu bar select '**Pick Distance**'.
19. Selecting two different atoms for which the distance has to be measured. The distance in Å unit will be displayed both in the terminal and command line.
20. To display the distance between two atoms only in the **Terminal window** (**Monitor**), under '**Setting**' in the menu bar select '**Pick Monitor**'.

### Displaying the Bond Angle Between Atoms

21. To display the **bond angle**, which defined by three atoms, under '**Setting**' in the menu bar select '**Pick Angle**' (Fig. 5.43).

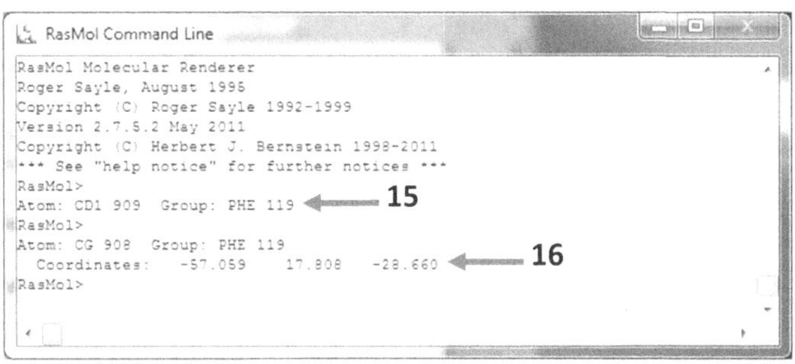


Fig. 5.42 The command line interface displaying the details of selected atom

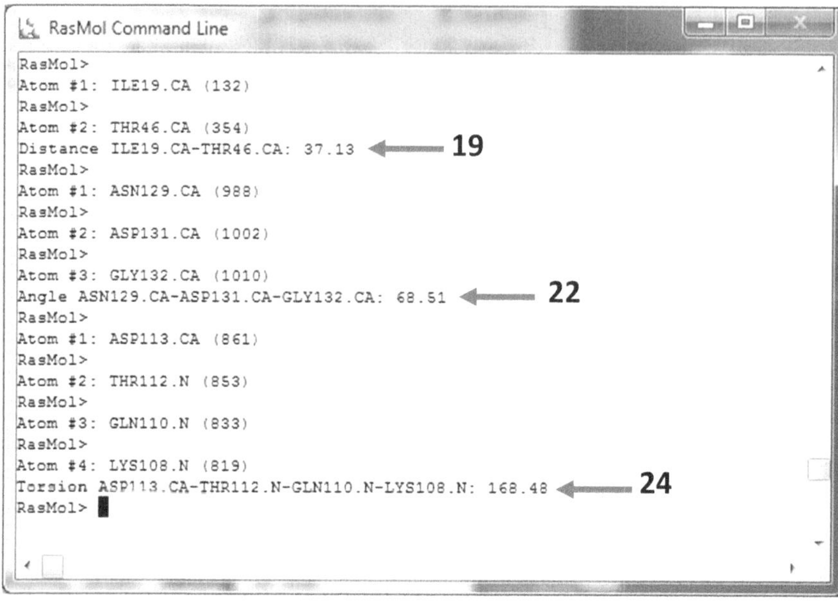


**Fig. 5.43** The command line interface displaying the details of distance, bond angle and torsion angle between selected atoms

22. Click on three different atoms for the bond angle is to be measured. The angle in degrees (°) will be displayed both in the terminal and command line (Fig. 5.43).

### Displaying the Torsion Angle Between Atoms

23. To display the **torsion angle**, which is defined by four atoms, under '**Setting**' in the menu bar select '**Pick Angle**'.
24. Click on four different atoms for the bond angle is to be measured. The angle in degrees (°) will be displayed both in the terminal and command line (Fig. 5.43).

## Creating Publication-Quality Images

### Setting the View

25. Once a desired angle of view is determined then it can transferred to a script by selecting '**Export**' from menu bar and clicking '**RasMol Script**'.

### Saving Molecules and Images

26. Saving the molecules or a view is straightforward. This can be done selecting '**Export**' from menu bar and clicking '**BMP/GIF**' and specifying the location of the file to be stored.

### Advance Manipulation Using Command Line Interface

#### Command Line Interface

RasMol can also be controlled by using interactive commands typed at the RasMol prompt in the terminal window. Every command should be typed on a separate line followed by entering the return key on the keyboard to execute the command. Commands are not case-sensitive and may be entered in lower or upper case letters. White spaces and blank lines are ignored, except that separates the keyword and the arguments in commands. RasMol command lines support a maximum of 256 characters per line, beyond which is ignored by a warning message. Strings can be entered within single or double quotation marks. Comments can be inserted by placing a hash '#' character and RasMol will ignore the rest of the line following it. In case of any syntax errors or the command is not recognised by RasMol, the program returns a warning message.

#### Editing in the Command Line

Some basic editing in the command line is supported in RasMol like pressing the backspace, delete or ^H (Control-H) keys can delete the previous character, and pressing the key ^D (Control-D) can delete the character next to the cursor. Use characters ^B, ^F, ^A and ^E for moving the cursor back a single character, forward a single character, to the beginning of the line and to the end of the line, respectively. Since RasMol maintains a history of recently used commands, the user need not type the same commands again and again. To display the previous command in the history type ^P (Control-P) and to display the following command, type ^N (Control-N).

Some other commonly used commands in RasMol are listed below. For more compressive list refer RasMol's quick reference card.

Commands in RasMol (from RasMol's quick reference card).

**General Commands**

| | |
|---|---|
| **load [format] <filename>** | To load a molecule |
| **zap** | To delete molecule |
| **exit** | To exit from RasMol |
| **help [topic [subtopic]]** | To display on-line help topic |
| **script<filename>** | To run a RasMol scipt |

**Selection**

| | |
|---|---|
| **select<expression>** | Update part of molecule |
| **restrict<expression>** | Display only part of mol |

**Expression**

| | |
|---|---|
| * | All atoms |
| cys | Atoms in cysteines |
| hoh | Atoms in water molecules |

| | |
|---|---|
| as? | Atoms in asparagine or aspartic acid |
| *120 | Atoms at residue 120 of all chains |
| *p | Atoms in chain P |
| *.n? | Nitrogen atoms |
| cys.sg | Sulphur atoms in cysteine residues |
| ser70.c? | Carbon atoms in serine-70 |
| hem*p.fe | Iron atoms in the Heme of chain P |

**Display Commands**
**thickness values are in 1/250 Å units**

| | |
|---|---|
| **wireframe [boolean]** | Display wireframe |
| **wireframe<value>** | Display stick bonds |
| **spacefill [boolean]** | Display spacefill spheres |
| **spacefill<value>** | Specify atom sphere radius |

| | |
|---|---|
| **spacefill temperature** | |
| **backbone [boolean]** | Display alpha backbone |
| **backbone<value>** | Specify backbone radius |
| **ribbons [boolean]** | Display solid ribbons |
| **ribbons<value>** | Specify ribbon width |
| **cartoon [boolean]** | Display ribbon & direction |
| **cartoon<value>** | Specify width |
| **label [boolean]** | Draw default atom labels |
| **label<string>** | Label with arbitrary text |
| **set fontsize<value>** | Set label font height |
| **ssbonds [boolean]** | Display disulphide bonds |
| **ssbonds<value>** | Specify ssbond radius |
| **set ssbonds backbone** | SSBonds between alphas |
| **set ssbonds sidechain** | SSBonds between sulphurs |
| **hbonds [boolean]** | Display hydrogen bonds |
| **hbonds<value>** | Specify hbond radius |
| **set hbonds backbone** | HBonds between alphas |
| **set hbonds sidechain** | HBonds donor/acceptor |
| **monitor <#>, <#>** | Draw bond and distance between arbitrary atoms |
| **monitor [boolean]** | Turn monitor on and off |
| **set monitor [boolean]** | Turn monitor label on/off |
| **dots [boolean]** | Display dot surface |
| **dots<value>** | Specify dot density |
| **set solvent [boolean]** | VDW or solvent surface |
| **set radius <value>** | Specify probe sphere rad |

**Colour Commands**

| | |
|---|---|
| **colour [object] <colour>** | Colour representation |

**Objects:**
**Atoms|bonds|backbone|ribbons|labels|hbonds|ssbonds|dots|axes|ribbons1|ribbons2**

**Predefined Colours:**
**Blue|black|cyan|green|greenblue|magenta|orange|purple|red|redorange|violet white|yellow**

**Atom Colour Schemes:**
**Cpk|amino|shapely|group|chain|structure|temperature|charge|user**

**Manipulation Commands**

| | |
|---|---|
| **rotate<axis> [-] <value>** | Rotate molecule |
| **translate<axis> [-] <value>** | Translate molecule |
| **zoom [boolean]** | Scale molecule |
| **zoom<value>** | Specify magnification |
| **slab [boolean]** | Enable/disable slabbing |
| **slab<value>** | Move Z-clipping plane |
| **centre [expression]** | Set centre of rotation |
| **reset** | Initial transformation |

**Misc. Commands**

| | |
|---|---|
| **structure** | DSSP secondary structure |
| **connect [boolean]** | Recalculate connectivity |
| **renumber** | Sequentially number chains |
| **show information** | Display molecule statistics |
| **show sequence** | Display molecule sequence |
| **show symmetry** | Display crystal space group |

**Command Line Editing**
In addition to the cursor keys, the following 'emacs' control keys may be used to edit the command line.

Ctrl-H/Ctrl-D Delete previous/next character
Ctrl-B/Ctrl-F Move backward/forward a character
Ctrl-A/Ctrl-E Move to beginning/end of line
Ctrl-P/Ctrl-N Display previous/next history

**Exercise**

1. How do you measure the following:

   Distance between atoms
   Bond angle
   Torsion angle

2. How will you generate publication-quality images?

3. How will you get the information on the secondary structure?
4. How will you display the amino acid sequence of a particular structure?
5. How will show only a particular amino acid in different display mode?
   Hint: SELECT SER
   WIREFRAME 50

### 5.5.2 *PyMol*

**Introduction**

PyMOL is an open-source tool available from (www.pymol.org) to visualize molecules. It is available for platforms like Windows, Linux and MacOS. PyMOL is capable of producing high-quality images and animations from 3D structures. The program also has advanced functions for structure manipulation and analyzing their chemical properties. It also supports scripts and plugins.

PyMOL has been developed using Python (www.python.org) C language. Though free, commercial version (Incentive PyMOL) is also available that comes with maintenance and support.

**Exploration Activity**

1. The PyMOL graphical user environment has two windows, the '**Viewer**' and '**External GUI**',
2. The '**Viewer**' displays 3D models, and the user can interact and manipulate the model (Fig. 5.44).

**Fig. 5.44** The PyMOL viewer

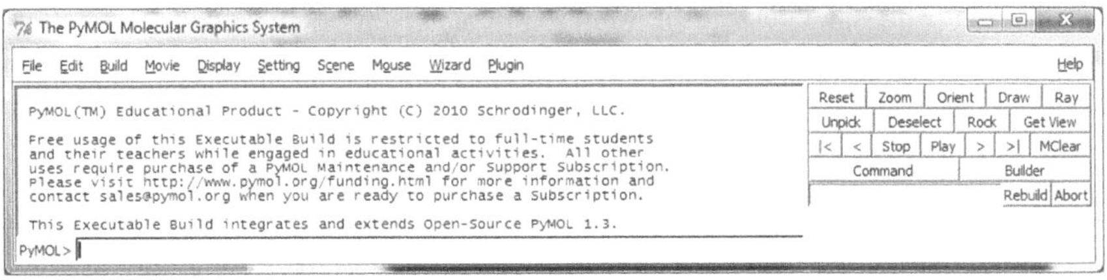


**Fig. 5.45** The PyMOL external GUI

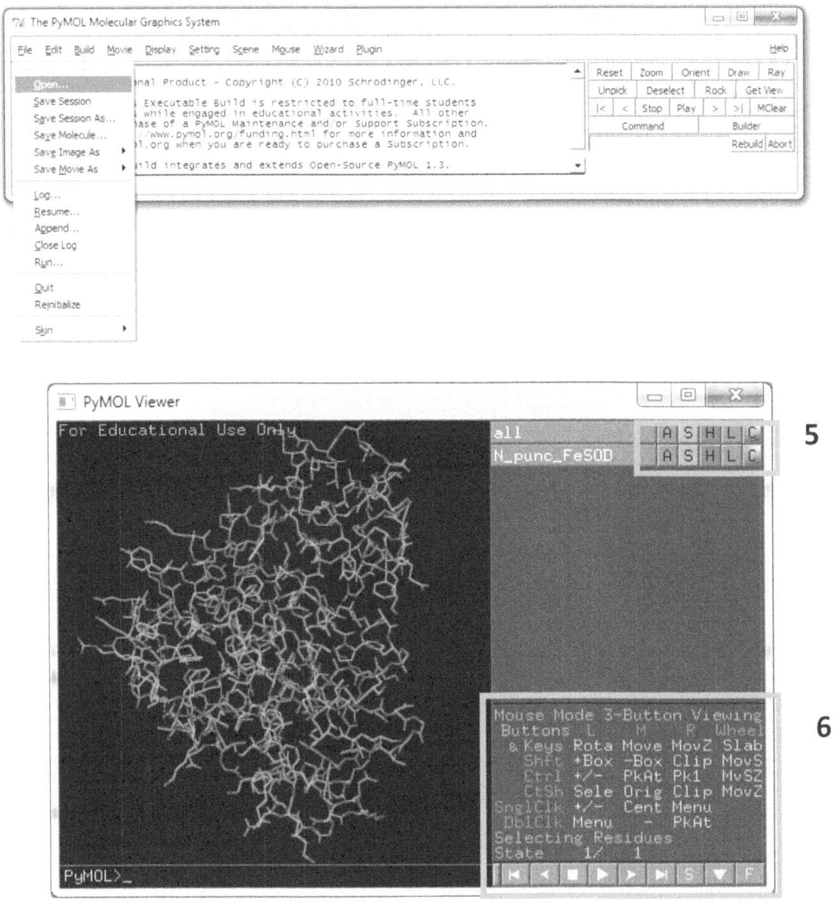


**Fig. 5.46** Loading PDB file from menu

3. The second window called '**External GUI**', is a small window having a menu bar, a display showing the latest action performed and command line. It also has buttons for controlling animations.

4. The command line interface allows the experienced users to gain control options that are not available with point and click graphical interface (Fig. 5.45).
5. The loaded molecules and user-defined selections appear at the **Object list (top right of viewer)**. Mere clicking the name can be toggled on and off. Moreover, display modes can be selected using the options like [A]ction, [S]how, [H]ide, [L]abel and [C]olor buttons.
6. For modelling purposes, the functions of the mouse buttons along with the keyboard are in **Mouse controls (bottom right of viewer)**. Clicking into this control box can toggle between editing and viewing atoms modes.

**Loading a Molecule**

7. You can open or load a PDB file using **File** > **Open**... from the menu bar and select the desired file. The protein's structure will appear as simple bonding lines (wireframe) by default (Fig. 5.46).

**Setting Display Modes**

8. The options include **lines**, **sticks**, **cartoon**, **spheres**, **surface**. In **Object list (top right of viewer)**, click '**S**' and choose desired representation. To represent in cartoon click [**S**]how and select '**Cartoon**' (Fig. 5.47).
9. To hide the lines in the viewer, click [**H**]ide and select '**Lines**'. This hides the side chains that are displayed earlier (Fig. 5.48).

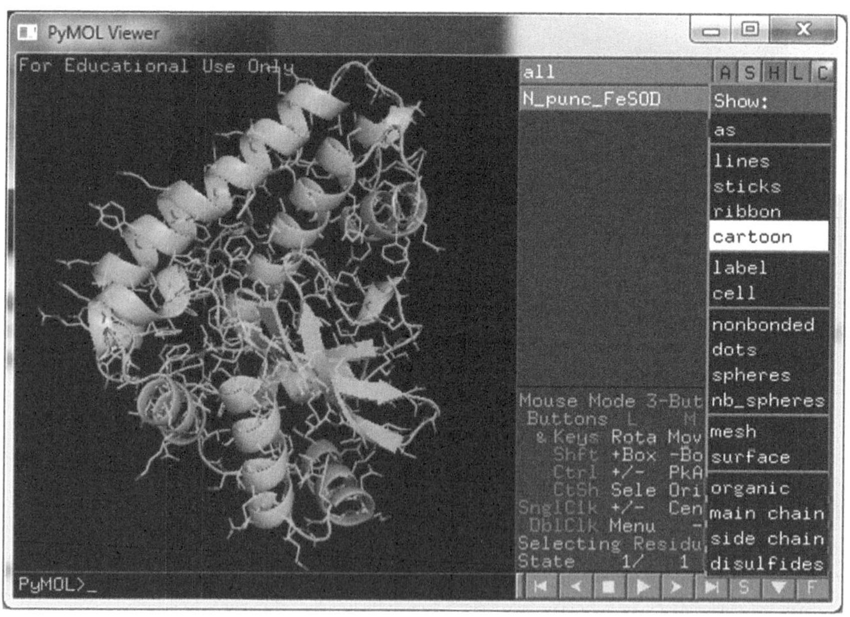


Fig. 5.47 Molecule represented as cartoon with side chains

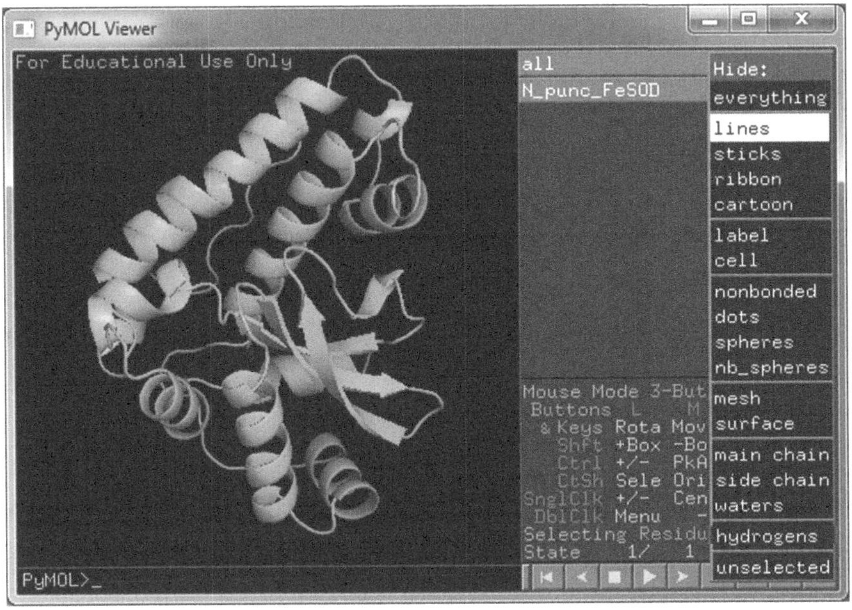


**Fig. 5.48** Molecule represented as cartoon without side chains

### To Reset a View

10. The molecule on the whole or even a single molecule can be brought to centre by right clicking the mouse button in the '**Viewer**' window, and selecting '**Reset**' option (Fig. 5.49).

### Change the Colour of the Molecule

11. The **color** command is used to change the colour of a given object. This can be done choosing [**C**]olor and selecting '**red**' (Fig. 5.50).

### Selecting Parts of Molecules

12. One or several molecules can be selected and represented in different colour or form or both. This is helpful especially for publishing images in journals.
13. First click '**Display**' and select '**Sequence**'. This displays the sequence of the protein in the '**Viewer**' (Fig. 5.51).
14. Select the desired amino acids to be highlighted by clicking on the sequence. Change the selected molecules colour by clicking [**C**]olor and selecting '**green**'. To change the selected molecules form click [**S**]how and select '**sticks**' (Fig. 5.52).

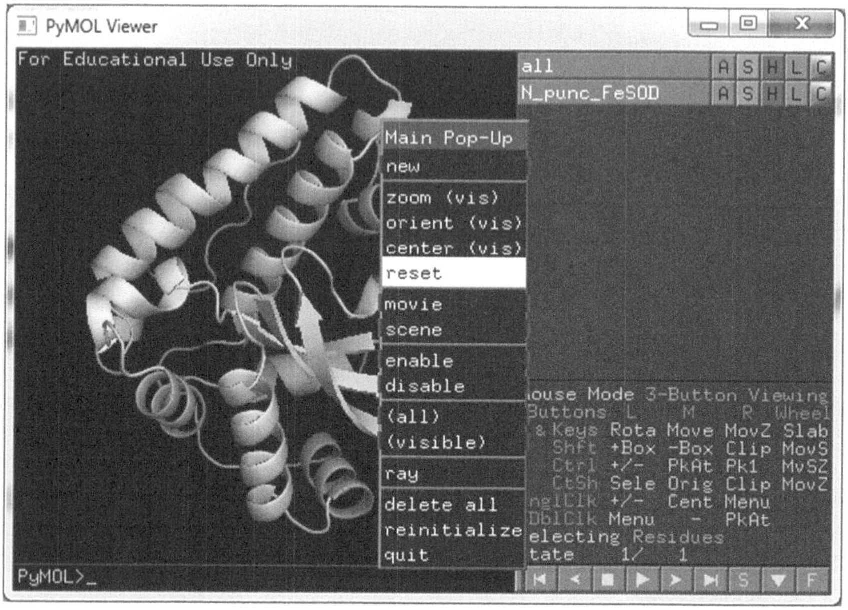


Fig. 5.49 Resetting molecule's view

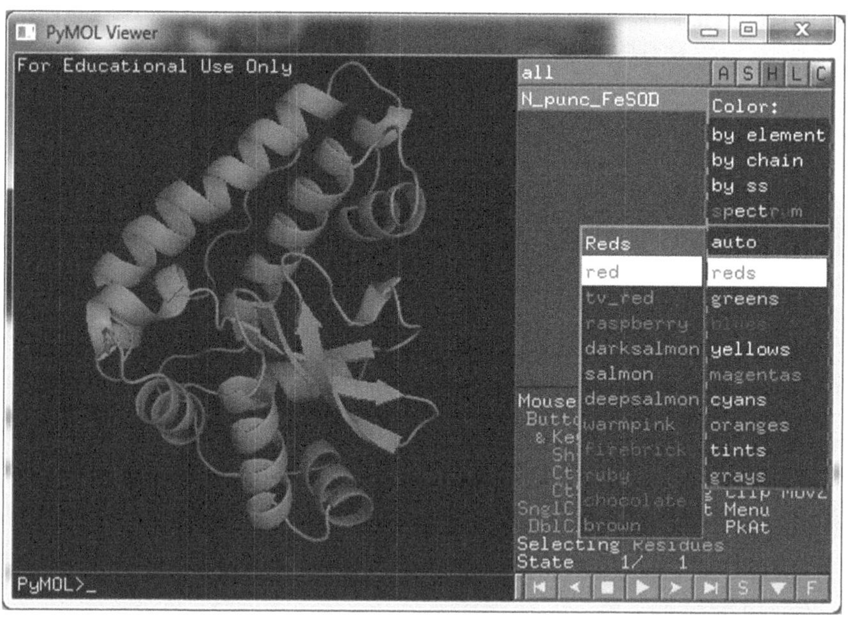


Fig. 5.50 Molecule represented in *red colour* (Color figure online)

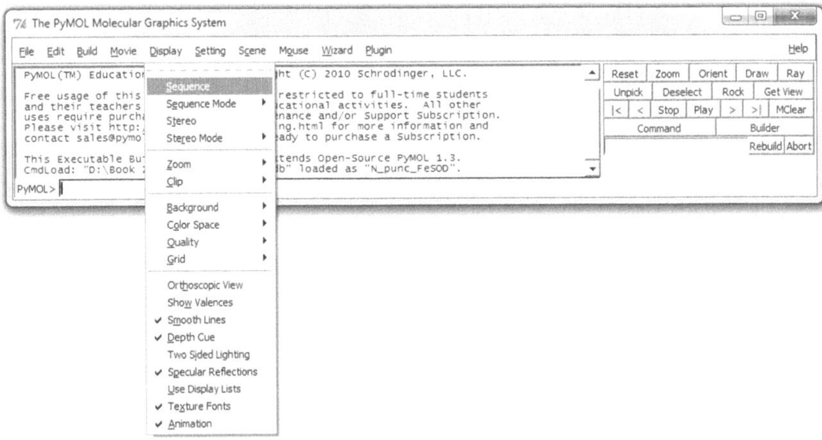


Fig. 5.51 Selecting option to display sequence in the viewer

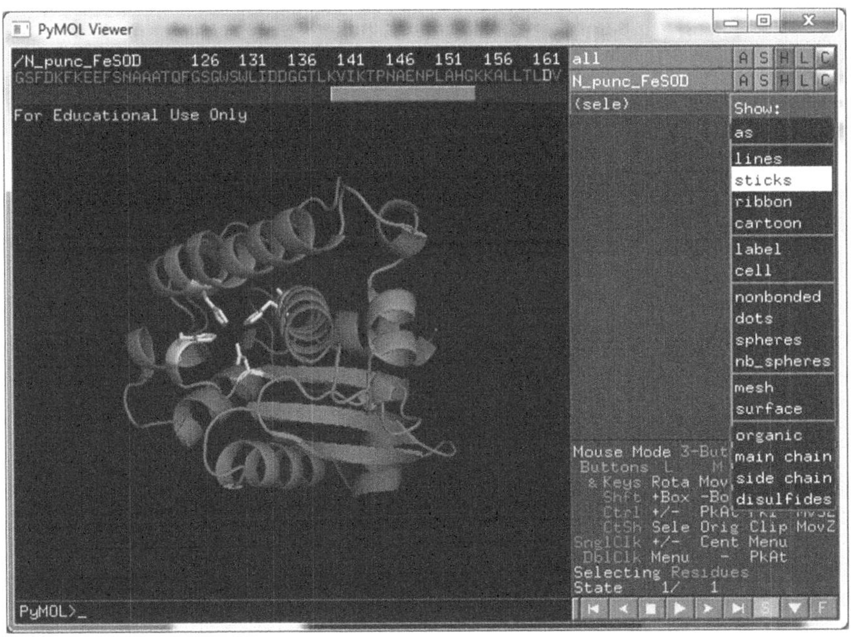


Fig. 5.52 Molecule represented as cartoon with active site molecules *highlighted*

## Background Colour

15. To change the background to white, which is often more appropriate for submitting to journals, this can be adjusted by clicking '**Display**' and selecting appropriate background option from '**Background**' (Fig. 5.53).

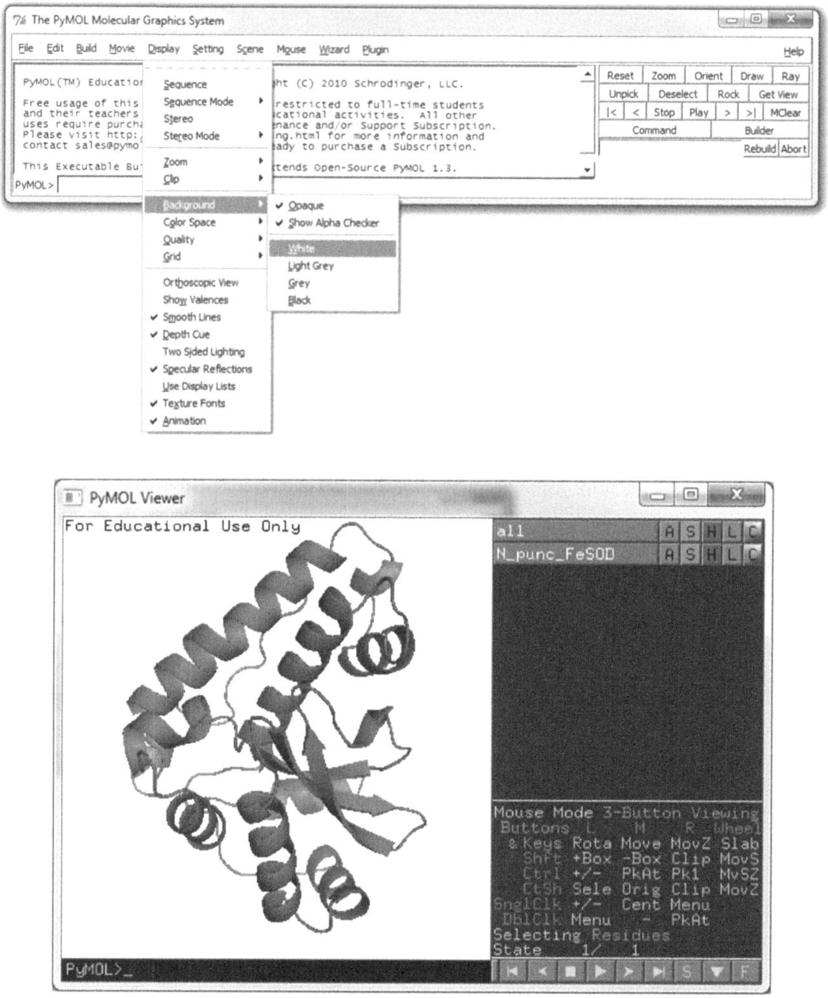


Fig. 5.53 Changing background of the image

## Saving Scenes and Molecules

PyMOL has provisions to save the work as Molecule, Image or as Movie. Saving is simple and can be done from using options from the menu bar.

## Saving Images

16. PyMOL allows the users to save the image in PNG, VRML @ or as POV Ray. The options are under **File > Save Image A** (Fig. 5.54).

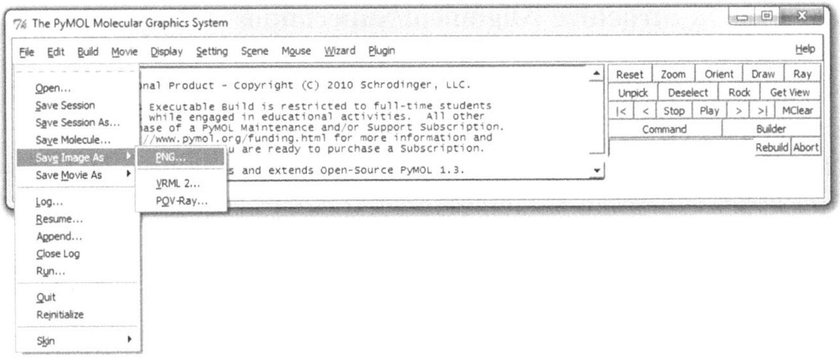


**Fig. 5.54** Saving the view as image

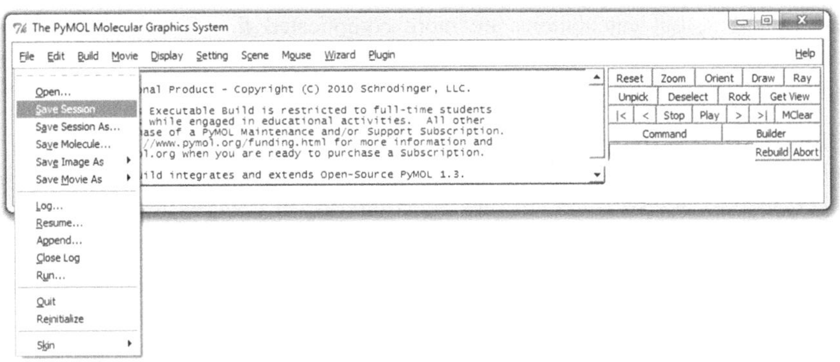


**Fig. 5.55** Saving molecule as session

## Raytracing

17. PyMOL's raytracer is efficient in rendering high-quality images.
18. You can save the whole PyMOL environment as PyMOL session file or as PyMOL Show file by choosing **File > Save Session** (Fig. 5.55).

## Exercise

1. For the models you generated in earlier exercises (for proteins **O81360** and **Q40412**) visualize the structure in different representation and export the image at publication quality.

## 5.6 Protein Structure Alignment/Superimpose Using SuperPose

**Introduction**

Superposition or superimposition is a method for comparing three-dimensional (3D) structures by rotating or orienting them. SuperPose, a freely accessible web server, takes PDB files or accession numbers as input and uses a modified quaternion eigenvalue approach for pairwise and multiple protein structure superpositions of 3D structures. Apart from generating structure alignments, the program provides sequence alignments, distance plots and RMSD statistics of the superimposed molecules (Maiti et al. 2004).

The root mean square deviation (RMSD) of two aligned structures indicates how for the two structures differ from one another. Distance RMSD is easy to calculate but it fails to distinguish between mirror images. Coordinate RMSD distinguishes mirror images, but calculations are more complicated first, both structures are translated so that their centroids are at the origin, and then one is rotated so that the squared deviation in corresponding coordinates is minimized.

**Exploration Activity**

In order to compare the modelled FeSOD structure of *Nostoc punctiforme* with the *E. coli* FeSOD, the two structures are superimposed to visualize the difference.

1. SuperPose can be accessed from http://wishart.biology.ualberta.ca/SuperPose/.
2. Upload the structures to be superpose in PDB format, under PDB entry A (*E.coli* FeSOD) and PDB entry B (*N. punctiforme* FeSOD) (Fig. 5.56).
3. Alternatively PDB accession numbers can also be provided.
4. If a user enters PDB accession numbers, the program retrieves the PDB and the necessary files and also allows users to select chains interactively in the PDB file.
5. SuperPose has options under '**Output Image**' for customizing the still image output, including choosing from Backbone or Ribbon styles, Greyscale or Colour, Mono or Stereo View and Black or White background (Fig. 5.57).
6. The program also allows the user to specify the residues to align by indicating under '**Alignment Options**' (Fig. 5.57).
7. Normally using default values works for many users.
8. Click '**Submit**' button.

**Interpreting Results**

9. The main results page display the superposed structure in WebMol viewer and links for other outputs under headings SuperPose Output Images and SuperPose Output Text Files (Fig. 5.58).

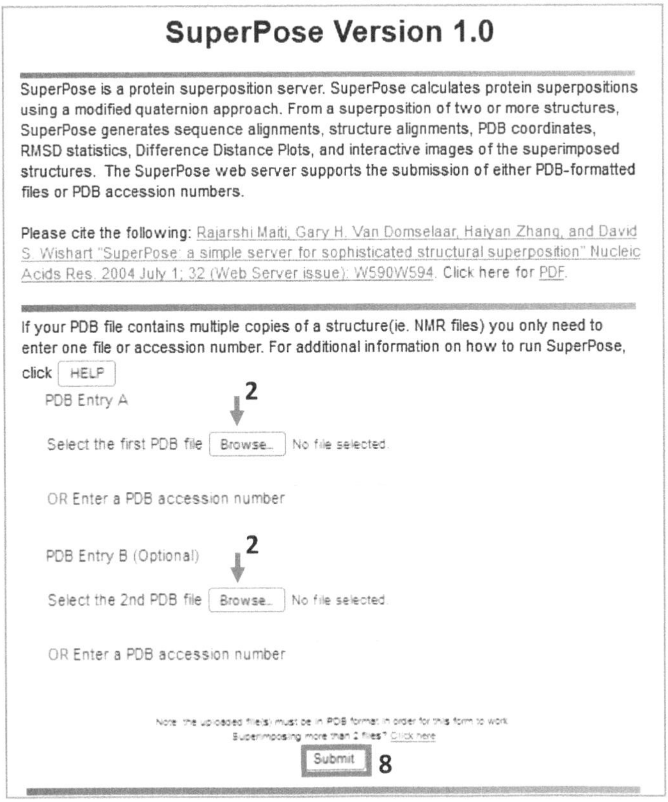


Fig. 5.56 The SuperPose interface by Wishart Research Group at University of Alberta

**SuperPose Output Images**

This section provides links for

10. **WebMol**: A Java based applet for the visualization of superimposed molecules. This requires Java runtime library to be installed (Fig. 5.58).
11. **MolScript Superposition Image**: A still image (in PNG format) of the superimposed molecules is generated using MolScript (Fig. 5.59).
12. **Difference Distance Matrix**: When two molecules are superimposed, a difference distance matrix is generated in the form of PNG image showing regions differences between two structures, like the lighter being more similar and darker representing more different (Fig. 5.60).

**SuperPose Output Text Files**

This section provides links for

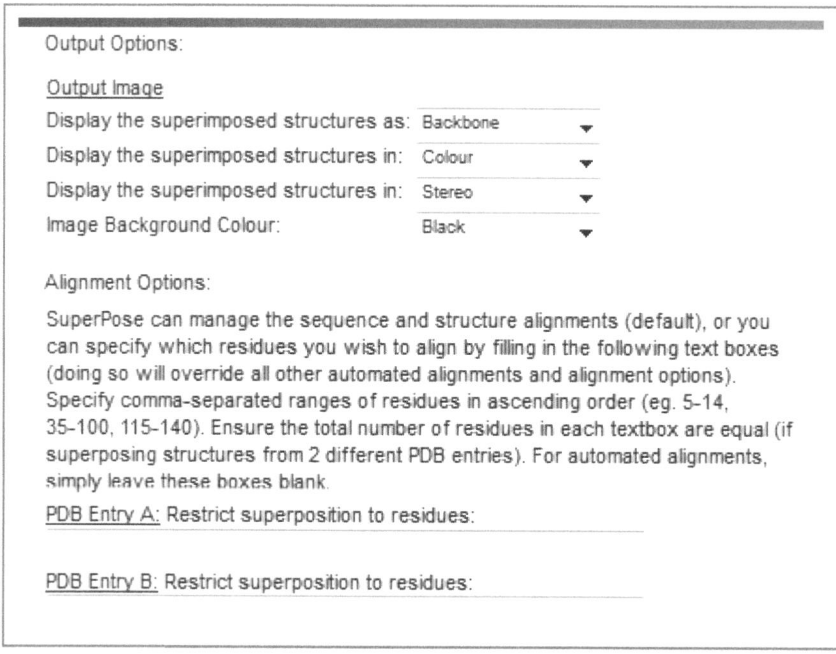


**Fig. 5.57** Advanced options in SuperPose

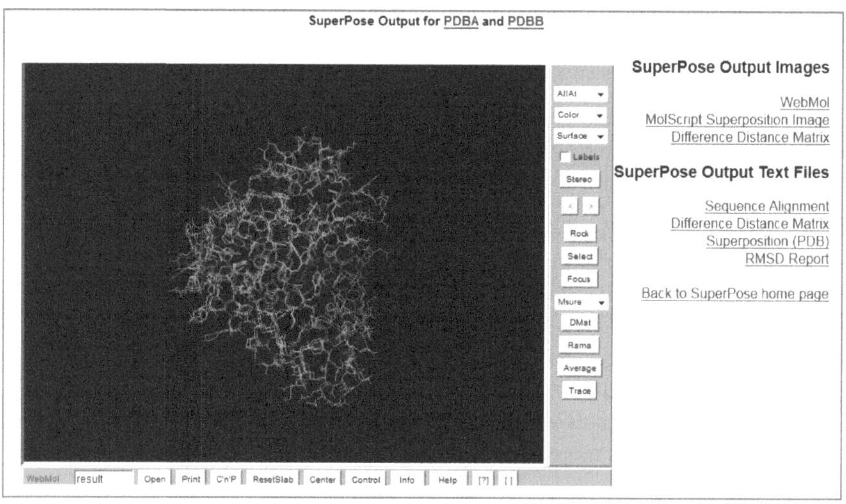


**Fig. 5.58** The SuperPose main result page

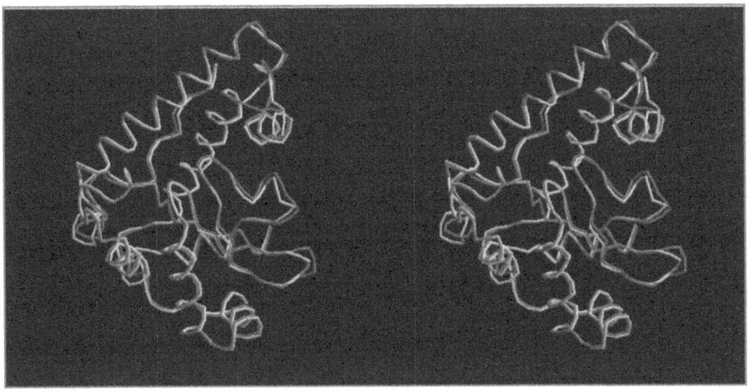

**Fig. 5.59** Snapshot showing the superposed structures in stereo

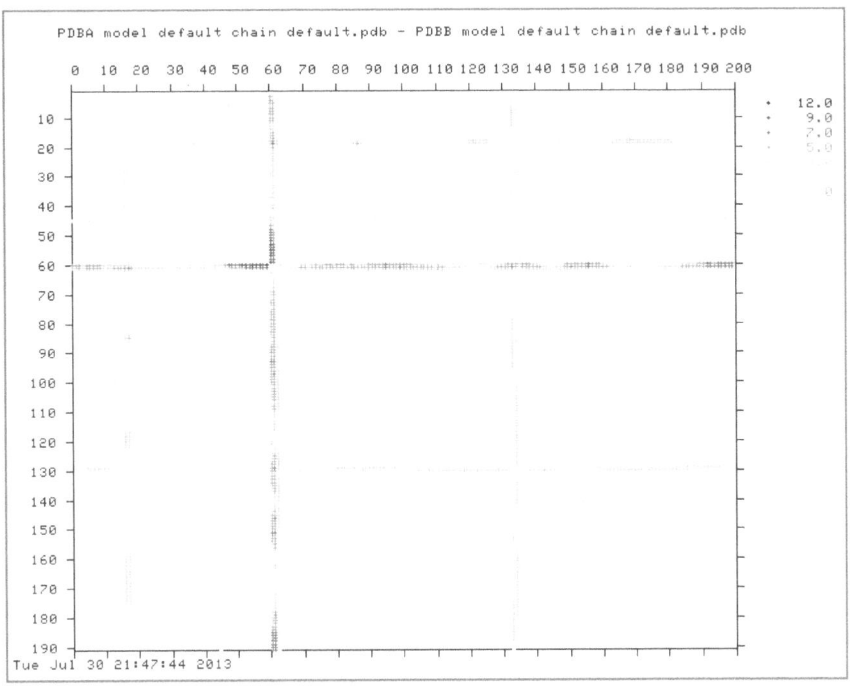


**Fig. 5.60** The plot of difference distance matrix for the two structures

```
#=======================================
#
# Aligned_sequences: 2
# 1: PDBA_model_default_chain_default
# 2: PDBB_model_default_chain_default
# Matrix: EBLOSUM62
# Gap_penalty: 10.0
# Extend_penalty: 0.5
#
# Length: 199
# Identity:     111/199 (55.8%)
# Similarity:   138/199 (69.3%)
# Gaps:           8/199 ( 4.0%)
# Score: 586.0
#
#
#=======================================

PDBA_model_de      1 SFELPALPYAKDALAPH-ISAETIEYHYGKHHQTYVTNLNNLIKGTAFEG     49
                     .|..|.||:.|:||.|: :.|||.||||||||:.||.|||.|.:||....
PDBB_model_de      1 VFVQPPLPFDKNALEPNGMKAETFEYHYGKHHKAYVDNLNKLTEGTELAD     50

PDBA_model_de     50 KSLEEIIR-----SSEGGVFNNAAQVWNHTFYWNCLAPNAGGEPTGKVAE     94
                     |.|||:|:     ||:.|:||||||||||||:||.|.||.||.|||.:|.
PDBB_model_de     51 KPLEEVIKISFQDSSKAGIFNNAAQVWNHTFFWNSLKPNGGGAPTGDLAA    100

PDBA_model_de     95 AIAASFGSFADFKAQFTDAAIKNFGSGWTWLVKNSDGKLAIVSTSNAGTP    144
                     .|...||||..||.:|::||...|||||:||: :..|.|.::.|.||..|
PDBB_model_de    101 RIDKDFGSFDKFKEEFSNAAATQFGSGWSWLI-DDGGTLKVIKTPNAENP    149

PDBA_model_de    145 LTTDATPLLTVDVWEHAYYIDYRNARPGYLEHFW-ALVNWEFVAKNLAA    192
                     |......|||:||||||||||||||||.::::|. .||||:|.|:|||.
PDBB_model_de    150 LAHGKKALLTLDVWEHAYYIDYRNARPAFIKNFLDNLVNWDFAAENLAK    198
```

**Fig. 5.61** Sequence alignment for the superposed structures

13. **Sequence Alignment**: This shows the alignment of sequences (pairwise or multiple) (Fig. 5.61).
14. **Difference Distance Matrix**: A difference distance matrix in text format.
15. **Superposition (PDB)**: The structure (superimposed molecules) as PDB files containing the coordinates that can be used directly in modeling or viewing software like RasMol (Fig. 5.62).
16. **RMSD Report**: This shows the local and global RMSD values (in Angstroms) calculated between the two superimposed molecules (Fig. 5.63).

From the superimposition of the two structures, it is evident that the modeled FeSOD of *N. punctiforme* is globally near equal to *E. coli* FeSOD. This is evident from the observed local and global RMSD values that are equal.

**Exercise**

1. For the models you generated for the proteins with Uniprot id **O81360** and **Q40412** (used in earlier exercise) superpose and compare the proteins.

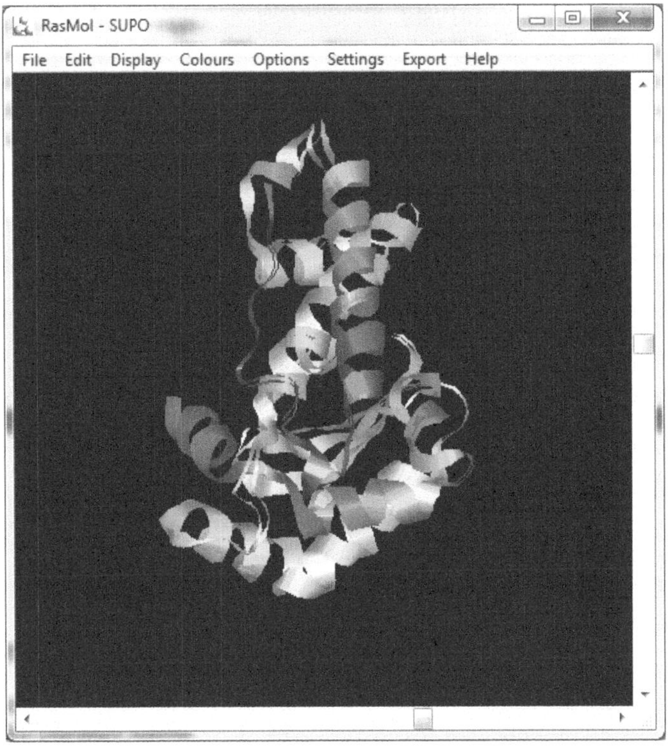


**Fig. 5.62** The PDB file of the superimposed molecules visualized using RasMol

**Fig. 5.63** RMSD report for the superposed structures

Local RMSD

| | Alpha Carbons | Back Bone | Heavy | All |
|---|---|---|---|---|
| RMSD | 1.34 | 1.35 | 1.52 | 1.52 |
| Atoms | 191 | 764 | 1362 | 1362 |

| Structure | Residues |
|---|---|
| PDBA | 2-18, 19-58, 59-127, 129-179, 180-193 |
| PDBB | 2-18, 20-59, 65-133, 134-184, 186-199 |

Global RMSD

| | Alpha Carbons | Back Bone | Heavy | All |
|---|---|---|---|---|
| RMSD | 1.34 | 1.35 | 1.52 | 1.52 |
| Atoms | 191 | 764 | 1362 | 1362 |

| Structure | Residues |
|---|---|
| PDBA | 2-18, 19-58, 59-127, 129-179, 180-193 |
| PDBB | 2-18, 20-59, 65-133, 134-184, 186-199 |

## 5.7 Protein Cleft Analysis

**Introduction**

Proteins interact with other molecules usually with the help of clefts observed on their surface. In case of enzymes the active site is mostly a large and deep cleft whereas in case of protein–protein interactions, it happens on the planar surfaces. It is observed in majority of cases the ligand binds with a single-chain enzyme in the largest cleft suggesting that active sites can be predicted using purely geometrical criteria alone (Lawkowki et al. 1996).

**PDBsum Generate**

PDBsum depicts information like ligands, protein chains, metal ions and their interactions as schematic representations structures deposited in the Protein Data Bank (PDB) and also user submitted structures.

**Exploration Activity**

1. PDBsum Generate can be accessed from http://www.ebi.ac.uk/thornton-srv/databases/pdbsum/Generate.html.
2. **Upload** your own PDB-format file in the submission form of PDBsum structural analyses (Fig. 5.64).

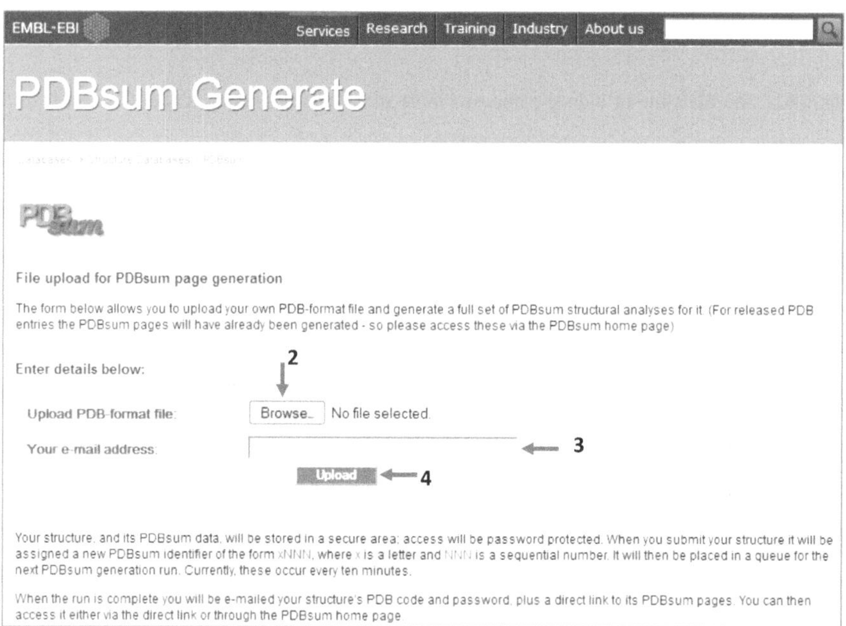


**Fig. 5.64** PDBsum Generate interface at EMBL-EBI

3. Provide **valid email address** so that when the run is complete, the user will be e-mailed with his structure's **PDB code**, a **password** and a **direct link** to its PDBsum pages so as to access the results.
4. Click **'Upload'** button.

**Interpreting Results**

5. For each entry, the first **'top page'** provides a summary of the entry's contents, its name, source authors and date. Links to pages containing PROCHECK analysis and the secondary structure is also provided (Fig. 5.65).
6. The second **'Protein'** tab displays the **thumbnail image(s)** generated with the help of **Raster3D**. The **Proteins** are depicted in terms of their secondary structure: **cylinders** corresponding to helices, **arrows** to beta strands and **threads** to random coil. SAS link to the related protein in PDB is also provided. Other links to view the Topology in postscript and PDF versions and view Hera diagram representing the main chain H-bonds as Postscript and PDF versions is also provided (Fig. 5.66).
7. In the **'Clefts'** tab, **Ligands** and **metals** are displayed (snapshots) as spacefill model (Fig. 5.67).

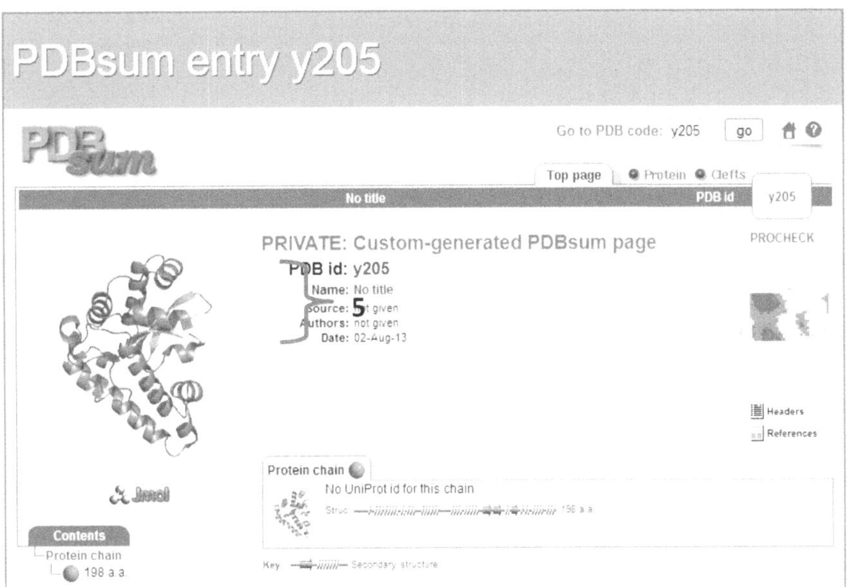


**Fig. 5.65** The main result page

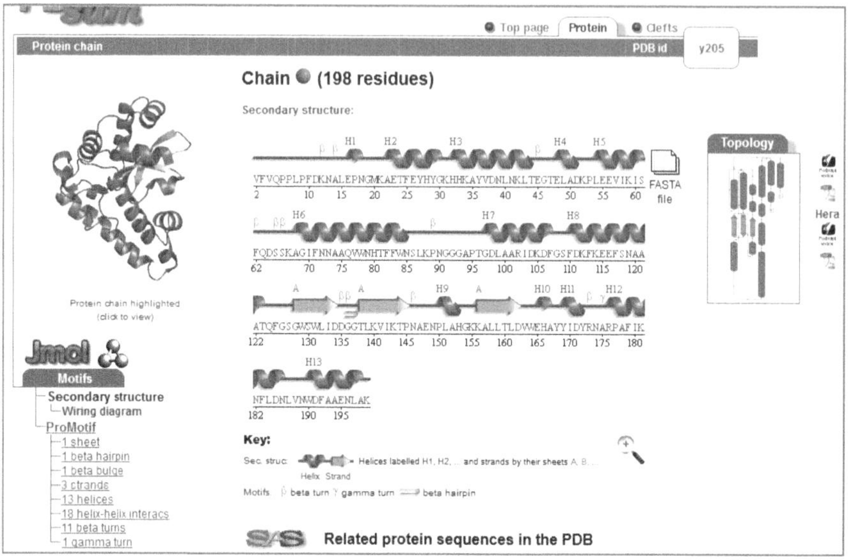


Fig. 5.66 The secondary structure representation of the protein

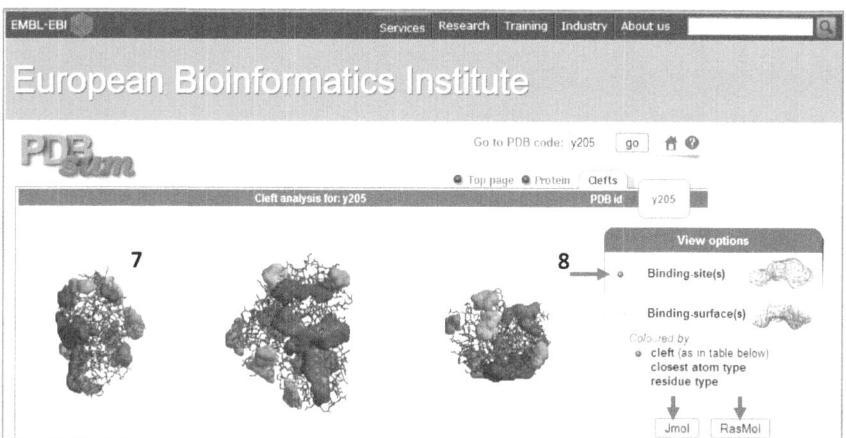


Fig. 5.67 The predicted clefts presented as snapshots in three different views

8. The binding site can be visualized in **Jmol** or **RasMol** by selecting **Binding-site(s)** and clicking '**Jmol**' or '**RasMol**' button (Fig. 5.67).
9. If the molecules are not too large, the molecular surface and clefts in that surface are also computed. These can be viewed using **Jmol** or **RasMol** (Figs. 5.67, 5.68, and 5.69).

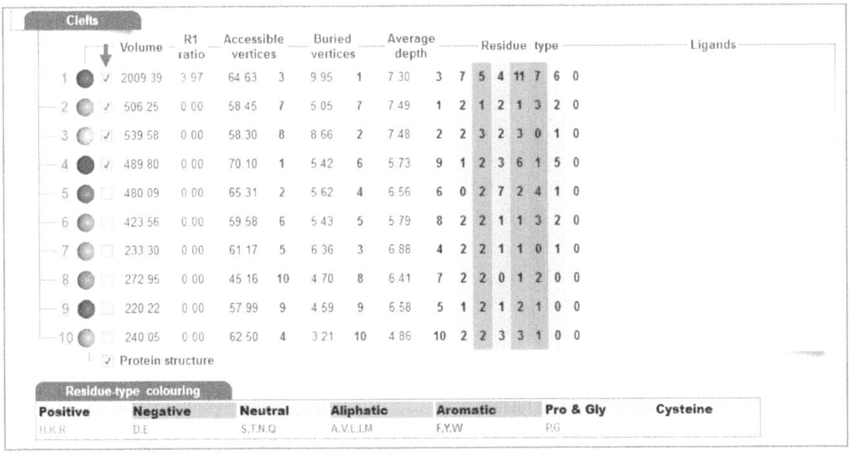

Clefts

| | Volume | R1 ratio | Accessible vertices | | Buried vertices | | Average depth | Residue type | | | | | | | | Ligands |
|---|---|---|---|---|---|---|---|---|---|---|---|---|---|---|---|---|
| 1 | 2009.39 | 3.97 | 64.63 | 3 | 9.95 | 1 | 7.30 | 3 | 7 | 5 | 4 | 11 | 7 | 6 | 0 | |
| 2 | 506.25 | 0.00 | 58.45 | 7 | 5.05 | 7 | 7.49 | 1 | 2 | 1 | 2 | 1 | 3 | 2 | 0 | |
| 3 | 539.58 | 0.00 | 58.30 | 8 | 8.66 | 2 | 7.48 | 2 | 2 | 3 | 2 | 3 | 0 | 1 | 0 | |
| 4 | 489.80 | 0.00 | 70.10 | 1 | 5.42 | 6 | 5.73 | 9 | 1 | 2 | 3 | 6 | 1 | 5 | 0 | |
| 5 | 480.09 | 0.00 | 65.31 | 2 | 5.62 | 4 | 6.56 | 6 | 0 | 2 | 7 | 2 | 4 | 1 | 0 | |
| 6 | 423.56 | 0.00 | 59.58 | 6 | 5.43 | 5 | 5.79 | 8 | 2 | 2 | 1 | 1 | 3 | 2 | 0 | |
| 7 | 233.30 | 0.00 | 61.17 | 5 | 6.36 | 3 | 6.88 | 4 | 2 | 2 | 1 | 1 | 0 | 1 | 0 | |
| 8 | 272.95 | 0.00 | 45.16 | 10 | 4.70 | 8 | 6.41 | 7 | 2 | 2 | 0 | 1 | 2 | 0 | 0 | |
| 9 | 220.22 | 0.00 | 57.99 | 9 | 4.59 | 9 | 6.58 | 5 | 1 | 2 | 1 | 2 | 1 | 0 | 0 | |
| 10 | 240.05 | 0.00 | 62.50 | 4 | 3.21 | 10 | 4.86 | 10 | 2 | 2 | 3 | 3 | 1 | 0 | 0 | |

Protein structure

Residue-type colouring

| Positive | Negative | Neutral | Aliphatic | Aromatic | Pro & Gly | Cysteine |
|---|---|---|---|---|---|---|
| H,K,R | D,E | S,T,N,Q | A,V,L,I,M | F,Y,W | P,G | |

**Fig. 5.68** The top ten largest clefts in the surface represented as table

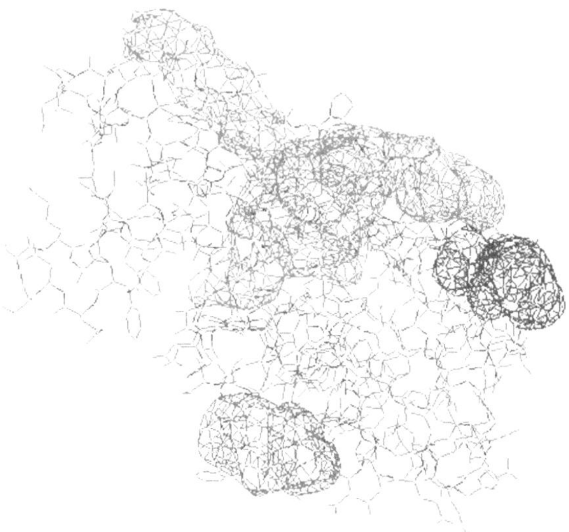

**Fig. 5.69** The top four largest clefts in the surface rendered in Jmol

10. The top ten largest clefts in the surface are depicted in different colours like red, purple, yellow, blue, green, brown, pink, greenblue, magenta, cyan in decreasing surface volume (Fig. 5.68).
11. The binding surface can be also visualized by choosing the option '**Binding Surface**' (Fig. 5.67) and it can be coloured on the basis of cleft (Fig. 5.70a), closest atom type (Fig. 5.70b) or by residue type (Fig. 5.70c).
12. For some enzymes the molecules involved (active site) in the reaction catalysed by the enzyme are also shown.

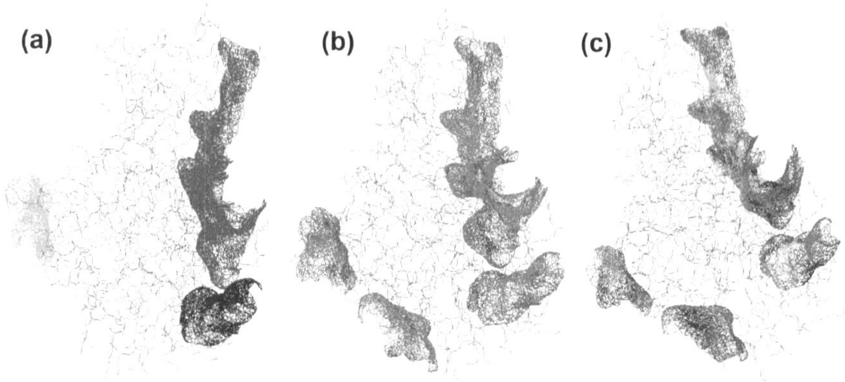


**Fig. 5.70** Binding surface coloured on the basis of cleft (**a**), closest atom type (**b**) or by residue type (**c**)

## Exercise

1. For the models you have generated for the proteins with UniProt id **O81360** and **Q40412** (used in earlier exercise) perform a cleft analysis and compare them.

# Chapter 6
# Protein–Ligand Interactions

## 6.1 Protein–Ligand Docking Using AutoDock4.1 and MGLTools

**Introduction**

Protein–ligand docking is a structural biology tool that predicts the possible binding modes of a ligand with protein for which three-dimensional structure is known. Docking can be applied in the field of pharma to perform virtual screening on large libraries of compounds and propose structural hypotheses of how the ligands inhibit the target, which is invaluable in drug discovery process (Morris and Lim-Wilby 2008).

AutoDock is collection of tools for docking studies that current distribution consists of AutoDock 4 and AutoDock Vina. **AutoDock 4** basically consists of two main programs: 'autodock' that performs docking of ligand to a defined location the target protein called grids, which is basically an interaction map, and 'autogrid' that pre-calculates these grids. Besides being used for docking, the atomic affinity grids can also be visualised thus helping, for example, synthetic chemists to guide organic design for better binders (Morris et al. 2009). These programs uses a graphical user interface called **AutoDock Tools (ADT)**, which helps for setting up the bonds that will treated as rotatable in the ligand and for analysis of dockings.

During these exercises, you will be guided to work on a macromolecule (the target protein), which will be kept fixed during the docking and the ligand being flexible.

In general, if you want to do protein–small molecule docking, you need to have information like:

a. Active site of the receptor,
b. Feasibility of binding of a particular small molecule to that receptor, the knowledge gained from literatures, and
c. Bond pattern of the ligand, mobility, flexibility, etc.

 
K. Syed Ibrahim et al., *Bioinformatics—A Student's Companion*,
DOI 10.1007/978-981-10-1857-2_6

**Exploration Activity**

The structure 1CPS from Protein Data Bank is used in the exercise. For our docking exercise, we shall try to separate the ligand that is already bound to this crystal structure (co-crystallized), prepare a separate file having only ligand and another file having target protein alone. We should also make sure that all the water molecules are also removed, while preparing the files. We will try to dock the separated ligand into the target protein using AutoDock. This process is called redocking. For our convenience, the files are name as Ligand.pdb and Target.pdb.

**Preparations Before Running AutoDock**

i. Download MGLTools for appropriate OS from http://mgltools.scripps.edu/downloads.
ii. Download AutoDock and AutoGrid for appropriate OS from http://autodock.scripps.edu/downloads after registration.
iii. Make a new working directory in the Desktop.
iv. Copy files like **ligand**, **target protein**, **AutoDock.exe**, **AutoGrid.exe** (four files) to the working directory.
v. Open ADT and click **File > Preferences > Set…** under Startup Directory… and locate the working directory you created in the step iii (Fig. 6.1).
vi. Click '**Set**' and then '**Dismiss**' button.

**Loading Ligand and Target Molecules**

1. **File > Read Molecule > select the' Ligand.pdb' and' Target.pdb' files and click 'Open'** (Fig. 6.2).
2. **Edit > Delete Water** (if water molecules are present).

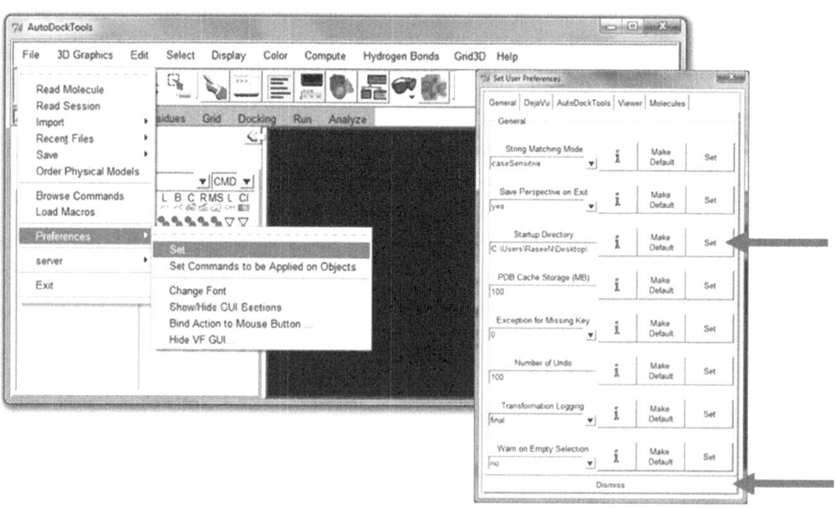


Fig. 6.1 Setting startup directory preferences

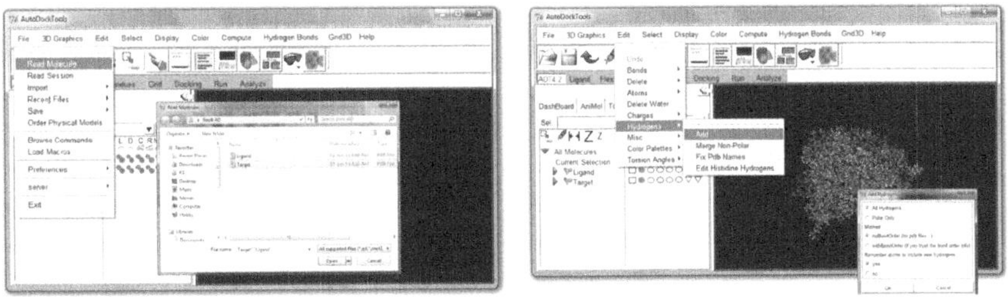

**Fig. 6.2** Loading target protein and ligand, deleting water molecules and adding hydrogen atoms

3. **Edit > Hydrogen > Add…** to add hydrogen atoms. Click '**OK**' in the widget. This is essential because the crystal structure will not have hydrogen atoms, as crystallographic experiments do not resolve hydrogen atoms (Fig. 6.2).

**Ligand Preparation**

4. In the Dashboard, select Ligand.
5. **Ligand > Input > Choose >** Select the ligand molecule and click '**Select molecule for AutoDock**' button (Fig. 6.3).
6. This summarizes the processes like type of charge used, the numbers of non-polar hydrogens merged, of aromatic carbons, of rotatable bonds found and the number of torsional degrees of freedom detected (TORSDOF). Gasteiger charges can be added only if the molecule has hydrogen atoms added earlier which includes both polar and non-polar ones (Fig. 6.3).
7. Click '**OK**'.

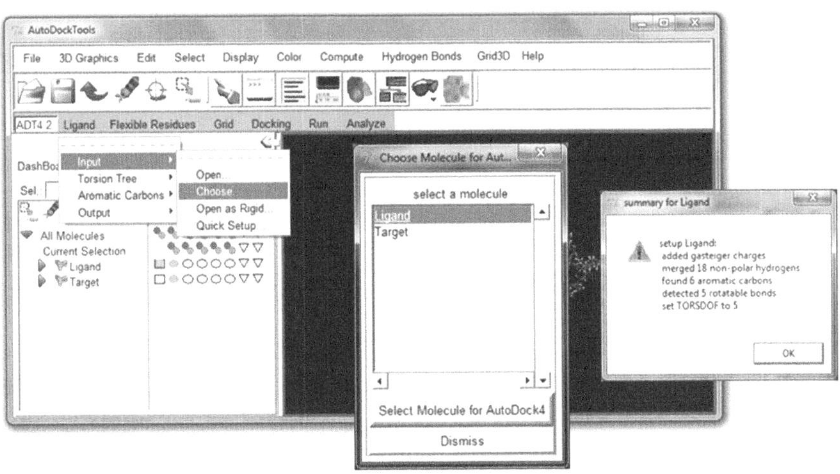


**Fig. 6.3** Choosing ligand and adding gasteiger charges to it

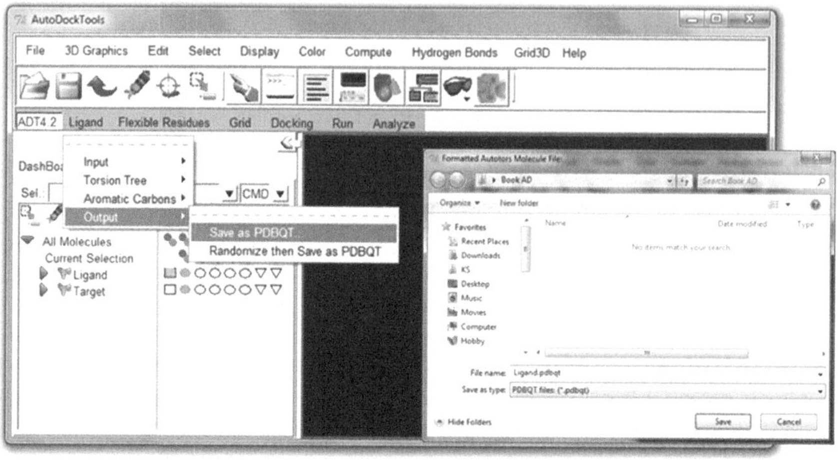


**Fig. 6.4** Saving the prepared ligand file in pdbqt format

8. Select **Ligand > Output > Save as PDBQT… to save the prepared ligand in PDBQT format** (Fig. 6.4). The pdbqt format usually contains atom types supported by AutoDock including other records that specify rotatable bonds.
9. In the Dashboard, deselect Ligand.

**Protein/Target Preparation**

10. In the Dashboard, select protein (target).
11. **Edit > Charges > Add Kollman Charges**. It shows the charge added. Click '**OK**' (Fig. 6.5).
12. **Grid > Macromolecule > Choose…** select the target protein. Click '**Select**' button. Click '**OK**' in the popup (Fig. 6.6).
13. When prompted for saving in PDBQT format after checking, save it (Fig. 6.6).
14. In the Dashboard, deselect protein (target).

Note: Adding charge usually assigns atomic partial charges. Charge assignment requires explicit hydrogen, which can be added beforehand. Structures often have problems such as truncated side chains leading to non-integral net charges.

**Grid Generation**

15. **Grid > Set Map Types > Choose Ligand…** to select the ligand and click 'Select Ligand' button (Fig. 6.7). AutoDock uses 'maps' generated by AutoGrid and it does not use the receptor directly. The maps include one map for each atom type in the ligand(s) including a map for desolvation and for electrostatics. The interaction energy is recorded by the AutoGrid program at each point in a 3D grid around the rigid receptor. During the docking process, the energetics is calculated by the AutoDock for a particular ligand

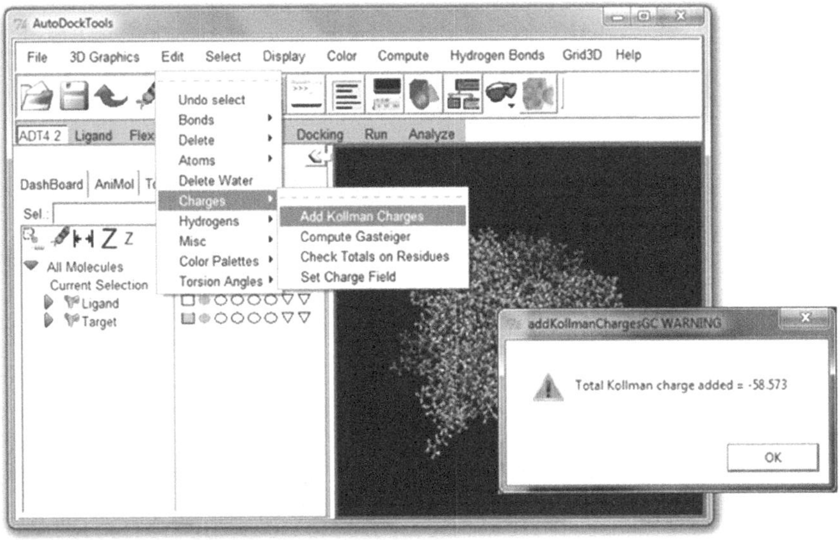


Fig. 6.5 Adding Kollman charges to the target protein

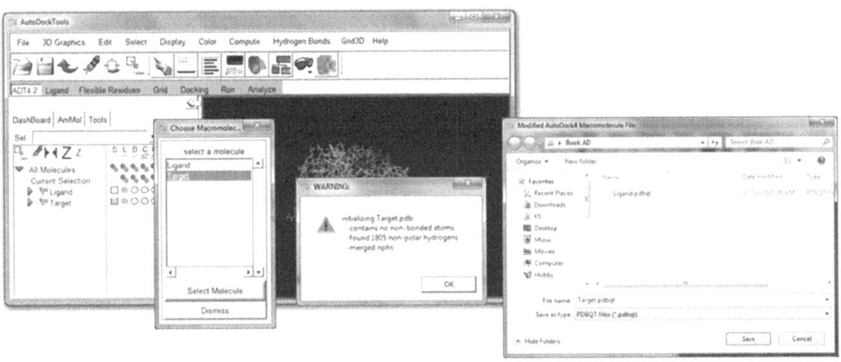


Fig. 6.6 Choosing the target protein, initialising and saving in pdbqt format

configuration which is evaluated using the values available from the gridmaps. It has to be noted that the types of maps depend on the types of atoms in the ligand(s). Choosing a ligand is one way to specify the types of maps.

16. Select **Grid > Grid Box…** and enter values of x, y and z to maximum (60) and spacing to 0.375 Å (Fig. 6.8). Optionally, user can directly type in the value by clicking on the thumbwheel widget with right mouse button, directly. The gridbox specifies the location and extent of the 3D area to be searched during the docking experiment.

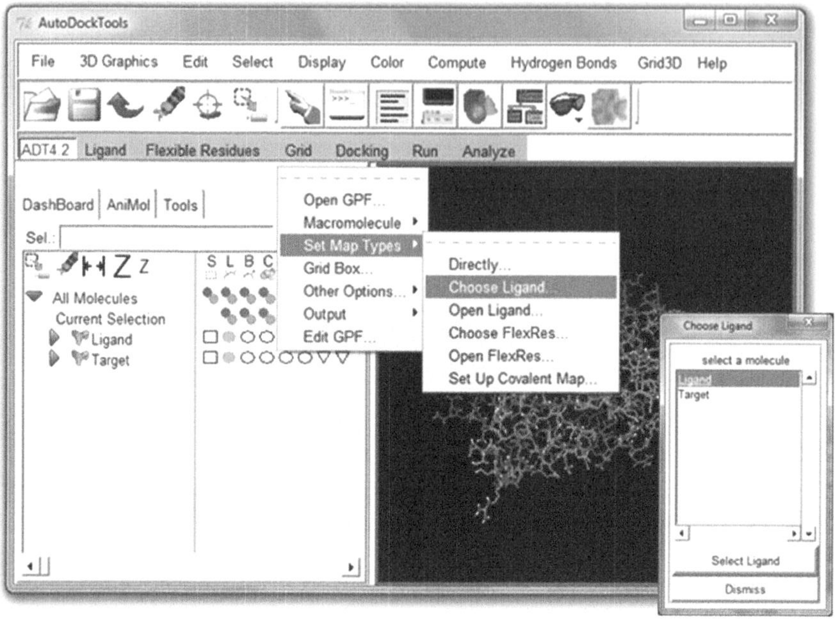


Fig. 6.7 Choosing ligand molecule for setting map type

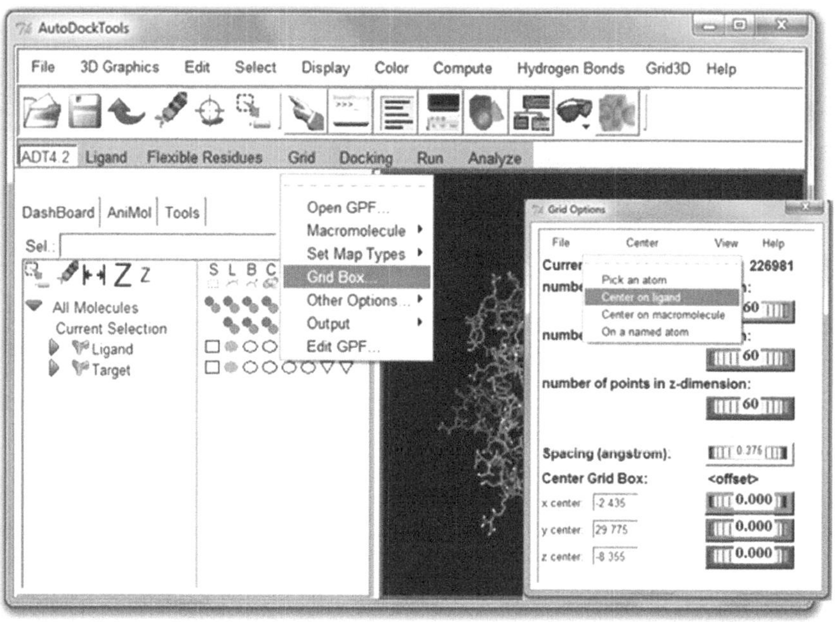


Fig. 6.8 Setting grid box parameters for specifying the search space

17. In the Grid options widget, click, **Center > Center on Ligand** to place the grid box's center at the ligand's centre (Fig. 6.8).
18. You can close the Grid options widget by selecting **File > Close Saving Current**.
19. Select **Grid > Output > Save GPF** to save the grid parameter file by providing a name like **grid.gpf**. Basically the grid parameter file will hold the information like filename of the rigid receptor, the search space location and its extent and the type of atoms in the flexible (ligand) molecule(s) that will be docked.
20. Select **Run > Run AutoGrid** (Fig. 6.9) to initiate AutoGrid.
21. Locate the file grid parameter file (grid.gpf) file you created earlier (step 19).
22. The program automatically creates the grid log file (grid.glg).
23. Click '**Launch**' button (Fig. 6.9).

### Docking Ligand with Target Protein

24. **Docking > Macromolecule > Set Rigid…** to open and specify the target protein as a rigid molecule (Fig. 6.10). Note: *This will not load a new molecule, rather it only specify the stem of the gridmap filenames.*
25. **Docking > Ligand > Choose…** to choose the Ligand file (Fig. 6.10).
26. A popup window opens up and lists the Ligand names. Select it and click '**Accept**' button.
27. **Docking > Search Parameters > Genetic Algorithm**.

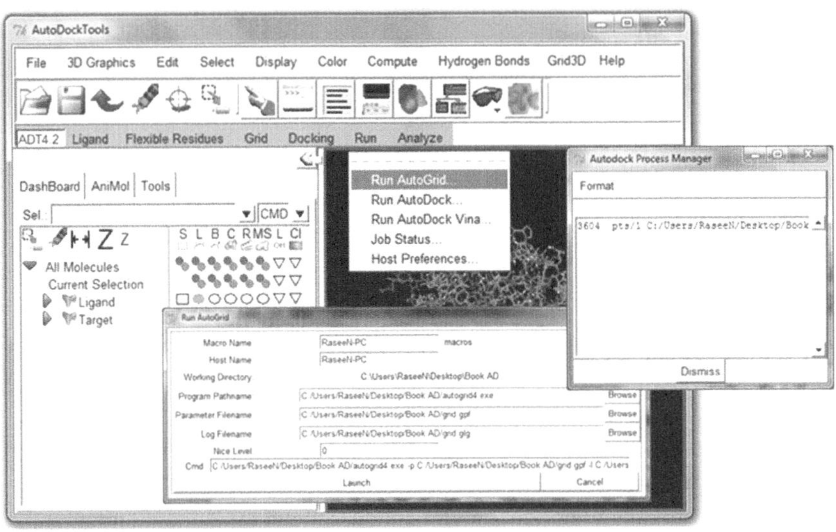


Fig. 6.9 Running AutoGrid

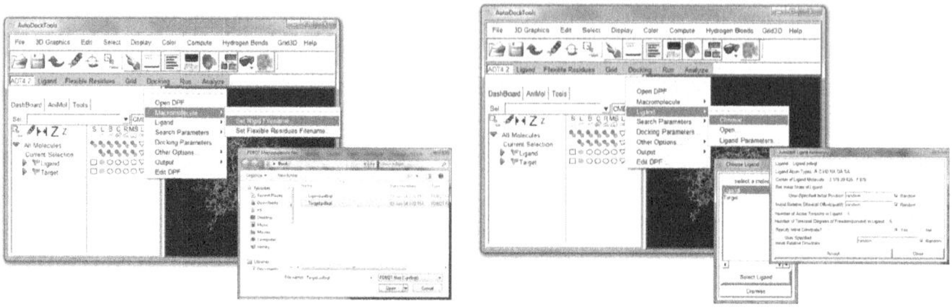

Fig. 6.10 Specifying target protein and ligand molecule for docking

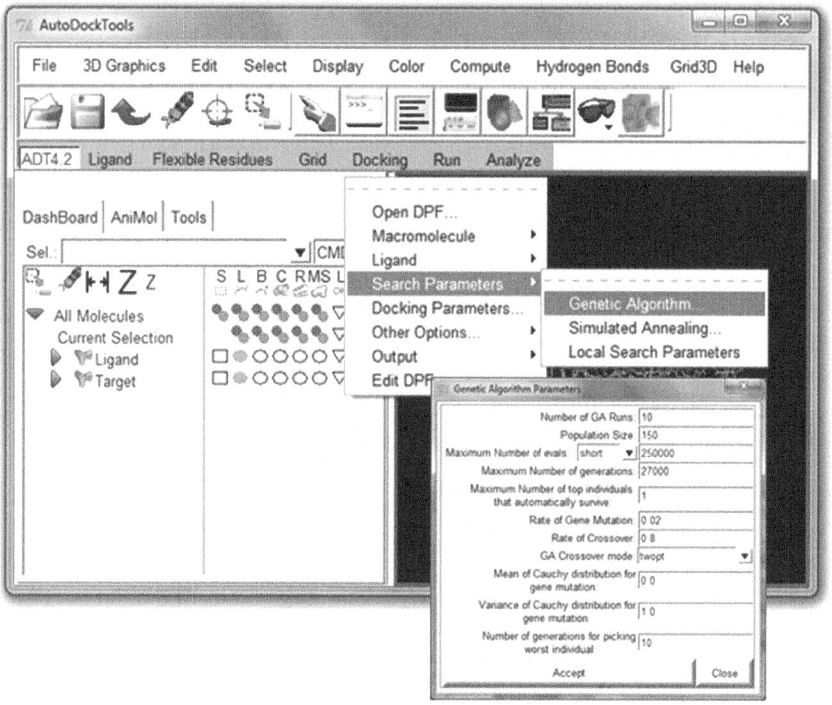


Fig. 6.11 Setting search parameters for AutoDock

28. A window opens up with the parameters. Change to '**Short**' under 'Maximum number of evals:' and click '**Accept**' button (Fig. 6.11). This is done to keep it short for the purpose of exercise.

**Table 6.1** Number of torsions and number of energy evaluations

| Number of torsions | ga_num_evals | ga_num_generations |
|---|---|---|
| 0 | 25,000–250,000 | 27,000 |
| 1–10 | 250,000–25,000,000 | 27,000 |
| >10 | >25,000,000 | 27,000 |

*Note* There are no hard-and-fast rules in setting up '**ga_num_evals**' and '**ga_num_generations**' parameters. Though there are recommendations to run at least 50 dockings (specified by the '**ga_run**' parameter), users are recommended to try a few varying of parameters before settling on your best values

**Docking Parameters to be Noted**

In any docking procedure, the number of energy evaluations depends on the number of torsions in the ligand. This applies to the receptor too, if flexible. Table 6.1 provides some general guidelines for rigid ligands.

29. Select **Docking > Docking Parameter** and click the '**Accept**' button to accept the default parameters.
30. To save the docking parameters file, select **Docking > Output > Lamarckian GA** and provide a file name say, **dock.dpf**. A docking parameter file contains instructions for a Lamarckian Genetic Algorithm (LGA) docking which is also known as a Genetic Algorithm-Local Search (GA-LS).
31. To view the content of the docking parameters file, select **Docking > Edit DPF**.
32. **Run > AutoDock** (Fig. 6.12) to initiate docking process.
33. Locate the file grid parameter file (dock.dpf) file you created earlier (step 30).
34. The program automatically creates the grid log file (grid.dlg).
35. Click 'Launch' button.

**Visualization and Analysis**

36. **Analyze > Dockings > Open** and locate the **dock.dlg** file to load the contents of dockings (Fig. 6.13).
37. Provide a name say, 'Results' and click OK.
38. **Analyze > Macromolecules > Choose** to load target protein and click 'Select Macromolecule' (Fig. 6.14).
39. Click **Analyze > Conformations > Load** to display all the docked conformations. This displays the scores of the top 10 dockings.
40. For better visualization the line representation of the protein molecule can be changed to ribbon. From the dashboard deselect the red dot in the under 'L' for the target protein and click on the circle under 'R' (Fig. 6.15).
41. **Analyze > Conformations > Play ranked by energy** to read 10 docked conformations the dock.dlg file. A player window will open now click on & **sign** then you can see details about docking conformation like binding energy, H bond, etc. (Fig. 6.16).

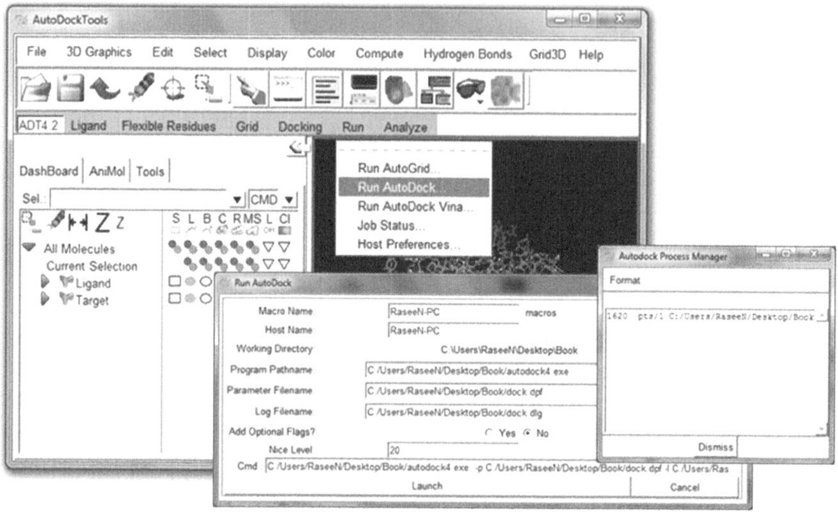


**Fig. 6.12** Running AutoDock

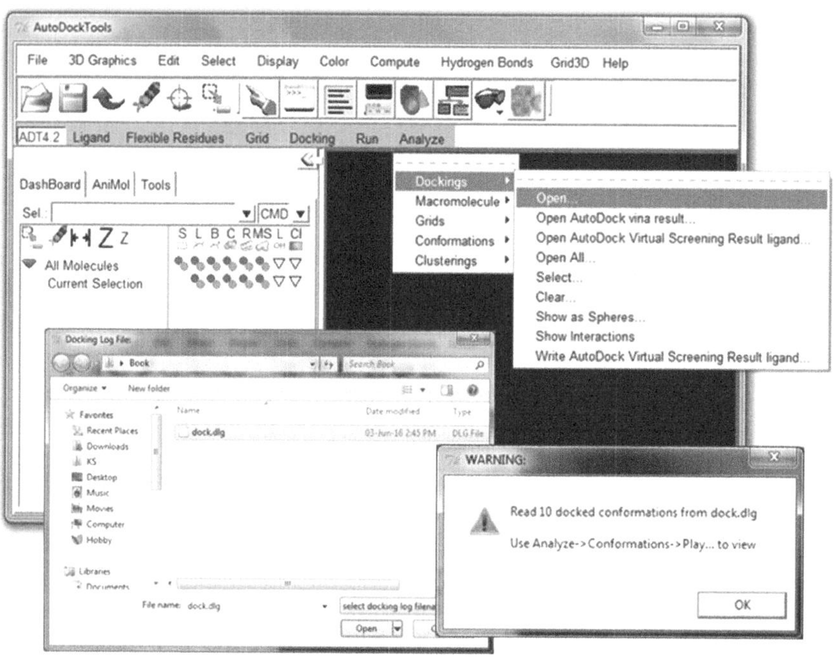


**Fig. 6.13** Loading the docking results for visualization

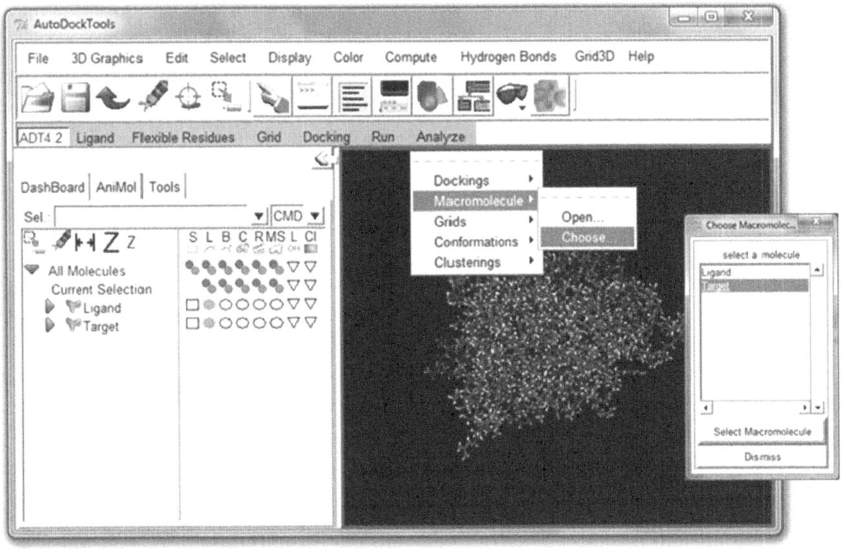


Fig. 6.14 Loading the target protein for visualising the docking poses of the ligand

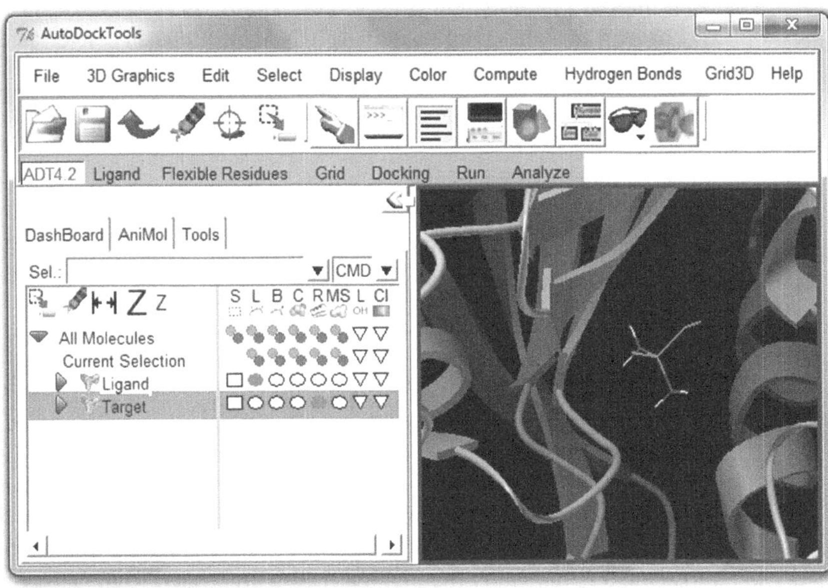


Fig. 6.15 Changing the representation of the target protein

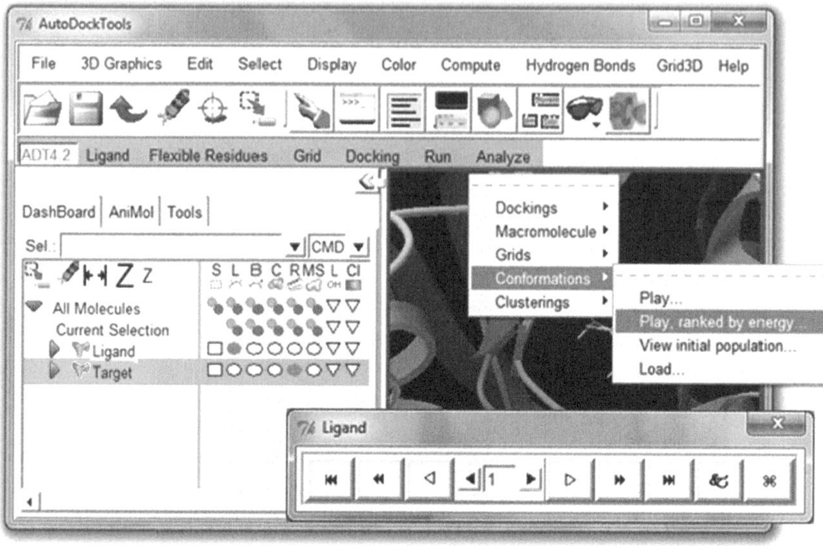


**Fig. 6.16** Visualising the docking based on the docking score rank

42. To write that docked conformation (pose) click '**Build current**' and '**Write Complex**'. This will save the complex in PDBQT format. You can convert later to PDB format.

**Exercise**

1. What are the applications of molecular docking?
2. What is an active site?
3. What is redocking and why it is recommended?
4. What is meant by energy minimization?

## 6.2 Protein–Protein Docking Using ClusPro2.0

**Introduction**

Protein-protein docking involves assigning a protein as target and the other as ligand and tries to predict the structure of the intermolecular complex formed by them. The ultimate goal is to predict the biological activity of the ligand. This can be achieved in two stages, namely posing and scoring. The former deals determination with the orientation and conformation that fits the activie site and the latter deals with the measurement of fit by performing simple energy calculations like van der Waals, electrostatic forces and much higher calculations like free energy binding and salvation.

ClusPro (Comeau et al. 2004) is a fully automated web-based program for the computational docking of protein structures that allows the users to upload the coordinate files of two protein structures or enter the PDB codes of the respective structures for docking through ClusPro's web interface. Since the output lists multiple potential complexes, the docking algorithms evaluate complexes with favourable surface complementarities by Fourier correlation method and finally select those with good electrostatic and desolvation free energies. They are further clustered and ranked according to their clustering properties. ClusPro has been reported to be successful in the blind **C**ritical **A**ssessment of **PR**edictions of **I**nteractions (**CAPRI**) (Camacho and Gatchell 2003) experiment for generating some of the best predictions for the given targets. ClusPro produces reliable high reproducible models for targets with qualities closer to the best human predictor groups, in less than 48 h.

**Exploration Activity**

Toxin–Antitoxin systems besides involving in normal bacterial physiology also participate in pathogenicity of bacteria. These toxins are normally produced but neutralized with their cognate antitoxins from a TA (toxin-antitoxin) operon in the growing cells (Yamaguchi et al. 2011). As an example, we shall use proteins from Toxin-Antitoxin (TA) system slr7041 (toxin) and ssr7040 (antitoxin) *Synechocystis* sp. PCC 6803. The protein sequences were retrieved from Toxin Antitoxin Database (TADB) (http://202.120.12.135/TADB2/index.php) and the protein structures were modelled using homology modelling. Here ssr7040 was used as target and slr7041 as ligand.

1. ClusPro 2.0 server page can be accessed from http://cluspro.bu.edu/login.php.
2. The server requires an academic or non-commercial email id for creating a login. However, it can also be used without an account with lesser benefits (Fig. 6.17).
3. Once logged, provide a job name. If no job name is given, by default a job id will be assigned.
4. You can specify the receptor and ligand for docking by providing PDB ID or by uploading one's own protein structure (in PDB format) by clicking on '**Upload PDB**' link (Fig. 6.18).
5. Optionally if the user wants to specify only one chain for docking, the information can be provided under '**Chains:**'. The chains should be designated as a single letter and if more than one chains are to be committed they should be separated by a space (white space). In case if no chains are mentioned, then all the chains in the structure will be considered (Fig. 6.18). Since in our exercise the files have single chains respectively, nothing is mentioned here.
6. In case if you wish to provide addition information like attraction and repulsion of residues or removing any unstructured terminal residues in the ligand or target, etc. can be provided with the 'Advanced Option' (Fig. 6.19).
7. Click on the agreement policy and the click on the 'Dock' button to start docking process (Fig. 6.18).

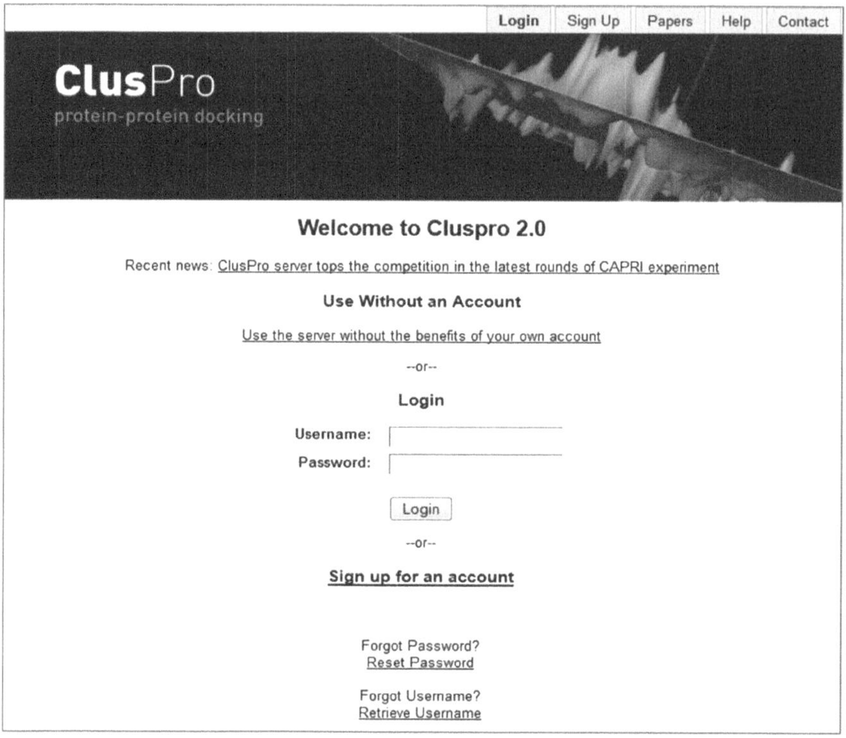


**Fig. 6.17** The Cluspro logging interface

8. Upon submission the program checks the PDB file for errors, processes it and puts in queue for docking. Every stage will be displayed under 'Running Jobs' (Fig. 6.20).
9. Docking time depends on the number of items in the queue and it also depends on the size of the protein. We can expect the process to be completed in few hours. Once done, results will be notified via. e-mail you have provided or you can click on the ID which is hyperlinked to see the status.

### Interpreting Results

10. Clicking on the id displays the results for that particular job. Here, the results of the docking can be viewed and can also be downloaded (Fig. 6.21).
11. The result page lists out the top 10 models by default using 'Balanced' force. If you have prior knowledge on the forces that dominate your complex, you can choose accordingly.
12. Individual models can also be downloaded by clicking on the number above the image for the model.

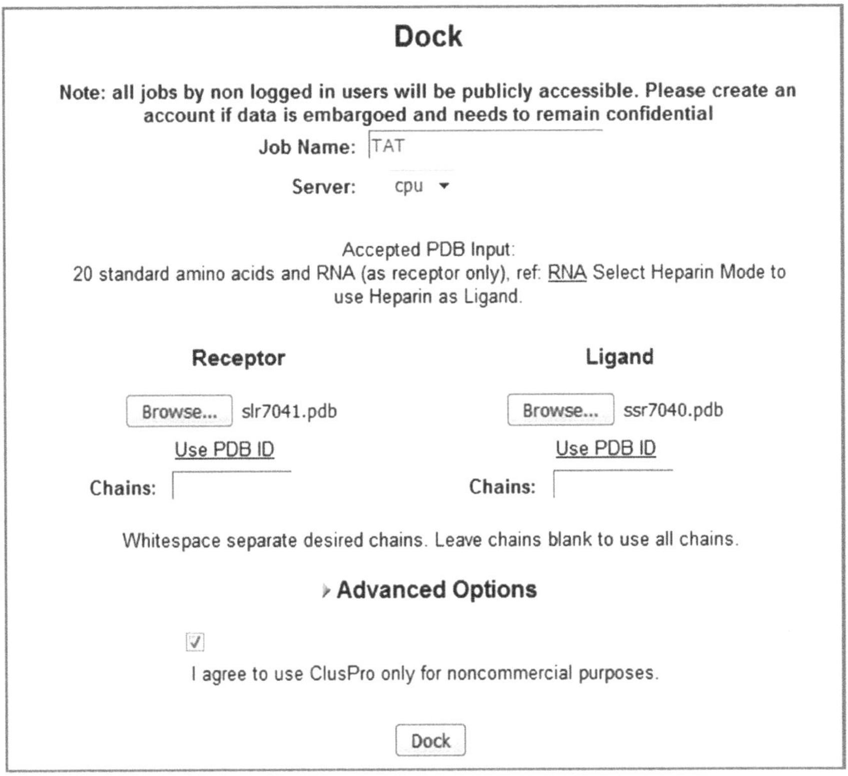


**Fig. 6.18** Specifying the receptor and ligand for docking

**Fig. 6.19** The advanced options for docking

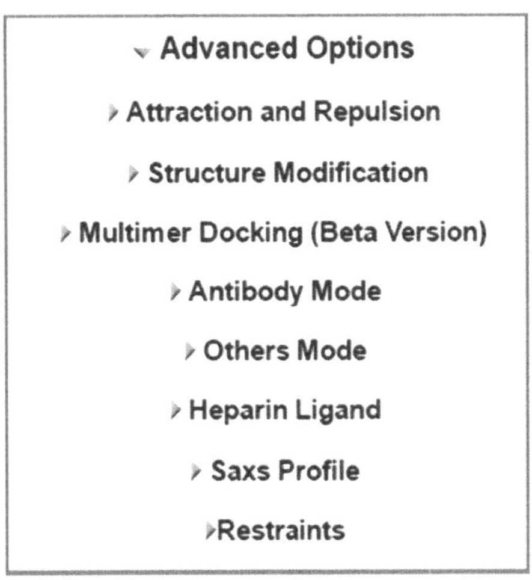

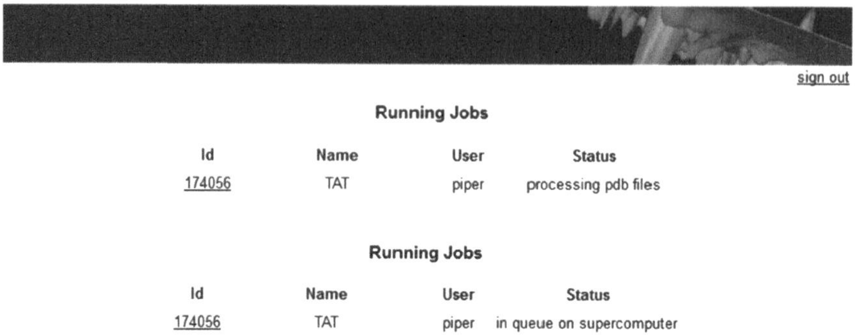


**Fig. 6.20** The different stages of processing in the server

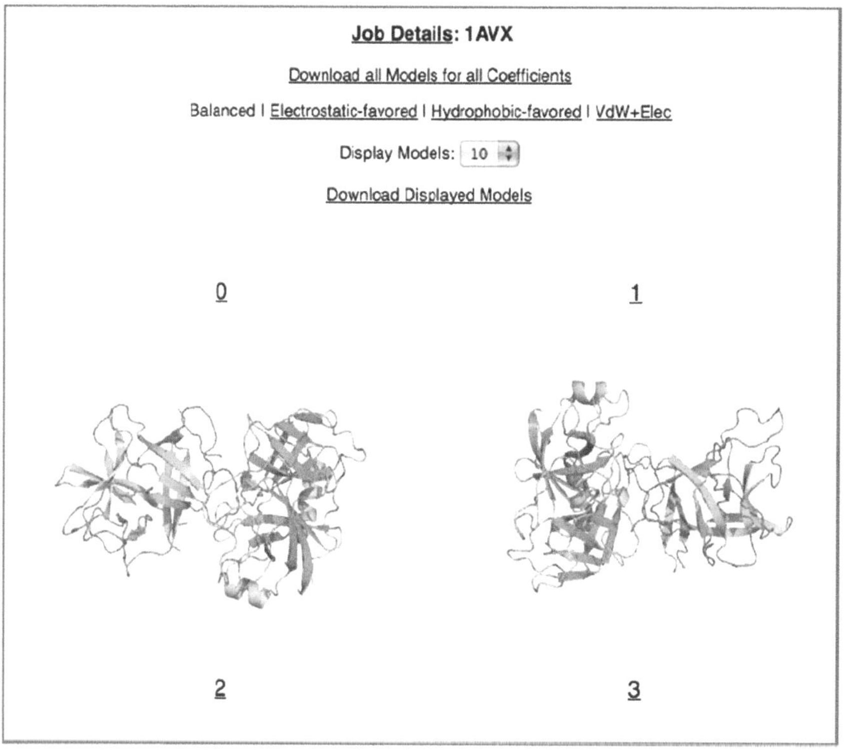


**Fig. 6.21** The results for the docking

13. There are provisions to change the number if there no satisfactory models are observed. There is also a provision to display the models based on the forces (Fig. 6.22). Links to download all models are also provided.
14. To view the scores of each docking, click on the 'View Model Scores' link. These scores come from 'Piper', which is a FFT-based rigid docking program, used in current versions (Fig. 6.23).

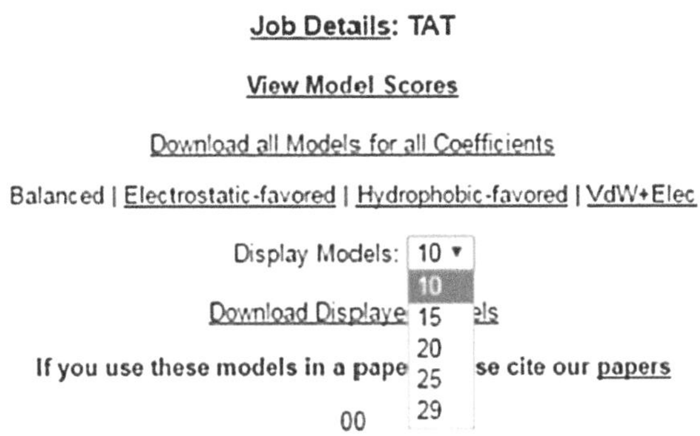


Fig. 6.22 The display options at results page

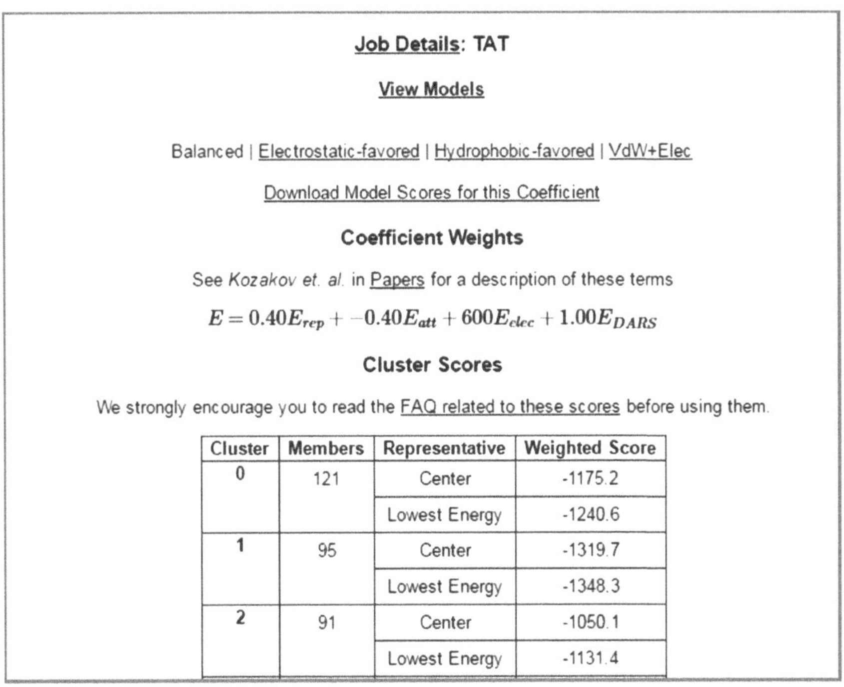


Fig. 6.23 The scores of the models

15. To analyse the model, it has to be downloaded and then loaded on to a different visualizing program like PyMOL. Some program might display only the target or the ligand file, as it may not support multiple files for loading. If original complex structure is known or available it can be loaded and superposed to the reliability of the docking.

## Exercises

1. What is docking scoring function?
2. What are the advanced options available in ClusPro?
3. If most of the docking poses are unsuitable or unsatisfactory, which coefficient (out of balanced, electrostatic-favoured, hydrophobic-favoured and VdW + Elec) do you choose to select an appropriate model?

# Appendix
# Online Resources on Bioinformatics

## (a) Tools

| Application | Description | URL |
|---|---|---|
| *DNA sequencing* | | |
| | Sequence scanner software from applied biosystems | https://products.appliedbiosystems.com/ab/en/US/adirect/ab?cmd=catNavigate2&catID=600583&tab=DetailInfo |
| | Technelysium | http://technelysium.com.au/ |
| | GENEWiz—sequence/chromatogram viewing software | http://www.genewiz.com/public/tools.aspx |
| | Nucleics—PeakTrace Basecaller | https://www.nucleics.com/peaktrace/ |
| | Geospiza—FinchTV | http://www.geospiza.com/Products/finchtv.shtml |
| *Sequence assembly* | | |
| | **CAP3** | http://doua.prabi.fr/software/cap3 |
| | Staden package—Gap4/Gap5 | http://sourceforge.net/projects/staden |
| | Phrap | http://www.phrap.org/phredphrapconsed.html#block_phrap |
| | ARACHNE | http://www.broad.mit.edu/crd/wiki/index.php/Main_Page |
| *Sequence retrieval* | | |
| | NCBI | http://www.ncbi.nlm.nih.gov/ |
| | EBI | http://srs.ebi.ac.uk/srsbin/cgi-bin/wgetz?-page+srsq2+-noSession |
| | UniProt | http://www.uniprot.org/ |
| *Database searching* | | |
| | 2013 NAR database summary paper alphabetic list | http://www.oxfordjournals.org/nar/database/a/ |
| | NCBI BLAST | http://blast.ncbi.nlm.nih.gov/ |
| | NCBI BLAST+ at EBI | http://www.ebi.ac.uk/Tools/sss/ncbiblast/nucleotide.html |
| | UniProt BLAST | http://www.uniprot.org/blast/ |

(continued)



K. Syed Ibrahim et al., *Bioinformatics—A Student's Companion*,
DOI 10.1007/978-981-10-1857-2

(continued)

| Application | Description | URL |
|---|---|---|
| *Dot plot analysis* | | |
| | Dot plot | http://www.vivo.colostate.edu/molkit/dnadot/ |
| | Dotlet | http://myhits.isb-sib.ch/cgi-bin/dotlet |
| | JDotter | http://athena.bioc.uvic.ca/virology-ca-tools/jdotter/ |
| | YASS | http://bioinfo.lifl.fr/yass/yass.php |
| *Pairwise alignment* | | |
| | BLAST2Seq at NCBI | http://blast.ncbi.nlm.nih.gov/Blast.cgi?PAGE_TYPE=BlastSearch&PROG_DEF=blastn&BLAST_PROG_DEF=megaBlast&BLAST_SPEC=blast2seq |
| | Pairwise sequence alignment at EBI | http://www.ebi.ac.uk/Tools/psa/ |
| | LALIGN | http://www.ch.embnet.org/software/LALIGN_form.html |
| *Multiple sequence alignment* | | |
| | MSA tools at EBI | http://www.ebi.ac.uk/Tools/msa/ |
| | PRALINE | http://www.ibi.vu.nl/programs/pralinewww/ |
| *MSA visualization* | | |
| | Boxshade | http://bioweb2.pasteur.fr/cgi-bin/MobylePortal/portal.py?form=boxshade |
| | ESPript | http://espript.ibcp.fr/ESPript/ESPript/ |
| *Phylogenetic analysis* | | |
| | MEGA | http://www.megasoftware.net |
| | PHYLIP | http://evolution.gs.washington.edu/phylip.html |
| | PAUP | http://users.iab.uaf.edu/ ~ derek_sikes/software2.htm. |
| | Collection of phylogenetic analysis software | http://evolution.genetics.washington.edu/phylip/software.html#methods |
| *Primer designing* | | |
| | Primer3 | frodo.wi.mit.edu/ |
| *Restriction analysis* | | |
| | NEBCutter | tools.neb.com/NEBcutter2/ |
| *Gene prediction* | | |
| | BLASTX | http://blast.ncbi.nlm.nih.gov/ |
| | ORFFinder | www.ncbi.nlm.nih.gov/gorf/gorf.html |
| *Protein primary structure analysis* | | |
| | ProtParam | web.expasy.org/protparam/ |
| | PROT-PROP | http://www.mzu.edu.in/index.php/biotech-facilities |
| *Protein secondary structure prediction* | | |
| | SOPMA | http://npsa-pbil.ibcp.fr/cgi-bin/npsa_automat.pl?page=/NPSA/npsa_sopma.html |
| | PSIPRED | bioinf.cs.ucl.ac.uk/psipred/ |
| | GOR V | gor.bb.iastate.edu/ |
| *Protein tertiary structure prediction* | | |
| Homology modeling | Swiss model | swissmodel.expasy.org/ |
| Threading | Phyre2 | www.sbg.bio.ic.ac.uk/phyre2/ |

(continued)

(continued)

| Application | Description | URL |
|---|---|---|
| *Protein tertiary structure analysis* | | |
| | RAMPAGE | mordred.bioc.cam.ac.uk/ ~ rapper/rampage.php |
| | PROCHECK | www.ebi.ac.uk/thornton-srv/software/PROCHECK/ |
| *Protein structure visualization* | | |
| | RASMOL | rasmol.org/ |
| | PyMOL | www.pymol.org/ |
| *Protein structure alignment* | | |
| Superimpose protein structures | SuperPose | http://wishart.biology.ualberta.ca/SuperPose/ |
| *Protein cleft analysis* | | |
| | PDBsum | http://www.ebi.ac.uk/thornton-srv/databases/pdbsum/Generate.html |
| *Transmembrane analysis* | | |
| | MEMSAT | http://bioinf.cs.ucl.ac.uk/psipred/ |
| *Signal peptide* | | |
| | SignalP | http://www.cbs.dtu.dk/services/SignalP/ |
| *Subcellular localization* | | |
| | TargetP | http://www.cbs.dtu.dk/services/TargetP/ |
| *RNA analysis* | | |
| RNA structure analysis | MFold | http://mfold.rna.albany.edu//?q=mfold/RNA-Folding-Form |
| Finding RNA in genome | RNAScan—SE | selab.janelia.org/tRNAscan-SE/ |
| Looking for RNA patterns | PatScan | www-unix.mcs.anl.gov/compbio/PatScan/HTML/scanner.html |

## (b) **Structural biology databases and software links**

| | |
|---|---|
| CADB: conformational angles database of proteins | http://cluster.physics.iisc.ernet.in/cadb/ |
| SSEP: secondary structural elements of proteins | http://cluster.physics.iisc.ernet.in/ssep/ |
| THGS: a web based database of transmembrane helices in genome sequences | http://pranag.physics.iisc.ernet.in/thgs/ |
| LySDB: lysozyme structural database | http://iris.physics.iisc.ernet.in/lysdb/ |
| SMS: sequence, motif and structure | http://pranag.physics.iisc.ernet.in/sms/ |
| MIPS: metal interactions in protein structures | http://dicsoft2.physics.iisc.ernet.in/mips/ |
| MLDB: macromolecule ligand database | http://dicsoft2.physics.iisc.ernet.in/mldb/ |
| PSST: protein sequence search tool—a web based interactive search engine | http://pranag.physics.iisc.ernet.in/psst/ |
| PDB goodies: a web-based GUI to manipulate the Protein Data Bank file | http://dicsoft2.physics.iisc.ernet.in/pdbgoodies/ |
| RP: Ramachandran plot on the web | http://dicsoft1.physics.iisc.ernet.in/rp/ |
| WAP: water analysis package—a web-based package to calculate geometrical parameters between water oxygen and protein atoms | http://dicsoft2.physics.iisc.ernet.in/wap/ |

(continued)

(continued)

| | |
|---|---|
| CAP: conformation angles package—displaying the conformation angles of side chains in proteins | http://dicsoft1.physics.iisc.ernet.in/cap/ |
| SEM: symmetry equivalent molecules—a web based GUI to generate and visualize the macromolecules | http://dicsoft2.physics.iisc.ernet.in/sem/ |
| BSDD: biomolecules segment display device—a web based interactive display tool | http://iris.physics.iisc.ernet.in/bsdd/ |
| FF: fragment finder—a web based software to identify similar three-dimensional structural motif | http://cluster.physics.iisc.ernet.in/ff/ |
| 3dSS: 3-dimensional structural superposition | http://cluster.physics.iisc.ernet.in/3dss/ |
| PSAP: structure analysis package | http://iris.physics.iisc.ernet.in/psap/ |
| RPMS: Ramachandran plot for multiple structures | http://cluster.physics.iisc.ernet.in/rpms/ |
| CSSP: consensus secondary structure prediction: a web based server for structural biologists | http://bioserver1.physics.iisc.ernet.in/cssp/ |
| SSMBS: a web-server to locate sequentially separated motifs in biological sequences | http://dicsoft1.physics.iisc.ernet.in/ssmbs/ |
| FAIR: a server for internal sequence repeats | http://bioserver1.physics.iisc.ernet.in/fair/ |
| ProSTRIP: a method to find similar structural repeats in three-dimensional protein structures | http://cluster.physics.iisc.ernet.in/prostrip/ |

## (c) Educational resources

| | |
|---|---|
| University of Missouri bioinformatics consortium | http://umbc.rnet.missouri.edu/resources/software/alphabetical.html |
| University of Colorado Bioinformatics | http://www.colorado.edu/chemistry/bioinfo/ |
| Train online—European Bioinformatics Institute | www.ebi.ac.uk/training/online/ |
| MIT libraries | http://libguides.mit.edu/bioinfo |
| Online lectures on bioinformatics | http://lectures.molgen.mpg.de/online_lectures.html |
| Digital world biology | http://digitalworldbiology.com |
| NIH library | http://nihlibrary.nih.gov/resourcetraining/Pages/ResourceTraining.asp#Bio |
| Network science | http://www.netsci.org/Resources/Software/index.html |
| Bio-soft | http://en.bio-soft.net/index.html |

# References

Altschul SF, Gish W, Miller W, Myers EW, Lipman DJ (1990) Basic local alignment search tool. J Mol Biol 215:403–410

Altschul SF, Madden TL, Schäffer AA, Zhang J, Zhang Z, Miller W, Lipman DJ (1997) Gapped BLAST and PSI-BLAST: a new generation of protein database search programs. Nucleic Acids Res 25:3389–3402

Apweiler R, Bairoch A, Wu CH (2004) Protein sequence databases. Curr Opin Chem Biol 8:76–80

Bachmair A, Finley D, Varshavsky A (1986) In vivo half-life of a protein is a function of its amino-terminal residue. Science 234:179–186

Bairoch A, Boeckmann B, Ferro S, Gasteiger E (2004) Swiss-Prot: juggling between evolution and stability. Brief Bioinform 5:39–55

Bao W, Kojima KK, Kohany O (2015) Repbase update, a database of repetitive elements in eukaryotic genomes. Mobile DNA 11. doi:10.1186/s13100-015-0041-9

Beaussart F, Weiner J 3rd, Bornberg-Bauer E (2007) Automated Improvement of Domain ANnotations using context analysis of domain arrangements (AIDAN). Bioinformatics 23:1834–1836

Benkert P, Tosatto SCE, Schomburg D (2008) QMEAN: a comprehensive scoring function for model quality assessment. Proteins 71:261–277

Boratyn GM, Schäffer AA, Agarwala R, Altschul SF, Lipman DJ, Madden TL (2012) Domain enhanced lookup time accelerated BLAST. Biol Direct 7:12

Bowie JU, Lüthy R, Eisenberg D (1991) A method to identify protein sequences that fold into a known three-dimensional structure. Science 253:70–164

Buchan DWA, Minneci F, Nugent TCO, Bryson K, Jones DT (2013) Scalable web services for the PSIPRED Protein Analysis Workbench. Nucleic Acids Res 41(W1):W340–W348

Camacho CJ, Gatchell D (2003) Successful discrimination of protein interactions. Proteins 52:92–97

Cochrane et al (2010) Access. Website: http://www.ncbi.nlm.nih.gov/VecScreen/UniVec.html

Colovos C, Yeates TO (1993) Verification of protein structures: patterns of nonbonded atomic interactions. Protein Sci 2:1511–1519

Comeau SR, Gatchell DW, Vajda S, Camacho CJ (2004) ClusPro: a fully automated algorithm for protein-protein docking. Nucleic Acids Res (Web Server issue):W96–W99

Dayhoff M, Schwartz R, Orcutt B (1978) A model of evolutionary change in proteins. In: Dayhoff MO (ed) Atlas of protein sequence and structure. National Biomedical Research Foundation, Washington, DC, pp 345–352

Edelhoch H (1967) Spectroscopic determination of tryptophan and tyrosine in proteins. Biochemistry 6:1948–1954

Edgar RC (2004) MUSCLE: multiple sequence alignment with high accuracy and high throughput. Nucleic Acids Res 32:1792–1797

Eisenberg D, Lüthy R, Bowie JU (1997) VERIFY3D: assessment of protein models with three-dimensional profiles. Methods Enzymol 277:396–404

 
K. Syed Ibrahim et al., *Bioinformatics—A Student's Companion*,
DOI 10.1007/978-981-10-1857-2

Emanuelsson O, Nielsen H, Brunak S, von Heijne G (2000) Predicting subcellular localization of proteins based on their N-terminal amino acid sequence. J Mol Biol 300:1005–1016
Fassler J, Cooper P (2011) BLAST glossary. In: BLAST® Help [Internet]. National Center for Biotechnology Information (US), Bethesda (MD). Available from: http://www.ncbi.nlm.nih.gov/books/NBK62051/
Fraser ID, Tavalin SJ, Lester LB, Langeberg LK, Westphal AM, Dean RA, Marrion NV, Scott JD (1998) A novel lipid-anchored A-kinase anchoring protein facilitates cAMP-responsive membrane events. EMBO J 17:2261–2272
French S, Robson B (1983) What is a conservative substitution? J Mol Evol 19:171–175
Gasteiger E, Hoogland C, Gattiker A, Duvaud S, Wilkins MR, Appel RD, Bairoch A (2005) Protein identification and analysis tools on the ExPASy server. In: Walker JM (ed) The proteomics protocols handbook. Humana Press, pp 571–607
Geourjon C, Deleage G (1994) Protein Eng 7:157–164
Geourjon C, Deléage G (1995) SOPMA: significant improvements in protein secondary structure prediction by consensus prediction from multiple alignments. Comput Appl Biosci 11:681–684
Gill SC, von Hippel PH (1989) Calculation of protein extinction coefficients from amino acid sequence data. Anal Biochem 182:319–326
Griffiths AJF, Miller JH, Suzuki DT et al (2000) An introduction to genetic analysis, 7th edn. W. H. Freeman, New York. Genetic code. Available from: http://www.ncbi.nlm.nih.gov/books/NBK21950/
Guruprasad K, Reddy BVB, Pandit MW (1990) Correlation between stability of a protein and its dipeptide composition: a novel approach for predicting in vivo stability of a protein from its primary sequence. Protein Eng 4:155–161
Henikoff JG (1992) Amino acid substitution matrices from protein blocks. Proc Natl Acad Sci USA 89:10915–10919
Hofacker IL, Fontana W, Stadler PF, Bonhoeffer S, Tacker M, Schuster P (1994) Fast folding and comparison of RNA secondary structures. Monatshefte f Chemie 125:167–188
Hooft RW, Vriend G, Sander C, Abola EE (1996) Errors in protein structures. Nature 381:272
Huang X, Madan A (1999) CAP3: a DNA sequence assembly program. Genome Res 9:868–877
Ikai AJ (1980) Thermostability and aliphatic index of globular proteins. J Biochem 88:1895–1898
Janusz M, Bujnicki (2008) Prediction of protein structures, functions, and interactions. Wiley, UK
Jin X (2006) Essential bioinformatics. Cambridge University Press, New York, USA
Johnson M, Zaretskaya I, Raytselis Y, Merezhuk Y, McGinnis S, Madden TL (2008) NCBI BLAST: a better web interface. Nucleic Acids Res 36(Web Server issue):W5–W9. doi:10.1093/nar/gkn201
Jones DT (1999) Protein secondary structure prediction based on position-specific scoring matrices. J Mol Biol 292:195–202
Junier T, Pagni M (2000) Dotlet: diagonal plots in a web browser. Bioinformatics 16:178–179
Jurka J, Kapitonov VV, Pavlicek A, Klonowski P, Kohany O, Walichiewicz J (2005) Repbase update, a database of eukaryotic repetitive elements. Cytogenet Genome Res 110:462–467
Käll L, Krogh A, Erik L, Sonnhammer L (2004) A combined transmembrane topology and signal peptide prediction method. J Mol Biol 338:1027–1036
Karlin S, Altschul SF (1990) Methods for assessing the statistical significance of molecular sequence features by using general scoring schemes. Proc Natl Acad Sci USA 87:2264–2268
Katoh K, Misawa K, Kuma K, Miyataa T (2002) MAFFT: a novel method for rapid multiple sequence alignment based on fast Fourier transform. Nucleic Acids Res 30:3059–3066
Kececioglu JD, Myers EW (1995) Combinatorial algorithms for DNA sequence assembly. Algorithmica 13:7–51
Kelley LA, Sternberg MJE (2009) Protein structure prediction on the web: a case study using the Phyre server. Nat Protoc 4:363–371
Kelley LA, Mezulis S, Yates CM, Wass MN, Sternberg MJ (2015) The Phyre2 web portal for protein modeling, prediction and analysis. Nat Protoc 10:845–858

Ko LJ, Engel JD (1993) DNA-binding specificities of the GATA transcription factor family. Mol Cell Biol 13:4011–4022
Kyte J, Doolittle R (1982) A simple method for displaying the hydropathic character of a protein. J Mol Biol 157:105–132
Laskowski RA, Chistyakov VV, Thornton JM (2005) PDBsum more: new summaries and analyses of the known 3D structures of proteins and nucleic acids. Nucleic Acids Res 33 (Database issue):D266–D268
Laskowski RA, Luscombe NM, Swindells MB, Thornton JM (1996) Protein clefts in molecular recognition and function. Protein Sci 5:2438–2452
Laskowski RA, MacArthur MW, Moss DS, Thornton JM (1993) PROCHECK: a program to check the stereochemical quality of protein structures. J Appl Cryst 26:283–291
Liu K, Muse SV (2005) PowerMarker: an integrated analysis environment for genetic marker analysis. Bioinformatics 21:2128–2129
Lorenz R, Bernhart SH, Höner Zu Siederdissen C, Tafer H, Flamm C, Stadler PF, Hofacker IL (2011) ViennaRNA package. Algorithms Mol Biol 6:26
Lowe TM, Eddy SR (1997) tRNAscan-SE: a program for improved detection of transfer RNA genes in genomic sequence. Nucleic Acids Res 25:955–964
Ma B, Tromp J, Li M (2002) PatternHunter: faster and more sensitive homology search. Bioinformatics 18(3):440–445
Magrane M, UniProt Consortium (2011) UniProt knowledgebase: a hub of integrated protein data. Database (Oxford) bar009. doi:10.1093/database/bar009
Maiti R, Van Domselaar GH, Zhang H, Wishart DS (2004) SuperPose: a simple server for sophisticated structural superposition. Nucleic Acids Res 32(Web Server issue):W590–W594
Mathews DH (2004) Using an RNA secondary structure partition function to determine confidence in base pairs predicted by free energy minimization. RNA 10:1178–1190
McCaskill JS (1990) The equilibrium partition function and base pair binding probabilities for RNA secondary structures. Biopolymers 29:1105–1119
Melo F, Feytmans E (1998) Assessing protein structures with a non-local atomic interaction energy. J Mol Biol 277:1141–1152
Melo F1, Feytmans E (1998) Assessing protein structures with a non-local atomic interaction energy. J Mol Biol 277(5):1141–1152
Mølhøj M, Degan FD (2004) Leader sequences are not signal peptides. Nat Biotechnol 22:1502
Morris GM, Lim-Wilby M (2008) Molecular docking. Methods Mol Biol 443:365–382
Morris AL, MacArthur MW, Hutchinson EG, Thornton JM (1992) Stereochemical quality of protein structure coordinates. Proteins 12:345–364
Morris GM, Huey R, Lindstrom W, Sanner MF, Belew RK, Goodsell DS, Olson AJ (2009) Autodock4 and AutoDockTools4: automated docking with selective receptor flexiblity. J Comput Chem 16:91–2785
Notredame C, Higgins DG, Heringa J (2000) T-Coffee: a novel method for fast and accurate multiple sequence alignment. J Mol Biol 302:205–217
Nugent T, Jones DT (2009) Transmembrane protein topology prediction using support vector machines. BMC Bioinformatics 10:159
Pace CN, Vajdos F, Fee L, Grimsley G, Gray T (1995) How to measure and predict the molar absorption coefficient of a protein. Protein Sci 11:2411–2423
Pauling L, Corey RB (1953) A proposed structure for the nucleic acids. Proc Natl Acad Sci USA 39:84–97
Pearson K (1901) On lines and planes of closest fit to systems of points in space. Philos Mag 2:559–572
Perrier X, Jacquemoud-Collet JP (2006) DARwin software. http://darwin.cirad.fr/

Perry JJ, Shin DS, Getzoff ED, Tainer JA (2010) The structural biochemistry of the superoxide dismutases. Biochim Biophys Acta 1804:245–262

Petersen TN, Brunak S, von Heijne G, Nielsen H (2011) SignalP 4.0: discriminating signal peptides from transmembrane regions. Nat Methods 8:785–786

Pontius J, Richelle J, Wodak SJ (1996) Deviations from standard atomic volumes as a quality measure for protein crystal structures. J Mol Biol 264:121–136

Ramachandran GN, Ramakrishnan C, Sasisekharan V (1963) Stereochemistry of polypeptide chain configurations. J Mol Biol 7:95–99

Rice P, Longden I, Bleasby A (2000) EMBOSS: the European molecular biology open software suite. Trends Genet 16:276–277

Rohlf FJ (2008) NTSYSpc: numerical taxonomy system, ver. 2.20. Exeter Publishing, Ltd. Setauket, NY

Rombel IT, Sykes KF, Rayner S, Johnston SA (2002) ORF-FINDER: a vector for high-throughput gene identification. Gene 282:33–41

Rutherford K, Parkhill J, Crook J, Horsnell T, Rice P, Rajandream MA, Barrell B (2000) Artemis: sequence visualisation and annotation. Bioinformatics 16:944–945

Sayle RA, Milner-White EJ (1995) RASMOL: biomolecular graphics for all. Trends Biochem Sci 20:374

Schwede T, Kopp J, Guex N, Peitsch MC (2003) SWISS-MODEL: an automated protein homology-modeling server. Nucleic Acids Res 31:3381–3385

Senthilkumar B, Sailo S, Guruswami G, Nachimuthu S (2003) Prot-Prop: J-tool to predict the subcellular location of proteins based on physiochemical characterization. Interdiscip Sci 4:296–301. doi:10.1007/s12539-012-0143-8

Sharma D, Issac B, Raghava GP, Ramaswamy R (2004) Spectral repeat finder (SRF): identification of repetitive sequences using Fourier transformation. Bioinformatics 20 (9):1405–1412

Sharp PM, Tuohy TMF, Mosurski KR (1986) Codon usage in yeast: cluster analysis clearly differentiates highly and lowly expressed genes. Nucleic Acids Res 14:5125–5143

Sokal RR, Sneath PHA (1963) Principles of numerical taxonomy. W.H. Freeman & Co, San Francisco

Spitale RC, Flynn RA, Torre EA, Kool ET, Chang HY (2014) RNA structural analysis by evolving SHAPE chemistry. Wiley Interdiscip Rev RNA 5(6):867–881. doi:10.1002/wrna.1253

Tamura K, Peterson D, Peterson N, Stecher G, Nei M, Kumar S (2011) MEGA5: molecular evolutionary genetics analysis using maximum likelihood, evolutionary distance, and maximum parsimony methods. Mol Biol Evol 28:2731–2739

Tatusov RL, Natale DA, Garkavtsev IV, Tatusova TA, Shankavaram UT, Rao BS, Kiryutin B, Galperin MY, Fedorova ND, Koonin EV (2001) The COG database: new developments in phylogenetic classification of proteins from complete genomes. Nucleic Acids Res 29:22–28

The PyMOL Molecular Graphics System, Version 1.5.0.4 Schrödinger, LLC

van Gunsteren WF, Billeter SR, Eising AA, Hünenberger PH, Krüger P, Mark AE, Scott WRP, Tironi IG (1996) Biomolecular simulations: the GROMOS96 manual and user guide. VdF Hochschulverlag ETHZ, Zürich

Vincze T, Posfai J, Roberts RJ (2003) NEBcutter: a program to cleave DNA with restriction enzymes. Nucleic Acids Res 31:3688–3691

Vriend G (1990) WHAT IF: a molecular modeling and drug design program. J Mol Graph 8:52–56

Waterhouse AM, Procter JB, Martin DMA, Clamp M, Barton GJ (2009) Jalview Version 2—a multiple sequence alignment editor and analysis workbench. Bioinformatics 25(9):1189–1191

Westbrook J, Feng Z, Chen L, Yang H, Berman HM (2003) The Protein Data Bank and structural genomics. Nucleic Acids Res 31:489–491

Wheeler DL, Barrett T et al (2006) Database resources of the National Center for Biotechnology Information. Nucleic Acids Res 34(Database Issue):D173–D180

Yamaguchi Y, Park JH, Inouye M (2011) Toxin-antitoxin systems in bacteria and archaea. Annu Rev Genet 45:61–79

Yuryev A (2007) PCR primer design. In: Methods in molecular biology, vol 402. Humana Press, Totowa, New Jersey, USA, p 431

Zhou H, Zhou Y (2002) Distance-scaled, finite ideal-gas reference state improves structure-derived potentials of mean force for structure selection and stability prediction. Protein Sci 11: 2714–2726

Zuckerkandl E, Pauling L (1965) Molecules as documents of evolutionary history. J Theor Biol 8:357–366

Zuker M, Stiegler P (1981) Optimal computer folding of large RNA sequences using thermodynamic and auxiliary information. Nucleic Acids Res 9:133–148

The manufacturer's authorised representative in the EU is Springer Nature Customer Service Centre GmbH, Europaplatz 3, 69115 Heidelberg, Germany. If you have any concerns regarding our products, please contact ProductSafety@springernature.com

Printed and bound by CPI Group (UK) Ltd, Croydon, CR0 4YY
22/09/2026
02220945-0009